Lecture Notes in Cc

Lecture Notes in Artificial Intelligence 15051

Founding Editor

Jörg Siekmann

Series Editors

Randy Goebel, *University of Alberta, Edmonton, Canada*
Wolfgang Wahlster, *DFKI, Berlin, Germany*
Zhi-Hua Zhou, *Nanjing University, Nanjing, China*

The series Lecture Notes in Artificial Intelligence (LNAI) was established in 1988 as a topical subseries of LNCS devoted to artificial intelligence.

The series publishes state-of-the-art research results at a high level. As with the LNCS mother series, the mission of the series is to serve the international R & D community by providing an invaluable service, mainly focused on the publication of conference and workshop proceedings and postproceedings.

M. Nazmul Huda · Mingfeng Wang ·
Tatiana Kalganova
Editors

Towards Autonomous Robotic Systems

25th Annual Conference, TAROS 2024
London, UK, August 21–23, 2024
Proceedings, Part I

Springer

Editors
M. Nazmul Huda
Department of Electronic and Electrical
Engineering
Brunel University London
London, UK

Mingfeng Wang
Department of Mechanical and Aerospace
Engineering
Brunel University London
London, UK

Tatiana Kalganova
Department of Electronic and Electrical
Engineering
Brunel University London
London, UK

ISSN 0302-9743   ISSN 1611-3349 (electronic)
Lecture Notes in Artificial Intelligence
ISBN 978-3-031-72058-1   ISBN 978-3-031-72059-8 (eBook)
https://doi.org/10.1007/978-3-031-72059-8

LNCS Sublibrary: SL7 – Artificial Intelligence

© The Editor(s) (if applicable) and The Author(s), under exclusive license
to Springer Nature Switzerland AG 2025

This work is subject to copyright. All rights are solely and exclusively licensed by the Publisher, whether the whole or part of the material is concerned, specifically the rights of translation, reprinting, reuse of illustrations, recitation, broadcasting, reproduction on microfilms or in any other physical way, and transmission or information storage and retrieval, electronic adaptation, computer software, or by similar or dissimilar methodology now known or hereafter developed.
The use of general descriptive names, registered names, trademarks, service marks, etc. in this publication does not imply, even in the absence of a specific statement, that such names are exempt from the relevant protective laws and regulations and therefore free for general use.
The publisher, the authors and the editors are safe to assume that the advice and information in this book are believed to be true and accurate at the date of publication. Neither the publisher nor the authors or the editors give a warranty, expressed or implied, with respect to the material contained herein or for any errors or omissions that may have been made. The publisher remains neutral with regard to jurisdictional claims in published maps and institutional affiliations.

This Springer imprint is published by the registered company Springer Nature Switzerland AG
The registered company address is: Gewerbestrasse 11, 6330 Cham, Switzerland

If disposing of this product, please recycle the paper.

Preface

This volume contains the papers presented at TAROS 2024, the 25th Towards Autonomous Robotic Systems (TAROS) Conference, held at the College of Engineering, Design and Physical Sciences, Brunel University London, UK, during August 21–23, 2024 (https://taros-conference.org/).

TAROS is the longest running UK-hosted international conference on robotics and autonomous systems (RAS), and is aimed at the presentation and discussion of the latest results and methods in autonomous robotics research and applications. The conference offers a friendly environment for robotics researchers and industry to take stock of current developments and plan for future progress. It welcomes senior researchers and research students alike, and specifically provides opportunities for research students and young research scientists to present their work to the scientific community.

TAROS 2024 was held at Brunel University London, a world-class university with an outstanding reputation for industry engagement in education, technology, research, and innovation. The papers in this volume were selected from 69 submissions under a single-blind peer review process. Out of these, 54 full papers and 11 short papers were selected for the conference. The conference programme included an academic conference, industry exhibitions, robot demonstrations, and other social events. TAROS 2024 covered topics of robotic systems, human–robot interaction, soft robotics, robot navigation and planning, robot control, and industrial robots, and highlights included – Keynote lectures by world-leading experts in robotics, including Robert Deaves from Dyson Ltd., UK; Ana Cavalcanti from University of York, UK; Asoke Nandi from Brunel University London, UK; Josie Hughes from École Polytechnique Fédérale de Lausanne, Switzerland; Kaspar Althoefer from Queen Mary University London, UK; Marc Hanheide from University of Lincoln, UK; Radhika Gudipati from Advanced Research and Invention Agency, UK; Cigdem Sengul from Brunel University London, UK; and Zakareya Hussein from Touchlab, National Robotarium, UK – Oral and Poster presentations of papers, covering various topics of robotics and autonomous systems. This year the conference also included contributions from Brunel's Robotics and AI for Healthcare group and the EPSRC Centre for Doctoral Training: AgriFoRwArdS, ALBERT, and FARSCOPE.

The TAROS 2024 Organizing Committee would like to thank all the authors, the reviewers, and the conference sponsors, including IET, MathWorks, RoboSavvy, Dobot, Unitree, Queen Mary University of London, UK-RAS Network, Springer, and MDPI for their support to the conference.

July 2024

M. Nazmul Huda
Mingfeng Wang
Tatiana Kalganova

Organization

General Chair

M. Nazmul Huda — Brunel University London, UK

Program Committee Chairs

Mingfeng Wang — Brunel University London, UK
Tatiana Kalganova — Brunel University London, UK

Programme Committee

Barry Lennox — University of Manchester, UK
Andrew Weightman — University of Manchester, UK
John Oyekan — University of York, UK
Marisé Galvez Trigo — Cardiff University, UK
Yohan Noh — Brunel University London, UK
Pedro Machado — Nottingham Trent University, UK

Web Chair

Farzana Sharmin Mou — Brunel University London, UK

Administrator

Surjeet Khurana — Brunel University London, UK

TAROS Steering Committee

Fumiya Iida — University of Cambridge, UK
Ana Cavalcanti — University of York, UK
Robert Richardson — University of Leeds, UK
M. Nazmul Huda — Brunel University London, UK

Mingfeng Wang	Brunel University London, UK
Chenguang Yang	University of Liverpool, UK
Antonia Tzemanaki	University of Bristol, UK
Farshad Arvin	Durham University, UK
Guodong Zhao	University of Manchester, UK
Barry Lennox	University of Manchester, UK
Marisé Galvez Trigo	Cardiff University, UK
Yang Gao	King's College London, UK

Invited Speakers

Robert Deaves	Dyson Ltd., UK
Ana Cavalcanti	University of York, UK
Asoke Nandi	Brunel University London, UK
Josie Hughes	École Polytechnique Fédérale de Lausanne, Switzerland
Kaspar Althoefer	Queen Mary University London, UK
Marc Hanheide	University of Lincoln, UK
Radhika Gudipati	Advanced Research and Invention Agency, UK
Cigdem Sengul	Brunel University London, UK
Zakareya Hussein	Touchlab, National Robotarium, UK

Additional Reviewers

Oneeba Ahmed	Brunel University London, UK
Sarfraz Ahmed	Coventry University, UK
Ibrahim Alispahic	University of Sarajevo, Bosnia-Herzegovina
Xiangyu An	Brunel University London, UK
Gerardo Aragon Camarasa	University of Glasgow, UK
Farshad Arvin	Durham University, UK
Kawsar Arzomand	Brunel University London, UK
Arjun Badyal	University of York, UK
Lingfan Bao	University of Leeds, UK
Joshua Bettles	University of Manchester, UK
Jordan Bird	Nottingham Trent University, UK
Pablo Borja	University of Plymouth
Matt Butler	Harper Adams University, UK
Daniele Cafolla	Swansea University, UK
Abu Bakar Dawood	Queen Mary University of London, UK
Sairaj R. Dillikar	Cranfield University, UK

Feliciano Domingos	Nottingham Trent University, UK
Ubada El Joulani	Brunel University London, UK
Laurenz Elstner	Western Norway University of Applied Sciences, Norway
Juan Pablo Espejel Flores	University of Lincoln, UK
Omar Faris	University of Lincoln, UK
Charles Fox	University of Lincoln, UK
Mohamed Gaballa	Brunel University London, UK
Ebony Harker-Bannister	Brunel University London, UK
Yuanzhi He	Cardiff University, UK
Henry Hickson	University of Bristol, UK
Ningzhe Hou	University of Oxford, UK
Junyan Hu	Durham University, UK
Marianne Huchard	University of Montpellier, France
Joseph Humphreys	University of Leeds, UK
Hoda Ibrahim	Brunel University London, UK
Kennedy Ihianle	Nottingham Trent University, UK
Katherine James	University of Lincoln, UK
Fahad Khan	Cranfield University, UK
Mohit Kumar	Software Competence Center Hagenberg, Germany
Seyonne Leslie-Dalley	University of Manchester, UK
Cunjia Liu	Loughborough University, UK
Haowen Liu	University of York, UK
Pengcheng Liu	University of York, UK
Qi Lu	University of Texas Rio Grande Valley, USA
Nan Ma	Northwestern Polytechnical University, China
Pedro Machado	Nottingham Trent University, UK
Soumya Kanti Manna	Canterbury Christ Church University, UK
Fraser McGhan	Cardiff University, UK
Nahal Memar Kocheh Bagh	University of York, UK
Hongying Meng	Brunel University London, UK
Dennis Monari	Nottingham Trent University, UK
Juan Sebastian Mosquera Maturana	Universidad Autónoma de Occidente, Colombia
Fama Ngom	University of Montpellier, France
Yohan Noh	Brunel University London, UK
Matthew Parris	University of Buckingham, UK
Tianhu Peng	University College London, UK
Vishnu Rajendran Sugathakumary	University of Lincoln, UK
Abhra Roy Chowdhury	Indian Institute of Science, India
Matteo Russo	University of Rome Tor Vergata, Italy

Philippa Ryan	University of York, UK
Weiyong Si	University of Essex, UK
Robert Stevenson	University of Lincoln, UK
Kedar Suthar	Brunel University London, UK
Yael Sznaidman	Ben-Gurion University of the Negev, Israel
Markus Weißflog	Chemnitz University of Technology, Germany
Emlyn Williams	University of Lincoln, UK
Salisu Yahaya	Nottingham Trent University, UK
Yanqiu Yang	Penn State University, USA
Ahmet Serhat Yildiz	Brunel University London, UK
Abdurrahman Yilmaz	University of Lincoln, UK
Yang You	RACE-UKAEA, UK
Kaiqiang Zhang	UK Atomic Energy Authority, UK
Ketao Zhang	Queen Mary University London, UK
Yulin Huaxi Zhang	University of Picardie Jules Verne, France
Zhixin Zhang	University of Manchester, UK
Hui Zhou	Coventry University, UK
Liyou Zhou	University of Lincoln, UK
Farbod Zorriassatine	Nottingham Trent University, UK

Contents – Part I

Robotic Learning, Mapping and Planning

A Performance Comparison of SLAM-Algorithms for Formula Student
Autonomous Driving 3
    *Kai Lascheit, Joseph Agrane, Oliver Neill, Jim Carty, Jamie Robb,
    and Gerardo Aragon-Camarasa*

Comparative Analysis of Unity and Gazebo Simulators for Digital Twins
of Robotic Tomato Harvesting Scenarios 15
    *Juan Pablo Espejel Flores, Abdurrahman Yilmaz,
    Luis Arturo Soriano Avendaño, and Grzegorz Cielniak*

Supporting Explainable Planning and Human-Aware Mission Specification
for Underwater Robots 28
    *Alan Lindsay, Andrés A. Ramírez-Duque, Bart Craenen,
    Andrea Munafò, Laurence Boé, Adam Campbell, and Ronald P. A. Petrick*

Improved Computation Efficiency 2D Visual SLAM Based on Particle
Filter With Distance Sliding Window 41
    Zhixin Zhang, Yichen Liang, and Alexandru Stancu

The Benefits of Ordinal Regression Under Domain Shift 53
    Andy Perrett, James M. Brown, and Petra Bosilj

Pretrained Visual Representations in Reinforcement Learning 60
    Emlyn Williams and Athanasios Polydoros

IntelliMove: Enhancing Robotic Planning with Semantic Mapping 72
    *Fama Ngom, Huaxi Yulin Zhang, Lei Zhang, Karen Godary-Dejean,
    and Marianne Huchard*

Localisation-Aware Fine-Tuning for Realistic PointGoal Navigation 84
    Fraser McGhan, Ze Ji, and Raphael Grech

Towards Revisiting Visual Place Recognition for Joining Submaps
in Multimap SLAM* 94
    Markus Weißflog, Stefan Schubert, Peter Protzel, and Peer Neubert

Consulting an Oracle; Repurposing Robots for the Circular Economy 107
    Helen McGloin, Matthew Studley, Richard Mawle, and Alan Winfield

"Incomplete Without Tech": Emotional Responses and the Psychology
of AI Reliance 119
   Mriganka Biswas and John Murray

Suspicious Activity Detection for Defence Applications 132
   *Matthew Marlon Gideon Parris, Hisham Al Assam,
and Mohammad Athar Ali*

Participatory AI: A Method for Integrating Inclusive and Ethical Design
Considerations into Autonomous System Development 144
   Christina E. Stimson and Rebecca Raper

Sampling-Based Motion Planning for Guide Robots Considering User
Pose Uncertainty 155
   *Juan Sebastian Mosquera-Maturana, Juan David Hernández Vega,
and Victor Romero Cano*

Safety Assurance Challenges for Autonomous Drones in Underground
Mining Environments 169
   *Philippa Ryan, Arjun Badyal, Samuel Sze, Benjamin Hardin,
Hasan Bin Firoz, Paulina Lewinska, and Victoria Hodge*

Who is the Chameleon? A Party Game to Explore Trust and Biases
Towards Alexa, Pepper and ChatGPT 182
   Charlotte Jones, Darren Reed, and Fanta Camara

Investigation of Gated-CNN and Self-Attention Mechanism for Historical
Handwritten Text Recognition 194
   Jizhang Li, Sarfraz Ahmed, and Md Nazmul Huda

Robotic Modeling, Sensing and Control

Enabling Tactile Feedback for Robotic Strawberry Handling Using AST
Skin 209
   *S. Vishnu Rajendran, Kiyanoush Nazari, Simon Parsons,
and E. Amir Ghalamzan*

Open Source Hardware Whisker Sensor 222
   *Robert Stevenson, Dimitris Paparas, Omar Faris, Xiaoxian Xu,
Catherine Merchant, Elliot Smith, Benjamin Nicholls, and Charles Fox*

Resonant Inductive Coupling Power Transfer for Mid-Sized Inspection
Robot 234
   Mohd Norhakim Bin Hassan, Simon Watson, and Cheng Zhang

3D Printer Based Open Source Calibration Platform for Whisker Sensors 249
   Liyou Zhou, Omar Ali, Soumo Emmanuel Arnaud, Eden Attenborough,
   Jacob Swindell, George Davies, and Charles Fox

Mitigating the Time Delay and Parameter Perturbation by a Predictive
Extended State Observer-Based Active Disturbance Rejection Control 256
   Syeda Nadiah Fatima Nahri, Shengzhi Du, Barend J. van Wyk,
   and Tawanda Denzel Nyasulu

Accessibility Framework for Determining Collisions and Coverage
for Radiation Scanning 270
   Joshua Bettles, Andrew West, Jeremy Andrew, Iain Darby,
   and Barry Lennox

A New Hybrid Teleoperation Control Scheme for Holonomic Mobile
Manipulator Robots Using a Ground-Based Haptic Device 283
   Bandar Aldhafeeri, Joaquin Carrasco, Bruno V. Adorno,
   and Erwin Jose Lopez Pulgarin

Variable Stiffness & Dynamic Force Sensor for Tissue Palpation 296
   Abu Bakar Dawood, Zhenyu Zhang, Martin Angelmahr,
   Alberto Arezzo, and Kaspar Althoefer

Real-World Testing of Ultrasonic Beacons for Mobile Robot Radiation
Emulation 302
   David Batty, Andrew West, Ipek Caliskanelli, and Paolo Paoletti

Robotic Tight Packaging Using a Hybrid Gripper with Variable Stiffness 313
   Michele Moroni, Ana Elvira Huezo Martin, Leonard Klüpfel,
   Ashok M. Sundaram, Werner Friedl, Francesco Braghin,
   and Máximo A. Roa

Design, Fabrication and Calibration of an Embroidery Textile Tactile
Sensor Array 327
   Ningzhe Hou, Marco Pontin, Leone Costi, and Perla Maiolino

SmartAntenna: Enhancing Wireless Range with Autonomous Orientation 339
   Michael Swann, Pedro Machado, Isibor Kennedy Ihianle,
   Salisu Yahaya, Farbod Zorriassatine, and Andreas Oikonomou

Machine Vision

Unsupervised Clustering with Geometric Shape Priors for Improved
Occlusion Handling in Plant Stem Phenotyping 355
   Katherine Margaret Frances James and Grzegorz Cielniak

Depth Priors in Removal Neural Radiance Fields 367
  Zhihao Guo and Peng Wang

YOLOv8-LiDAR Fusion: Increasing Range Resolution Based on Image
Guided-Sparse Depth Fusion in Self-Driving Vehicles 383
  Ahmet Serhat Yildiz, Hongying Meng, and Mohammad Rafiq Swash

Efficient 2D and 3D Corresponding Object Identification Using Deep
Learning Models 397
  Haowen Liu, Mark Post, and Andy Tyrrell

WeedScout: Real-Time Autonomous Blackgrass Classification
and Mapping Using Dedicated Hardware 409
  *Matthew Gazzard, Helen Hicks, Isibor Kennedy Ihianle, Jordan J. Bird,
  Md Mahmudul Hasan, and Pedro Machado*

What Criteria Define an Ideal Skeletonisation Reference in Object Point
Clouds? 422
  Qingmeng Wen, Seyed Amir Tafrishi, Ze Ji, and Yu-Kun Lai

Advancements in 3D X-Ray Imaging: Development and Application
of a Twin Robot System 434
  *Seemal Asif, Yuliya Hryshchenko, Martin Holden, Matteo Contino,
  Ndidiamaka Adiuku, Bryn Hughes, Angelos Plastropoulos,
  Nico Avdelidis, and Phil Webb*

Author Index 447

Contents – Part II

Human-Robot Interaction/Collaboration

Online Robust Robot Planning for Human-Robot Collaboration 3
   Yang You, Vincent Thomas, Francis Colas, Robert Skilton, and Olivier Buffet

The Dilemma of Decision-Making in the Real World: When Robots Struggle to Make Choices Due to Situational Constraints 14
   Khairidine Benali and Praminda Caleb-Solly

Controlling a Robotic Arm Through Neural Activity 27
   Hannah Gofton, Daniel H. Baker, and Fanta Camara

Human Facial Emotion Recognition for Adaptive Human Robot Collaboration in Manufacturing 33
   Fahad Khan, Seemal Asif, and Phil Webb

Do People Ascribe Similar Emotions to Real and Robotic Dog Tails? 48
   Alexandra Lee and Matthew Studley

An Perception Enhanced Human-Robot Skill Transfer Method for Reactive Interaction 58
   Weiyong Si, Jiale Dong, Ning Wang, and Chenguang Yang

Locomotion and Manipulation

Advancing Robotic Jumping with CVT Enhanced SEA 73
   Jingcheng Sun and Chengxu Zhou

Few-Shot Transfer Learning for Deep Reinforcement Learning on Robotic Manipulation Tasks 85
   Yuanzhi He, Christopher D. Wallbridge, Juan D. Hernndez, and Gualtiero B. Colombo

A ROS-Based Control Framework for Simulating Locomotion of a Multi-arm Space Assembly Robot 93
   Sairaj R. Dillikar, Cameron Leslie, Saurabh Upadhyay, Leonard Felicetti, and Gilbert Tang

Size Matters: Exploring the Impact of Scaling on Quadruped Robot
Dynamics 106
    Faraz Rahvar and Abdurrahman Yilmaz

Learning Bipedal Walking on a Quadruped Robot via Adversarial Motion
Priors 118
    *Tianhu Peng, Lingfan Bao, Joseph Humphreys,
    Andromachi Maria Delfaki, Dimitrios Kanoulas, and Chengxu Zhou*

Ground-Truth Based Calibration for a Two Wheel Differential Drive Robot 130
    Youssef G. Alboraei, Chayada Thidrasamee, and Paul O'Dowd

Control Strategies Using the Extended Cooperative Dual Task-Space 143
    Seyonne Leslie-Dalley, Bruno Vilhena Adorno, and Keir Groves

Mechanism Design

Design of a Robotic Trainer for Upper Body Physical Therapy 157
    Yael Sznaidman, Maya Krakovski, Shirley Handelzalts, and Yael Edan

A Bio-inspired Jumping Robot: Design, Modelling and Experimental Tests 164
    Harrison Elliott, Xiangyu An, and Mingfeng Wang

BlimpleBee: The Helium Assisted Indoor Inspection Drone 171
    Henry Hickson, Andrew T. Conn, and Hemma Philamore

Effects of Passive Ankle on Bipedal Robot Walking Locomotion 184
    Xiangyu An, Mayo Adetoro, and Mingfeng Wang

Development of Underactuated Geometric Compliant (UGC) Module
with Variable Radial for Robotic Applications 195
    Mark Krysov and Seyed Amir Tafrishi

Design and Motion Analysis of a Reconfigurable Pendulum-Based Rolling
Disk Robot with Magnetic Coupling 208
    Ollie Wiltshire and Seyed Amir Tafrishi

Soft Robotics

Contact-Rich Task Learning on an Articulated Soft Robot Arm Through
Simulation 223
    Laurenz Elstner, Erik Kyrkjebø, and Martin F. Stoelen

Using Barometric Pressure Sensors in Soft Robots 236
 Kedar Suthar, Chapa Sirithunge, and Mingfeng Wang

Modelling of Buoyancy Based Actuation of an Inflatable Underwater Soft
Robot 242
 Danyaal Kaleel, Benoit Clement, and Kaspar Althoefer

Designing and Manufacturing Low-Cost, Tendon-Driven Soft Robots 254
 Ted Winter-Glasgow and Pablo Borja

Bio-Inspired Soft Pneumatic Gripper for Agriculture Harvesting 266
 *Alex Clark, Liam Goodsell-Carpenter, Pia Buckow, Daniel Hewett,
 Francis White, Adil Imam, Nabila Naz, Breeshea Robinson,
 Soumya K. Manna, and Abdullahi Ahmed*

Swarms and Multi-Agent Systems

A Leader-Follower Collective Motion in Robotic Swarms 281
 *Mazen Bahaidarah, Ognjen Marjanovic, Fatemeh Rekabi-bana,
 and Farshad Arvin*

Duckling Platooning - Safety Guarantees Through Controlled Information
Disclosure 294
 James R. Heselden and Gautham P. Das

Robust Mitigation Strategy for Misleading Pheromone Trails in Foraging
Robot Swarms 307
 Ryan Luna and Qi Lu

Predator-Prey Q-Learning Based Collaborative Coverage Path Planning
for Swarm Robotics 320
 Michael Watson, Hanchi Ren, Farshad Arvin, and Junyan Hu

Path Planning for Multi-agent Systems Using Deep Q-Networks
Reinforcement Learning 333
 Ibrahim Alispahić and Adnan Tahirović

Author Index 345

Robotic Learning, Mapping and Planning

A Performance Comparison of SLAM-Algorithms for Formula Student Autonomous Driving

Kai Lascheit, Joseph Agrane, Oliver Neill, Jim Carty, Jamie Robb, and Gerardo Aragon-Camarasa(✉)

School of Computing Science, University of Glasgow, Scotland G12 8QQ, UK
{eng-ugracing,gerardo.aragoncamarasa}@glasgow.ac.uk

Abstract. This paper provides a comparative analysis of SLAM algorithms applied at Formula Student AI (FS-AI) using data collected by the University of Glasgow's FS-AI team, UGRacing, at the 2023 competition. As part of the international autonomous driving competition, UGRacing have developed a software capable of racing a Formula Student car through unknown circuits. An essential part of the software of UGRacing and other FS-AI teams across the globe is SLAM, which allows the car to memorise track boundary positions and estimate the car's position on the track. Optimising this component enables a vehicle to calculate and follow an ideal racing line, leading to significantly improved lap times. Hence, this paper examines the performance of three state-of-the-art SLAM approaches: EKF-SLAM, FastSLAM and Graph-SLAM. Experiments suggest that EKF-SLAM outperforms FastSLAM and GraphSLAM in terms of car position and landmark position estimates and computational efficiency. The dataset is publicly available at https://gitlab.com/ugrdv/fs-ai-datasets.

Keywords: Autonomous Racing · SLAM · Autonomous Driving

1 Introduction

Autonomous Racing has experienced an increase in popularity during the past decade. Recent developments include the introduction of Amazon's Deep-Racer platform which allows developers to apply Reinforcement Learning to autonomous driving [4] and the launch of the Abu Dhabi Autonomous Racing League where autonomous cars will race around the Abu Dhabi Formula 1 circuit [2]. As one of the world's biggest student engineering competitions, Formula Student (FS) [1] aims to provide university students with unique opportunities to gain practical experience in developing automotive technologies, from designing

---

We thank Mapix Technologies Ltd for providing the Velodyne Puck LiDAR, and Thales UK Ltd and the School of Computing Science, University of Glasgow, for their support in developing UGRacing's autonomous driving software.

Fig. 1. Autonomous Vehicle on Track at the 2023 FS-AI Competition

car components to implementing autonomous driving software. The challenge of the autonomous racing part of the competition, termed FS-AI, is to develop software that can safely and reliably navigate a real car through unseen circuits at racing speeds (see Fig. 1). The limiting factors for the software are the computational resources onboard the car and the inaccuracies of sensor measurements. Therefore, the software needs to be designed with close attention to computational efficiency and robustness to noise introduced by the sensors. Furthermore, the software must combine all required components for autonomous racing, from vision and motion estimation to localisation mapping, path planning and control.

In the context of FS-AI, SLAM is used to produce a map of the race track boundaries denoted by blue and yellow cones – as shown in Fig. 1. This map is created based on sensor measurements of the cones and car pose. The inputs to SLAM are cone position and colour estimates extracted from LiDAR point clouds and stereo camera images, as well as GPS measurements of the car pose. Additionally, UGRacing utilises motion information about the car based on dead reckoning. The information used for this comes from wheel speed and steering angle measurements provided by the vehicle control unit and, hence, are available to every FS-AI team. Sensor measurements are passed in together with estimates of the associated noise levels, allowing SLAM algorithms to weight measurements accordingly. The created map is used to improve the accuracy of car pose estimates and to calculate an ideal racing line once the full layout of the track has been memorised, which is essential for reaching top speeds. Therefore, since SLAM is a crucial component of autonomous driving, this paper addresses the question: *Which SLAM algorithm is most suited for the FS-AI competition?*

Formula Student is a worldwide engineering competition for university students. However, there is a scarcity of scientific literature on related methods and algorithms, such as SLAM, in the context of this competition. As a result, university student teams must implement and compare algorithms independently. Due to a lack of available resources, this is usually done in simulations where sensor faults and inaccuracies can only be approximated. Additionally, since each team develops the software from scratch, individual components are initially tested in isolation without indicating how they would perform once integrated with the complete autonomous software. To address these issues, this paper aims *to provide a rigorous analysis of three highly performing SLAM algorithms executed on real-world data collected at the FS-AI competition.* Additionally, SLAM is tested as a fully integrated component of the driverless software of the UGRacing[1] team such that sensor inputs are pre-processed by the vision system, giving a realistic indication of the accuracy of the measurements passed into SLAM.

The three algorithms chosen for this paper are GraphSLAM [8], FastSLAM [16], and EKF-SLAM [26]. These are widespread, landmark-based approaches to SLAM which are commonly applied by leading FS-AI teams such as KA-RaceIng (Karlsruhe Institute of Technology) [19] and AMZ Racing (ETH Zurich) [11]. For this paper, SLAM algorithms have been evaluated on input data collected by UGRacing at the Formula Student UK competition in July 2023, comprising cone position estimates based on a Velodyne Puck LiDAR and a ZED2 stereo camera as well as wheel speed and steering angle measurements provided by the car's vehicle control unit. Additionally, GPS measurements have been simulated with estimated noise levels based on tests carried out with UGRacing's GPS sensor as described further in Sect. 3.4. Ground truth car poses and landmark positions have been estimated precisely using offline Iterative Closest Points and manual post-processing to evaluate the accuracy of the algorithms' car pose and landmark position estimates.

2 Related Work

Various approaches to SLAM could be applied to FS-AI. Modern techniques include visual feature matching and landmark-based SLAM, which focuses on distinct objects in the robot's environment. Mono-SLAM [5] was one of the first feature-based approaches which track distinct visual features in camera images to recover the robot's trajectory. ORB-SLAM [17], an extension to Mono-SLAM, employs ORB keypoints to detect distinctive image features at high precision [22]. ProSLAM [24] provides an alternative feature-based approach with a lower computational load; however, it is considered to be less robust due to simplifications which are made in favour of a more lightweight solution such as the removal of bundle adjustment optimisation [7].

Another class of SLAM algorithms are based on detecting specific types of objects (i.e. landmarks) in the robot's environment, which can be tracked to deduce the robot's relative motion. GraphSLAM, FastSLAM, and EKF-SLAM are considered among the most popular landmark-based approaches [3] and are commonly employed at Formula Student. A major benefit of these algorithms is that they create a map which only contains the desired types of landmarks, i.e. traffic cones in the context of FS-AI. Furthermore, visual feature matching approaches are expected to produce unsatisfactory results on the particularly featureless visual landscape of a Formula Student track, consisting mostly of grey tarmac, as shown in Fig. 1. Finally, the computational resources onboard the ADS-DV are strictly limited, as described in Sect. 3.2, favouring lightweight

---

[1] https://ugracing.co.uk/.

approaches which do not require, for example, visual feature matching. Therefore, only the landmark-based approaches GraphSLAM, FastSLAM and EKF-SLAM are explored in this paper.

The three chosen SLAM approaches employ different concepts to fuse measurements from sensors and optimise the accuracy of their predictions. GraphSLAM creates a connected graph consisting of two nodes: one representing car poses based on GPS measurements and one representing landmark positions. Car pose nodes are connected using the dead reckoned trajectory in the period between two GPS measurements. The created graph is optimised regularly by running the Levenberg-Marquardt least squared-based optimisation algorithm [15]. In comparison, EKF-SLAM employs an Extended Kalman Filter [10] to calculate weighted means of disagreeing sensor measurements based on uncertainty estimates provided by the sensors. Finally, FastSLAM uses a particle-filter approach where each landmark is assigned its individual EKF and particles are continuously resampled based on their congruence with new measurements.

Due to their differing underlying concepts, the three algorithms provide varying advantages and disadvantages with respect to their localisation and mapping accuracy as well as their computational efficiency. GraphSLAM's optimisation step produces highly accurate results but requires large computational resources, as reported in [12]. Furthermore, the optimisation step achieves its highest accuracy when the car reaches a point it has already seen, a process known as loop closure [13]. In the context of FS-AI, this happens for the first time at the end of the first lap, resulting in a large delay between the initiation point of the map and the first occurrence of loop closure.

EKF-SLAM's accuracy is not dependent on an optimisation step (as opposed to GraphSLAM). Still, it remains approximately constant throughout the lap due to the continuous fusion of previous and new sensor measurements. Nevertheless, EKF-SLAM's accuracy is highly dependent on the estimated noise levels. If estimates of noise in sensor data are imprecise, this can significantly impair the effectiveness of the Kalman Filter. This problem is addressed in FastSLAM, whose particle filter approach makes it less vulnerable to imprecise noise estimates, leading to more accurate outputs in less predictable environments. However, FastSLAM's particle filter imposes high computational demands as numerous estimates of the car pose and landmark positions must be updated and evaluated simultaneously.

3 Materials and Methods

3.1 Software Stack

The implemented SLAM algorithms were integrated with UGRacing's autonomous racing software. For completeness, a brief description of the software stack is described below, while Fig. 3 shows a high-level overview of this software stack. The perception module uses YOLO [21] and OpenCV to extract cone positions and colours from stereo images (Fig. 2(left)). Additionally, a combination of noise removal [9] and K-D tree-based clustering [20] is used to localise cones

Fig. 2. Left: Left Camera Image from the ZED2 mounted on the ADS-DV. Right: Example of a LiDAR Point Cloud from the Dataset captured at the 2023 FS-AI Competition.

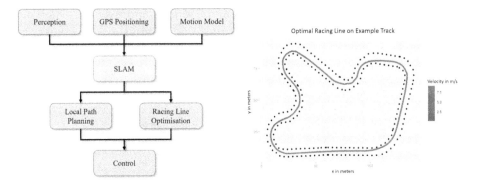

Fig. 3. Left: High-Level Control Flow of Autonomous System. Right: Example Output of the Racing Line Optimisation

in LiDAR point clouds (Fig. 2(right)) to increase the accuracy and robustness of the vision system. Simultaneously, wheel speed and steering angle measurements approximate the car's trajectory. GPS measurements are converted to a local Cartesian coordinate system to estimate the car's position relative to its starting point. SLAM takes the resulting cone positions and colours along with the absolute and relative information about the car's position and uses these to create a map of the race track and determine the car's pose within the map. Positions of cones close to the car are used immediately to calculate a path for the car using a cost function. Additionally, the map created by SLAM is used to compute a racing line, taking advantage of the knowledge of the complete track layout. The racing line is optimised using a neuroevolutionary approach, NEAT [25], and an example of this optimisation can be found in Fig. 3. Finally, control outputs are calculated using a PID controller and sent to the car's control unit. The autonomous software has been built using ROS2 [14], Python and C++.

3.2 Hardware Setup

The vehicle which was used to collect the Formula Student dataset in this paper was the *Autonomous Driving Systems – Dedicated Vehicle* (ADS-DV) [18] provided by the Institute of Mechanical Engineers (IMechE). The ADS-DV has

Fig. 4. ADS-DV collecting Data at the FS-AI Competition

an InCarPC CQ67G computer comprising an Intel i7 processor, an NVIDIA GeForce GTX 1050Ti GPU, and 16GB RAM. Additionally, the ADS-DV carries an Inertial Measurement Unit (IMU) and a GPS sensor [18]. The vehicle is the standard platform used by most FS-AI teams in the UK and, hence, the data collected by it is particularly representative of the data FS-AI teams have at their disposal during the competition. The SLAM algorithms were tested offline using the collected data as inputs. This experiment was conducted on a Fourth Generation Microsoft Surface Laptop with an Intel Core i5 CPU, 8 GB RAM, and an AMD Ryzen 5 4680U onboard graphics card. The underlying operating system was Ubuntu 22.04 LTS.

3.3 Ground Truth Reconstruction and Dataset Acquisition

The dataset used for this paper comprises one lap (approx. 250 m) around a previously unknown track, completed using UGRacing's autonomous driving software at the FS-AI competition in July 2023. It includes the stereo camera images, LiDAR point clouds, and all-wheel speed and steering angle information collected during the lap. Furthermore, the cone positions and colours, as estimated by the vision system, have been logged and added to the dataset. Each measurement has a timestamp. To evaluate the performance of the SLAM algorithms, ground truth car poses and cone positions were reconstructed after the test run by using the iterative closest point approach KISS-ICP [27]. KISS-ICP deskews and downsizes consecutive LiDAR scans and then iteratively estimates the transformation between them. KISS-ICP was chosen as it provides accurate results with minimal parameter tuning and does not rely on a vehicle model. Due to the high accuracy of LiDAR measurements and the volume of information available in the point clouds, ground truths were reconstructed based on LiDAR data only. Car poses were estimated using pairs of consecutive LiDAR scans, limiting the frequency of ground truth car pose estimates to 10 Hz. Figure 4 shows the ADS-DV during the lap from which the dataset was obtained.

3.4 GPS Noise Modelling

While UGRacing has collected camera, LiDAR, wheel speed and steering angle measurements, GPS measurements are missing in the dataset and have been simulated from the estimated ground truth car poses by adding artificial noise. Noise in GPS data is complex and influenced by many factors, such as satellite positions, obstructing buildings and atmospheric effects causing signal delay. Since these conditions are difficult to predict, we have adopted a simplified model, tested under varying conditions and proven to provide accurate estimates of the measurement noise, which can be adjusted to our case using a cost parameter [6]. According to [6], the Gaussian normal function can approximate the distribution probability,

$$P(x) = \frac{1}{2\pi\sigma} e^{-\frac{(x-\sigma)^2}{2\sigma^2}}; \sigma = 0.787c + 2.192 \quad (1)$$

where c is a cost parameter. The constants used in this function result from the experiment detailed in [6] and adjust the normal distribution to optimise GPS noise. Tests with UGRacing's Ellipse-2N IMU and GPS sensor [23] indicate an average error of 0.91 m with a standard deviation of 0.85 m. The aforementioned paper suggests a cost parameter of -1.7 in this case and estimates that the resulting GPS noise is accurate up to between 10 and 30 cm. This is a sufficient approximation for our purpose as the exact noise levels are difficult to predict, and discrepancies of the given magnitude (i.e. 10 cm to 30 cm) will likely balance out throughout the lap.

4 Experiments

A set of quantitative metrics was required to evaluate the performances of the SLAM algorithms and identify their advantages and disadvantages. The aim is to measure the performance on three levels: the map's accuracy, car pose estimates and computational efficiency. [12] has suggested a set of metrics, from which the following were selected: *Landmark Mean Error, Percentage of Landmark Measurements above an Error Threshold of 1.5 m, Update Rate* and *RAM Usage*. These summarise the quality of landmark positioning and the algorithms' efficiencies. We refer the reader to [12] for their definitions. Since the proposed metrics only evaluate the computational efficiency and map accuracy, we additionally measured the accuracy of car pose estimates and included the standard deviations of all accuracy measurements to capture the spread of the error levels.

GraphSLAM, EKF-SLAM and FastSLAM have been tested independently on the real-world dataset described in Sect. 3.3. FastSLAM has been run using 50 particles and initial test runs with the algorithm indicate that a higher particle count would exceed the capacity of the limited resources on the car's onboard PC. After the algorithms terminated, the estimated car poses and landmark positions were collected and compared to the ground truths.

4.1 Car Pose Accuracy

As Fig. 5(left) shows, EKF-SLAM produced the lowest mean error in car position estimates at 0.94 m with a standard deviation of 0.77 m, indicating that approximately 68% of the estimates had an error lower than 1.71 m. GraphSLAM produced slightly less accurate results, with an average error of 1.24 m and a standard deviation of 0.82 m. Meanwhile, FastSLAM's estimates were significantly less precise, with an average error of 2.70 m with a standard deviation of 1.81 m. The high quality of EKF-SLAM's estimates could be the result of an accurate sensor noise model, optimising the weighting of measurements in accordance to their reliability as described in Sect. 2. This has been achieved through testing with the sensors prior to the FS-AI competition and takes advantage of the predictability of sensor noise at FS-AI. GraphSLAM's high accuracy can be attributed to the effectiveness of the graph optimisation algorithm as described in Sect. 2. However, the algorithm appears to suffer from the lack of loop closure occurrences. FastSLAM's comparatively low accuracy indicates that more particles would be needed to increase precision, leading to higher computational demands.

Figure 5(right) observes similar trends for the algorithms' estimates of the car's heading. While FastSLAM produced the least accurate results with a mean error of 4.12°, GraphSLAM and EKF-SLAM's predictions were more precise, with mean errors of 1.81 and 1.22°, respectively. Notably, EKF-SLAM's estimates carry the highest standard deviation of 7.91°, implying that while a large portion of estimates are precise, a significant number of heading estimates also carry a large error, even exceeding that of FastSLAM. Upon further investigation, EKF-SLAM heavily miscalculated the car's heading at a particular turn, leading to a disproportionate increase in error at this part of the track. The danger of this type of outliers resulting from GraphSLAM is lower as GraphSLAM connects all poses and optimises them simultaneously, keeping the error low consistently. Nevertheless, further optimisations applied to the EKF-SLAM implementation might also mitigate these outliers such that EKF-SLAM achieves heading estimates that are consistently precise such as those of GraphSLAM.

4.2 Landmark Accuracy

Figure 6 shows the errors in landmark position estimates and the percentage of landmarks for which estimates exceeded the error threshold of 1.5 m, indicating that they will notably impact the car's trajectory. In contrast to the car pose estimates, the landmark estimates show a similar error magnitude between the three algorithms, where EKF-SLAM's estimates are the most precise, followed by FastSLAM and GraphSLAM with mean errors of 1.03 m, 1.15 m and 1.19 m, respectively. The standard deviations of the errors lie at 0.47 m for FastSLAM and EKF-SLAM and 0.46 m for GraphSLAM. As expected, EKF-SLAM's and GraphSLAM's landmark position estimates observe a slightly higher error than their car position estimates as imprecisions in the localisation of the car are transferred onto the landmarks. In particular, GraphSLAM's landmark estimates were notably less precise than those of EKF-SLAM. We hypothesize that this is a direct consequence of GraphSLAM's car pose estimates which show a higher error than those of EKF-SLAM, as described above. In contrast, FastSLAM achieves a higher accuracy on its landmark position estimates than its car position estimates. This is potentially due to FastSLAM's particle-filter nature, which offers increased flexibility to correct faulty landmark estimates despite errors in the car pose. EKF-SLAM produces the lowest number of landmarks above the error threshold at 20%. In comparison, 28.7% of FastSLAM's and 30.4% of GraphSLAM's landmark position estimates exceed the threshold. These landmarks are particularly problematic as their error is significant enough to cause the car to leave the track. Hence, this statistic implies that EKF-SLAM will be the most reliable algorithm with regard to mapping out the circuit and high errors can be compensated by other components in the software stack.

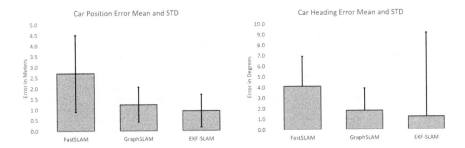

Fig. 5. Left: Error in Car Position Estimates. Right: Error in Car Heading Estimates

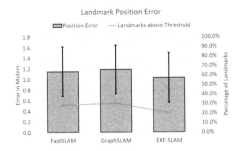

Fig. 6. Error in Landmark Position Estimates

4.3 Computational Efficiency

Figure 7(left) showcases the update rates of the three algorithms. Car pose updates based on GPS measurements and dead reckoning happen almost instantaneously and are therefore not considered in this metric. Instead, the overall runtime of the algorithm is measured and divided by the number of updates from the vision subsystem. To process measurements in real-time, SLAM's rate needs to be higher than that of the vision system, i.e. higher than 30 Hz in this paper. As Fig. 7(left) shows, GraphSLAM achieved the highest update rate at

127.68 Hz, followed by EKF-SLAM with 50.45 Hz. The high rate of both algorithms can be attributed to the comparatively small number of landmarks on the map, as an FS-AI track rarely contains more than 250 cones. Contrary to expectations, GraphSLAM was faster than EKF-SLAM, potentially because its optimisation step was not carried out often due to the absence of loop closures. In comparison, FastSLAM achieved a rate of 31.32Hz, marginally exceeding the speed of the vision system. Considering that FastSLAM would be expected to run in parallel with the remainder of the autonomous system onboard the car, its rate might drop further, falling below the minimum rate requirement.

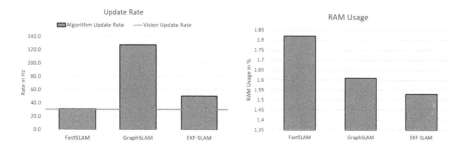

Fig. 7. Update Rates (left) and RAM Usage (right) for FastSLAM, GraphSLAM and EKF-SLAM

Figure 7(right) shows the amount of RAM taken up by the algorithms. As can be seen, EKF-SLAM performed best in terms of RAM usage, however, by a small absolute difference compared to the other two algorithms. While memory efficiency is critical given the limited resources onboard the car, the performance difference of the three algorithms with respect to this metric can be seen as negligible, as they are all within 0.3% points of each other. Furthermore, a RAM usage of less than 2% can be classified as highly memory-efficient compared to the other components of the software, considering the SLAM's central role in the autonomous racing software.

5 Conclusions

This paper compared FastSLAM, EKF-SLAM and GraphSLAM in the Formula Student AI autonomous racing competition context. These three SLAM algorithms have been chosen as they have been deemed particularly suitable to FS-AI due to being landmark-based and not relying on prominent visual features which are sparse on an FS-AI race track. In order to predict the algorithms' performances at Formula Student as accurately as possible, they have been tested on a dataset collected by the University of Glasgow's FS-AI team at the 2023 competition using the team's autonomous racing software. The algorithms have been evaluated with respect to their accuracy in estimating car poses and track boundary positions, as well as their computational efficiency.

As discussed in Sect. 4, EKF-SLAM predicted the car poses most accurately with an average positional error of 0.94 m, an average heading error of 1.22° and a notably high error spread in the heading component. GraphSLAM produced slightly less accurate estimates in terms of the car position. FastSLAM's car pose results were significantly less precise, with a mean error more than twice as large as that of GraphSLAM. When estimating the track boundary positions, the performance of the algorithms was more consistent. EKF-SLAM again produced the most accurate results with a mean error of 1.03 metres, closely followed by FastSLAM and GraphSLAM with mean errors of 1.15 and 1.19 metres, respectively. Using EKF-SLAM, only 20% of landmark position errors exceeded the significance threshold of 1.5 metres, compared to 28.7% for FastSLAM and 30.4% for GraphSLAM, indicating that EKF-SLAM's estimates were most reliable. When evaluating the algorithms' update rates, GraphSLAM was by far the fastest with an update rate of 127 Hz, followed by EKF-SLAM with 50 Hz. While both of these figures are competitive at FS-AI, FastSLAM's update rate of 31 Hz lies only marginally above the minimum required rate of 30 Hz and is at risk of falling below this threshold when run alongside the entire autonomous racing software at the competition. Based on these results, we believe EKF-SLAM will be the highest-performing algorithm and consider GraphSLAM a competitive alternative for FS-AI. Future work consists of including real GPS measurements in the input data and to run SLAM onboard the car during testing sessions and at the 2024 FS-AI competition.

References

1. Formula student – institution of mechanical engineers (2023). https://www.imeche.org/events/formula-student
2. Abu dhabi autonomous racing league (2024). https://a2rl.io/
3. Abaspur Kazerouni, I., Fitzgerald, L., Dooly, G., Toal, D.: A survey of state-of-the-art on visual slam. Expert Syst. Appl. **205** (2022)
4. Balaji, B., et al.: Deepracer: autonomous racing platform for experimentation with sim2real reinforcement learning. In: 2020 IEEE International Conference on Robotics and Automation (ICRA), pp. 2746–2754 (2020). https://api.semanticscholar.org/CorpusID:219683891
5. Davison: Real-time simultaneous localisation and mapping with a single camera. In: Proceedings Ninth IEEE International Conference on Computer Vision, vol. 2, pp. 1403–1410 (2003)
6. El Abbous, A., Samanta, N.: A modeling of GPS error distributions. In: 2017 European Navigation Conference (ENC), pp. 119–127 (2017). https://doi.org/10.1109/EURONAV.2017.7954200
7. Gao, B., Lang, H., Ren, J.: Stereo visual slam for autonomous vehicles: a review. In: 2020 IEEE International Conference on Systems, Man, and Cybernetics (SMC), pp. 1316–1322 (2020)
8. Grisetti, G., Kümmerle, R., Stachniss, C., Burgard, W.: A tutorial on graph-based slam. IEEE Intell. Transp. Syst. Mag. **2**(4), 31–43 (2010)
9. Himmelsbach, M., Hundelshausen, F.v., Wuensche, H.J.: Fast segmentation of 3D point clouds for ground vehicles. In: 2010 IEEE Intelligent Vehicles Symposium, pp. 560–565 (2010)

10. Julier, S., Uhlmann, J.: A new extension of the Kalman filter to nonlinear systems. In: Proceedings of AeroSense: The 11th International Symposium on Aerospace/Defense Sensing, Simulations and Controls (1997)
11. Kabzan, J., et al.: AMZ driverless: the full autonomous racing system. arXiv:1905.05150 (2019)
12. Large, N., Bieder, F., Lauer, M.: Comparison of different slam approaches for a driverless race car. tm - Technisches Messen **88** (2021)
13. Latif, Y., Cadena, C., Neira, J.: Robust loop closing over time for pose graph slam. The Int. J. Robot. Res. **32**(14), 1611–1626 (2013). https://doi.org/10.1177/0278364913498910
14. Macenski, S., Foote, T., Gerkey, B., Lalancette, C., Woodall, W.: Robot operating system 2: design, architecture, and uses in the wild. Sci. Robot. **7**(66), eabm6074 (2022). https://doi.org/10.1126/scirobotics.abm6074
15. Marquardt, D.W.: An algorithm for least-squares estimation of nonlinear parameters. J. Soc. Ind. Appl. Math. **11**(2), 431–441 (1963). https://doi.org/10.1137/0111030
16. Montemerlo, M., Thrun, S., Koller, D., Wegbreit, B.: FastSLAM: a factored solution to the simultaneous localization and mapping problem. In: Eighteenth National Conference on Artificial Intelligence (2002)
17. Mur-Artal, R., Montiel, J., Tardos, J.: Orb-slam: a versatile and accurate monocular slam system. IEEE Trans. Robot. **31**, 1147 – 1163 (2015). https://doi.org/10.1109/TRO.2015.2463671
18. Murphy, I.: FS-AI ADS-DV AI base compute and sensors hardware information (2020). https://github.com/FS-AI/FS-AI_Compute/issues/1
19. Nekkah, S., et al.: The autonomous racing software stack of the KIT19d. arXiv:2010.02828 (2020)
20. Panigrahy, R.: An improved algorithm finding nearest neighbor using KD-trees. In: LATIN 2008: Theoretical Informatics, pp. 387–398 (2008)
21. Redmon, J., Divvala, S., Girshick, R., Farhadi, A.: You only look once: unified, real-time object detection. In: Proceedings of the IEEE Conference on Computer Vision and Pattern Recognition, pp. 779–788 (2016)
22. Rublee, E., Rabaud, V., Konolige, K., Bradski, G.: Orb: an efficient alternative to sift or surf. In: International Conference on Computer Vision, pp. 2564–2571 (2011)
23. SBG-Systems: Minature high-performance inertial sensors (2023). https://www.sbg-systems.com/wp-content/uploads/Ellipse_Series_Leaflet.pdf
24. Schlegel, D., Colosi, M., Grisetti, G.: ProSLAM: graph slam from a programmer's perspective. In: 2018 IEEE International Conference on Robotics and Automation (ICRA), pp. 3833–3840 (2018)
25. Stanley, K.O., Miikkulainen, R.: Evolving neural networks through augmenting topologies. Evol. Comput. **10**(2), 99–127 (2002)
26. Thrun, S., Burgard, W., Fox, D.: Probabilistic Robotics. MIT Press (2005)
27. Vizzo, I., Guadagnino, T., Mersch, B., Wiesmann, L., Behley, J., Stachniss, C.: Kiss-ICP: in defense of point-to-point ICP simple, accurate, and robust registration if done the right way. IEEE Robot. Autom. Lett. **8**, 1029–1036 (2023). https://doi.org/10.1109/LRA.2023.3236571

Comparative Analysis of Unity and Gazebo Simulators for Digital Twins of Robotic Tomato Harvesting Scenarios

Juan Pablo Espejel Flores[1,2], Abdurrahman Yilmaz[1(✉)], Luis Arturo Soriano Avendaño[2], and Grzegorz Cielniak[1]

[1] University of Lincoln, Lincoln, UK
28658568@students.lincoln.ac.uk, {ayilmaz,gcielniak}@lincoln.ac.uk
[2] Universidad Autónoma Chapingo, Texcoco, Mexico
lsorianoa@chapingo.mx

Abstract. Robotic simulators play a crucial role in agricultural robotics research, enabling preliminary validations, digital twin experiments, and AI-based sim-to-real methodologies. However, creating realistic farming environments in simulators is challenging and often requires significant manual effort. This paper addresses this challenge by proposing a semi-automated method for creating realistic tomato farm environments compatible with two popular simulators Gazebo and Unity. We conduct a comparative analysis of these simulators in terms of visual fidelity and integration effort, with a focus on their compatibility with the ROS 2 framework. Our contributions include the development of a realistic and parameterisable digital twin of a tomato farm, a detailed comparative analysis of Gazebo and Unity frameworks, and the publication of open-source code integrated into ROS 2. Through our study, we identify limitations in current digital twins, such as their static nature and limited realism, and suggest future enhancements, including the integration of physics for improved fidelity and the deployment of robots for practical simulated tasks. Our findings provide valuable insights for researchers and practitioners in the field of agricultural robotics.

Keywords: ROS 2 · Gazebo · Unity · Agriculture · Field Robots · Tomato Glasshouse

1 Introduction

Robotic simulators have been gaining increasing significance recently due to their utility in conducting preliminary validations before transitioning to real-world applications [15], facilitating digital twin experiments [3,8], and serving as training grounds for artificial intelligence-based sim-to-real methodologies [7]. Consequently, the frequency of their use, the diversity of available simulators,

---

Work supported by the InnovateUK-funded project Agri-OpenCore #10041179.

and the size of the user base opting for simulation tools continue to increase steadily [10]. For agricultural robotics, developing virtual scenarios that realistically replicate the variability of conditions in modern farms is very challenging. Such simulations require significant manual effort in creating realistic key components of the world such as the field robot, crop plants, and the crop growing infrastructure. Additionally, these key components are highly unique to specific farm and crop types. Therefore, there is a growing need for automating the creation of digital twins in this domain.

Currently, there are several robotic simulators available of varying characteristics, which have been deployed and evaluated in various robotic scenarios. One of the most popular choices is Gazebo, a default option for ROS users for many years, especially those interested in mobile robotic applications [2,13]. Other options include Unity, a game engine renowned for its realistic scene rendering and advanced character animations, which has recently expanded its focus to include robotics applications [9], and NVidia's Isaac Sim and Gym, which offer efficient graphics and physics processing tailored for specialist GPU hardware [5]. The use of these simulators for agricultural robotics scenarios, however, has been relatively unexplored [6].

This paper proposes a semi-automated method for creating a realistic farming environment compatible with the two popular robotic simulators, Gazebo and Unity, along with their subsequent evaluation targeting the simulator capability, realism, and ease of integration into the robotic ecosystem (i.e. ROS 2). The key contributions of the paper include:

- Development of a realistic and parametrisable digital twin of a tomato glasshouse farm for robotics research;
- Comparative analysis and evaluation of the Unity and Gazebo simulators for the proposed scenario;
- Publication of the open source code as reusable components integrated into ROS 2 framework[1].

2 Related Work

A digital twin is a virtual environment mirroring the appearance and behaviour of a real-world scenario [11]. In the smart farming domain, agricultural experts, researchers, and farmers can utilise digital twins to simulate various scenarios, test different strategies, and predict future outcomes [8]. A particular use of digital twins for simulating agricultural robotic fleets and testing a multi-agent framework has been presented in [3]. The paper describes practical considerations for integrating ROS with Gazebo and Unity simulators and highlights Unity's high visual fidelity together with simple and easily scalable features, whilst contrasting Gazebo's extensive support and documentation, minimal resource consumption, and seamless integration with ROS which improve development workflows.

---

[1] https://github.com/LCAS/aoc_tomato_farm.

The integration of Unity with ROS has been addressed by several authors previously. In [2], Unity is compared against other robotic frameworks such as Gazebo and V-REP for their suitability in robotic simulations. The studies concluded that, although feasible, the integration of different robot models in Unity lacks features and options when compared to Gazebo and V-Rep. A similar sentiment is expressed in [9], in which several practical robotic scenarios involving the integration of ROS with Gazebo and Unity have been undertaken. The study highlights Unity's superior performance in large and dynamic scenarios, in which realistic shadows could significantly affect the robot's performance, whilst Gazebo excelled in smaller and static environments. The strengths and weaknesses of both simulation frameworks for integration with ROS have been assessed also in [14] reinforcing claims about Unity's graphical quality and Gazebo's ease of integration. Moreover, the integration of ROS with Unity has been explored in intelligent manufacturing demonstrating a digital twin of a robotic manufacturing station [4].

The majority of state-of-the-art techniques consider manually created assets and their arrangements which require significant skills and resources and limit the potential for re-deployment in scenarios of a different scale. In contrast, our work presents a method for semi-automated digital asset creation and configuration of realistic digital twin environments for a unique environment of tomato glasshouses.

3 Tomato Glasshouse Simulation Model

In this section, the key assets, general infrastructure, workspace, and capabilities of the proposed tomato glasshouse farm environments are explained in detail together with implementation details in the Unity and Gazebo simulators.

3.1 Glasshouse Components

The development of our tomato glasshouse models was informed by a dedicated data collection session undertaken during a visit to one of the real tomato glasshouse farms in the UK in Feb 2024. The data collected include the visual and structural details of tomato plants, their spatial layout and key infrastructure components of the glasshouse. The actual metric measurements were accompanied by visual references in the form of images which helped the asset creation and to inform design and parameter choices (see Figs. 2 and 4 for reference).

The following key components were identified as essential for the realistic appearance of a glasshouse environment: tomato plants, plant pots, soil beds, lamps, metal frames and glasshouse outer glazing structures (see Fig. 1 for visual reference). Whilst the real environments include many more of the finer details, limiting the core to six components enables the storage efficiency and re-useability of the implementation. All models were designed in Blender, an open-source 3D modelling software, using the visual references and measurements collected from the real farm.

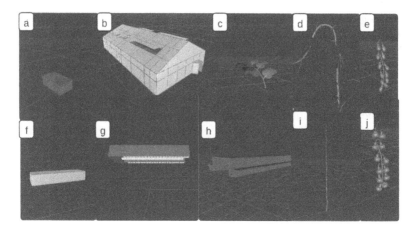

Fig. 1. Key tomato glasshouse components: a) plant pot, b) glasshouse outer glazing structure, c) plant leaf, d) branch type I, e) tomato type I, f) soil bed, g) industrial lamp, h) metal frames, i) branch type II, j) tomato type II.

The tomato plant models are configurable and allow for a parameterised generation of plants with a desired number of tomato fruit, leaves, and branches. The plant generator creates a final model which can then be incorporated as an asset into the glasshouse generator. Example plant models can be seen in Fig. 4.

3.2 Farm Layout

In addition to customizable tomato plants, our digital twin features an automated layout generation enabling scalability and easy customisation. A glasshouse can feature a different number of rows and plants per row with customisable spacing between the rows and plants. A single glasshouse can then be replicated into a larger, rectangular farm by specifying the number of rows and glasshouses per row.

The spatial arrangement of key assets such as soil containers, lamps, and metal frames is adjusted automatically based on the spatial layout of tomato plants. Figure 2 illustrates a real tomato glasshouse farm for visual reference together with a snapshot of the simulated environments in Unity and Gazebo.

3.3 Software Implementation

To facilitate the generation and simulation of glasshouse environments, three distinct software packages were developed: two glasshouse generators creating key file structures for Unity and Gazebo simulators and a platform-independent glasshouse simulator package integrating both simulation frameworks with ROS. The Unity generator is developed in C# whilst the Gazebo generator is developed in Python. The common simulator package contains a parameterisable ROS launch file enabling environment customisation for both frameworks. An illustrative example of the framework in a particular scenario is depicted in Fig. 3.

Fig. 2. A tomato glasshouse farm: a real tomato farm (a) and corresponding digital twin implemented in Unity (b) and Gazebo (c).

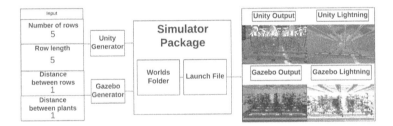

Fig. 3. A digital twin of a single tomato glasshouse in a 5 plants per row and 5 rows arrangement.

Both generator packages allow the generation of lighting conditions. To accomplish this, a light point with a range of 10 m is linked to each displayed industrial lamp model. The brightness of this light can be adjusted according to the input parameters selected by the user. The output of the lighting feature can be seen in Fig. 3.

3.4 Complexity of the Models

Table 1 summarises the complexity of the key tomato glasshouse components as presented in Fig. 1. These data were collected by displaying the Collada models through the Blender software (faces, vertices, edges, triangles) and Unity (size). The placed size represents the dimensions of the object's bounding box, with W,

H, and D representing the width, height, and depth of the box, respectively. The number of faces, vertices, edges, and triangles in a model significantly impacts the simulation performance and memory usage in both Unity and Gazebo. Both rendering engines need to process each vertex and face to generate the visuals on the screen. More faces and vertices increase the computational workload, potentially leading to lower frame rates. Additionally, complex models with a high number of these elements consume more memory during simulation, emphasizing the importance of efficient memory management in both simulators to maintain performance, particularly with large and detailed models.

4 Evaluation Criteria

In this section, we present the evaluation scenarios and criteria used for comparisons of the two simulation environments. Both digital twins are analysed in terms of realism and visual fidelity, ease of integration with ROS 2, customization extensibility, and computational efficiency.

Table 1. The complexity of the models employed in the digital environment

Model	Number of				Size in m (W × H × D)
	Faces	Vertices	Edges	Triangles	
Pot	13	20	30	30	0.113×0.07×0.04
Glasshouse Structure	13,711	13,107	27,453	27,124	21.6×6×11.4
Soil Bed	1,350	1,352	2,700	2,700	0.28×0.18×1.0
Light	8,730	11,216	19,768	17,468	57.3×30.5×203
Frames	252	128	378	252	81×11.7×33.7
Leaf	741	496	1,224	741	0.36×0.12×0.3
Branch I	14,460	7,976	22,409	14,460	1×1×0.48
Branch II	575	576	1,150	1,127	0.024×1.3×0.036
Tomato I	19,247	19,238	38,447	38,158	0.14×0.36×0.07
Tomato II	19,247	19,238	38,447	38,158	0.13×0.303×0.078

Visual fidelity directly affects the realism of simulations which are critical for robotics applications employing computer vision techniques. To evaluate this aspect, we collect screenshot images from both simulators and subsequently make direct visual comparisons between the simulators and against the visual reference from the real tomato farm. We employ a comparison protocol similar to the one used for the comparisons between the Unity and Unreal game engines in [12].

The **ease of integration with ROS 2** is a key consideration for employing the digital twins in robotics applications. There is direct support for ROS integration in both Gazebo and Unity, and in our comparisons, we highlight the key differences, strengths, and limitations of both platforms in this aspect.

We refer to **customization and extensibility** as the capability provided by the frameworks to add new models or functionalities, as well as to modify the existing features. In particular, we analyse the entire development process from generating the models with particular modelling software to integrating them into virtual environments. In our case, we rely on models in the Collada format (i.e., .dae) which can be exported from the Blender modelling software.

To assess the **computational efficiency** of simulations, three tomato farm scenarios of different complexity measurements were generated and used for collecting the performance measures including the model loading time before the simulation is ready, frame rate (FPS), processor and memory usage (in percentage values compared to the full physical capabilities of the hardware). The considered three scenarios include:

- **small-size farm:** a single glasshouse with 4 rows of 3 plants per row. The distance between plants is 4 m and between rows 6 m;
- **medium-size farm:** a single row of 3 glasshouses each featuring 8 rows of 4 plants per row. The distance between plants is 2 m and between rows 2 m;
- **large-size farm:** 3 rows of 2 glasshouses in each row with 6 units in total each featuring 6 rows of 7 plants per row. The distance between plants is 1 m and between rows is 1 m.

In scenarios, plants typically have between 12 and 20 leaves and between 2 and 8 branches. Consequently, there can be between 24 and 96 tomatoes per plant, as each branch supports 12 tomatoes. Considering all components in the simulation field, the overall complexity of the scenarios can be approximated in terms of the complexity of the models shown in Sec. 3.4. The small-size farm has more than 1.6 million faces and vertices, 3.3 million edges, and roughly 3.2 million triangles. The medium-size farm has approximately 37.8 million faces, 36.9 million vertices, 74.5 million edges, and around 71.5 million triangles. On the other hand, the large-size farm has roughly 195.4 million faces, 190 million vertices, approximately 384.6 million edges, and about 369.7 million triangles.

All experiments were undertaken with the following setup: Gazebo (v. 11.10.2) and Unity (ver. 2022.3.17.f1) running on a PC with Ubuntu 22.04.4 OS and ROS 2 Humble Hawksbill release. The PC was equipped with an Intel Core i7-6500U CPU (4 cores, 2 threads/core) and an Intel HD Graphics 520 graphics card.

5 Results and Discussions

In this section, we present the results of a comparative assessment of Unity and Gazebo simulators for our use case. The key points from individual assessments presented below are summarised in Table 2 demonstrating associated performance scores for each category and for each simulator.

Table 2. Summary comparison scores for Unity and Gazebo digital twins, indicating the perceived level of support for individual features, from the lowest indicated by 1 and the highest by 5.

Feature	Unity	Gazebo
Realism and visual fidelity	4	2
Integration with ROS 2	3	5
Customization and extensibility	5	3
Computational performance	3	2

5.1 Realism and Visual Fidelity

The tomato farm implemented in Unity demonstrates a comprehensive representation of the key farm components (see Figs. 4 and 5). We can observe clear colour and texture detail of the tomato models including the individual fruit and leaves. It is possible to distinguish soft orange-coloured tomatoes from those with a stronger orange hue. In addition, the materials exhibit more vibrant colours when exposed to sunlight with full intensity and the glasshouse, despite lacking transparency, stands out prominently. Moreover, the simulated shadows further increase the visual fidelity.

The colour rendering in the tomato farm generated in Gazebo is less vibrant and textures are less prominent. There is a lack of distinct definition in the tomato fruit material, which resembles a flat colour without depth. Additionally, there is no observable interaction of the material with sunlight in the glasshouse model, showing a more uniform colour. This behaviour is likely linked with the implementation of lighting in Gazebo.

Compared to real photographs, environments implemented in Unity look closer than those in Gazebo. We particularly note the improved definition of materials, providing a better representation of textures as well as the lighting effects present in the environment. Unity's capabilities in this regard are showcased effectively. As expected, the Gazebo tomato farm still accurately represents the key elements of the glasshouse, and its fidelity is acceptable if we are not concerned with high graphic quality.

5.2 Ease of Integration with ROS 2

To facilitate this aspect of the comparisons, we have integrated a custom-made ROS 2 package supporting a mobile robot, AgileX Hunter 2.0, into both simulation environments. Unity provides official support for integration with ROS[2] and the community has generated open-source alternatives as in [1]. There is partial support for robot controllers and a URDF importer, simplifying robot modelling in Unity. The URDF importer enables the re-use of the existing library of robot

---

[2] https://blog.unity.com/engine-platform/advance-your-robot-autonomy-with-ros-2-and-unity.

Fig. 4. An image depicting real tomatoes (a) together with digital models implemented in Unity (b) and Gazebo (c).

models available in numerous ROS packages already. The process of integrating a specific robot into Unity requires additional manual effort in creating dedicated components written in C#. This process is somewhat simplified by a dedicated package called ROS 2 for Unity module[3], which provides a communication layer and template scripts that need to be adjusted for a particular robot platform. To implement our farm, we needed to import the robot's URDF file and establish the connection between the simulation scenario and ROS 2 using that package.

Gazebo is free simulation software (Apache 2.0 licence) that supports many types of robots and, by default, is released together with ROS 2, enabling seamless integration. There is a vast amount of documentation and additional projects demonstrating the integration of these two frameworks in numerous use cases, which is very helpful as reference material. The implementation of our glasshouse in Gazebo required only a single package containing the robot's information (model, controls, programs), and then a simple launch file which spawns the robot in the environment.

Developing ROS-based robotic simulation projects in Gazebo is easier thanks to its seamless integration and excellent support. In contrast, Unity still requires a significant manual effort in integrating simulated environments, even with the use of additional software packages. Furthermore, the Gazebo user community

---

[3] https://github.com/RobotecAI/ros2-for-unity.

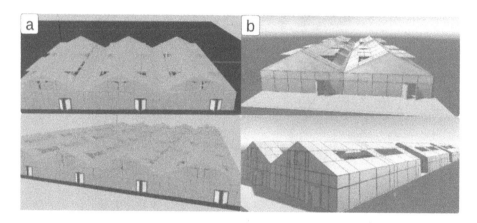

Fig. 5. Glasshouse models implemented in Gazebo (a) and Unity (b).

is larger and more active when compared to the robotic community using Unity, which often results in developing particular functionality or models from scratch without useful reference projects and support.

5.3 Customization and Extensibility

Figure 6 shows the process of generating asset models from a Collada mesh file into Unity and Gazebo. In Unity, the external asset models need to be imported into the asset folder of the project. Once the model resides in the asset directory, one can modify it through code or the scene view. The editor provides real-time visual feedback during the adjustments of the model, allowing modifications without the need for coding expertise, although it is possible to create and place the assets programmatically if required.

In Gazebo, a separate folder needs to be created within the package to store the models. Each model directory should contain a mesh folder to save the Collada files, a structured data file (.SDF) necessary for displaying the models in Gazebo, a configuration file (.config) enabling compatibility with our Jupyter Notebook file, which generates the entire digital environment, and a materials folder where are saved the materials that our model is using. The characteristics and parameters of the model can only be set programmatically without instant visual feedback. The simulator, however, allows for visualising changes in the environment by reloading the entire world file.

The tomato plant models for Gazebo were generated using a specialised Blender script, which is automatically invoked by the Jupyter Notebook file, generating custom tomato plants based on farm specifications. While Gazebo could potentially generate the plants independently, the development process would be more extensive due to the challenges of model management within the environment, and therefore, usually, third-party tools such as Blender are being used. In contrast, all tomato plant generation in Unity was solely accomplished within the simulator without the need for additional tools. Implementing the plant models in Unity was simpler than in Blender, offering better model management. Despite using different software pipelines, the process for generating the plants remained the same in both simulators.

The Unity asset editor offers a greater variety of elements and functionality compared to Gazebo, and visual feedback proves to be a valuable asset for development. Processes that require multiple lines of code in the Gazebo generator can be replicated with just a single press of the button in Unity's GUI. Unity's advantage becomes evident when calibrating parameters of virtual environments, particularly lighting. Visual feedback is essential for inspection, necessitating the generation of a new world in each iteration. Development time in Gazebo was considerably longer due to the absence of an interface for such tasks.

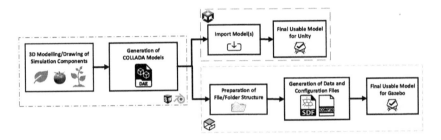

Fig. 6. Asset management in Unity and Gazebo with the aid of design tools.

5.4 Computational Efficiency

Table 3 presents the computational performance when using both simulators with the three farm models representing different complexity scales. The results represent average values from 100-second-long scenarios when the user performed a predefined virtual visit, looking at different directions of the virtual environment. Unity showed nearly constant loading times across all scenarios, while Gazebo's

Table 3. Computational performance comparison

Scenario	Simulator	Loading Time [s]	FPS #	CPU Usage [%]	Memory Usage [%]
Small-size	Unity	7	54.0	210	1.6
	Gazebo	36	45.1	145.4	6.9
Medium-size	Unity	7	28.625	220	2.1
	Gazebo	160	18.82	217.23	17.4
Large-size	Unity	7	14.8	235.4	3.1
	Gazebo	417	2.83	280.33	34.7

loading times increased linearly. FPS decreased linearly in both engines as scenario sizes increased, with Unity having a gentler decline than Gazebo. CPU usage in Unity rose moderately, whereas Gazebo's CPU usage increased more sharply. Memory usage remained constant in Unity but increased linearly in Gazebo. The results indicate the Unity model's superior performance in most metrics. Gazebo simulation excelled only in the CPU usage category. When exploring larger-scale scenarios beyond the scope of the large-size farm scenario, no significant issues arise in Unity. However, executing Gazebo simulations on available PC hardware is unfeasible due to Gazebo's limited capabilities, underscoring its computational inefficiency.

6 Conclusions and Future Work

This study addressed an important issue of integrating and evaluating robotic digital twins for agricultural scenarios. Firstly, we have developed a realistic and parameterisable digital twin of a tomato farm for robotics research. Through a comparative analysis of simulation platforms, we have identified the strengths and weaknesses of Unity and Gazebo in terms of visual fidelity and integration effort, providing valuable insights for researchers and practitioners. Furthermore, by publishing our open-source code as a reusable component integrated into the ROS 2 framework, we have facilitated further collaboration and innovation within the robotics community.

However, the study revealed limitations to be addressed in future research. The current digital twins generate static environments, limiting their use for tasks like harvesting. Future improvements should integrate physics for better user experience and fidelity, and refine plant models to simulate growth. Expanding comparisons to include different digital twins, such as more autonomous scenarios or other agricultural environments like strawberry polytunnels, would broaden the research scope.

Deploying robots to perform practical simulated tasks within the developed digital twins, such as crop monitoring or other agricultural operations, represents an exciting avenue for future exploration. Additionally, while Gazebo excels in robotic development, there is a need for enhancements to replicate the visual fidelity exhibited in Unity environments, including aspects such as shadow and lighting fidelity. Moreover, the integration of Unity and ROS 2 requires more manual effort compared to Gazebo, indicating a potential area for streamlining and automation in future developments. Similarly, conducting further comparisons with other simulation frameworks, such as NVidia's Isaac Sim, could provide valuable insights into their respective strengths and weaknesses.

References

1. Babaians, E., Tamiz, M., Sarfi, Y., Mogoei, A., Mehrabi, E.: ROS2Unity3D; high-performance plugin to interface ROS with Unity3D engine. In: 2018 9th Conference on Artificial Intelligence and Robotics and 2nd Asia-Pacific International Symposium, pp. 59–64. IEEE (2018). https://doi.org/10.1109/AIAR.2018.8769798

2. De Melo, M.S.P., da Silva Neto, J.G., Da Silva, P.J.L., Teixeira, J.M.X.N., Teichrieb, V.: Analysis and comparison of robotics 3D simulators. In: 2019 21st Symposium on Virtual and Augmented Reality (SVR), pp. 242–251. IEEE (2019). https://doi.org/10.1109/SVR.2019.00049
3. Gutiérrez Cejudo, J., Enguix Andrés, F., Lujak, M., Carrascosa Casamayor, C., Fernandez, A., Hernández López, L.: Towards agrirobot digital twins: agri-ro5–a multi-agent architecture for dynamic fleet simulation. Electronics **13**(1), 80 (2023). https://doi.org/10.3390/electronics13010080
4. Li, Y., Zhang, Q., Xu, H., Lim, E., Sun, J.: Virtual monitoring system for a robotic manufacturing station in intelligent manufacturing based on unity 3D and ROS. Mater. Today Proc. **70**, 24–30 (2022). https://doi.org/10.1016/j.matpr.2022.08.486
5. Makoviychuk, V., et al.: Isaac gym: high performance GPU-based physics simulation for robot learning. arXiv preprint arXiv:2108.10470 (2021). https://doi.org/10.48550/arXiv.2108.10470
6. Mansur, H., Welch, S., Dempsey, L., Flippo, D.: Importance of photo-realistic and dedicated simulator in agricultural robotics. Engineering **15**(5), 318–327 (2023). https://doi.org/10.4236/eng.2023.155025
7. Muratore, F., Ramos, F., Turk, G., Yu, W., Gienger, M., Peters, J.: Robot learning from randomized simulations: a review. Front. Robotics AI **9**, 799893 (2022). https://doi.org/10.3389/frobt.2022.799893
8. Peladarinos, N., Piromalis, D., Cheimaras, V., Tserepas, E., Munteanu, R.A., Papageorgas, P.: Enhancing smart agriculture by implementing digital twins: a comprehensive review. Sensors **23**(16), 7128 (2023). https://doi.org/10.3390/s23167128
9. Platt, J., Ricks, K.: Comparative analysis of ROS-Unity3D and ROS-gazebo for mobile ground robot simulation. J. Intell. Robot. Syst. **106**(4), 80 (2022). https://doi.org/10.1007/s10846-022-01766-2
10. Ramasubramanian, A.K., Mathew, R., Kelly, M., Hargaden, V., Papakostas, N.: Digital twin for human-robot collaboration in manufacturing: Review and outlook. Appl. Sci. **12**(10), 4811 (2022). https://doi.org/10.3390/app12104811
11. Rasheed, A., San, O., Kvamsdal, T.: Digital twin: values, challenges and enablers from a modeling perspective. IEEE Access **8**, 21980–22012 (2020). https://doi.org/10.1109/ACCESS.2020.2970143
12. Šmíd, A.: Comparison of unity and unreal engine. Ph.D. thesis, Czech Technical University in Prague (2017)
13. Tsolakis, N., Bechtsis, D., Bochtis, D.: AgROS: a robot operating system based emulation tool for agricultural robotics. Agronomy **9**(7), 403 (2019). https://doi.org/10.3390/agronomy9070403
14. Wijaya, G.D., Caesarendra, W., Petra, M.I., Królczyk, G., Glowacz, A.: Comparative study of Gazebo and unity 3D in performing a virtual pick and place of universal robot UR3 for assembly process in manufacturing. Simul. Model. Pract. Theory **132**, 102895 (2024). https://doi.org/10.1016/j.simpat.2024.102895
15. Yilmaz, A., Ervan, O., Temeltas, H., Akduman, I.: An autonomous robotic system for ground surface and subsurface imaging. In: 2022 International Conference on Engineering and Emerging Technologies (ICEET), pp. 1–6. IEEE (2022). https://doi.org/10.1109/ICEET56468.2022.10007262

Supporting Explainable Planning and Human-Aware Mission Specification for Underwater Robots

Alan Lindsay[1(✉)], Andrés A. Ramírez-Duque[1], Bart Craenen[1], Andrea Munafò[2], Laurence Boé[2], Adam Campbell[2], and Ronald P. A. Petrick[1]

[1] Heriot-Watt University, Edinburgh, UK
alan.lindsay@hw.ac.uk
[2] SeeByte, Edinburgh, UK

Abstract. The problem of effectively supporting human operators as they model missions, supervise their plans, and observe execution, combines several open questions in human-aware planning. These include how to provide support during the mission modelling and specification phase, and how to provide effective assistance during the planning phase. In this paper, we present ongoing work to create a tool for human operators, which aims to support effective interactions and improve operator understanding and awareness throughout autonomous underwater vehicle missions. We present an architecture which incorporates modelling assistance, plan generation, explanation, and plan space exploration. Our approach starts with a task-bot interface that provides modelling assistance during the mission specification step and comprehensible plan explanations during planning, while supporting the operator in exploring the plan space. In addition, we present an example walkthrough to showcase the human-machine interaction and explanation representations.

1 Introduction

The task of supporting human operators during the specification, planning, and monitoring of surveillance missions for autonomous underwater and surface robots is a challenging problem for human-machine interaction. In such missions, an operator must take into account various types of information, including topology, weather, robot capabilities, mission objectives and parameters, and safety requirements. This makes specifying the mission time-consuming and potentially error-prone. Furthermore, various competing factors must be considered when planning the mission, making it a process of trial and error.

Mission specification is time-consuming as there are various types of information that must be captured in the model, and specification typically requires carefully describing each aspect. This includes aspects such as mission objectives, launch points, and available assets, but also includes potentially dangerous areas such as sandbanks and exclusion zones. However, the various existing information sources provide rich information relating to the mission, which could be used to provide assistance in specification (e.g., exploiting existing known positions), and even in anticipating potential structures of the mission.

The mission planning process we consider uses the SeeTrack [19] and Neptune [18] software, which involves a two-phase approach. The first phase involves allocating resources by selecting behaviours to meet the objectives, arranging objectives in order of priority, and allocating assets to objectives. The second phase, called rehearsal, is similar to path planning. Here, the operator plays an active role in modifying the allocation to explore the plan space and discover alternative paths and optimal trade-offs. However, neither SeeTrack nor Neptune provide a specialized tool to support the operator's plan space understanding.

Within the Explainable Planning (XAIP) field, techniques have been developed that help the user understand generated plans [14] and support exploration of the plan space [11]. However, these techniques require the use of the planning model, and they do not consider how an operator will naturally interact with the concepts: how they will express queries, what model structures the queries will relate to, and how they will understand the generated explanations.

In this paper, we discuss the development of a system with the purpose of supporting operators through each of the stages of a mission involving underwater vehicles. The system's development so far has been guided by SeeByte, one of the project partners. SeeByte is a company specialized in command and control end-to-end mission software. With this guidance, we have developed a system prototype. Our starting point for the interaction is a task-bot, which the user can use to interact with the system during each of the three mission stages. Through the task-bot, the operator can build a mission, query the system about it, and lead the exploration through alternative plans. During execution, the operator can monitor the mission. The main contributions of this paper are a framework for supporting human-in-the-loop mission specification and planning, which aims to improve the interaction between the human operators and the planning system. The system combines and interprets several information resources in order to provide modelling assistance, reducing the amount of detail the operator needs to specify. This framework uses contrastive explanations in order to support operator understanding of the generated plans. It extends explainable planning approaches to support user queries for the underwater robot mission scenario. We also present a task-bot that combines textual and visual content in order to improve user interaction and comprehension.

2 Related Work

In automated planning, modelling has been identified as a bottleneck, due to the skills required to develop these models. This has inspired a variety of methods for supporting the authoring of domain models, including frameworks similar to Integrated Development Environments for use by software engineers [20,21], and approaches that build the planning model from ontologies [5,21]. Existing research has considered the planning problem for autonomous underwater vehicles, including problem formulation [17], search strategies [3], monitoring information opportunities during execution [4], and Human-Robot Teaming [15]. Within XAIP a growing number of approaches have used visualisations [6,12,16], and these have proven successful in creating explanations [12].

3 The Underwater Autonomous Vehicle Mission Scenario

The scenario we consider in this work revolves around one or more underwater autonomous vehicles (UAV), or *assets* for short [7,9]. Assets can have different configurations, and this determines what tasks they can perform and how they behave when performing them. For example, an asset with a side-scanning sonar can perform a sonar sweep of an area, while an asset with a camera can take underwater images or ascertain the exact location of a target.

In our scenario, a *mission* consists of assets performing tasks to achieve a number of objectives. The first and last objective of each asset during a mission are the *launch* and *recovery* objectives, respectively. In addition, there are two main objectives the assets can be asked to perform: to *survey* an area (for example, with a sonar); or to *acquire* (the exact location of) a target. Since objectives are usually situated in different locations, assets must also be able to travel from one location to another, i.e., perform a *transit* task. Transit takes place along a path of several locations, called *waypoints*. Finally, part of the area in which the mission takes place can be dangerous or unsuitable for assets to operate in, e.g., when a sandbank is present. These areas can be declared off-limit to assets by enclosing them in *exclusion zones*.

Tasks and objectives can be achieved in different ways, also depending on the configuration of the asset tasked to achieve them. This can be done by specifying, to the asset, a *behaviour* (or sequence of behaviours) for a task or objective. For example, an area can be surveyed using a lawnmower pattern behaviour, or by using spiral pattern behaviour. An asset can also be asked to exhibit certain behaviours at a waypoint, e.g., getting a fix on its exact location before attempting to acquire a target.

Given the above, an example scenario for a single asset could then be for it: to be launched; to transit to a survey area while avoiding an exclusion zone; to survey an area using the lawnmower pattern behaviour; to transit to a waypoint to get an exact fix of its location; to acquire the exact location of a target; to transit back to a recovery location; to finally be recovered there.

The problem of effectively supporting human operators as they plan and observe UAV missions, such as the ones described above, combines several open questions in human-aware planning. These include how to provide support during the mission specification phase, and how to provide effective assistance during the planning phase. In this work, we are developing a framework for supporting human operators throughout UAV missions.

4 System Architecture

The system architecture implementing this framework builds on proprietary software (SeeTrack [19] and Neptune [18]) and their components: *Wetside* and *Top-*

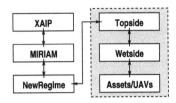

Fig. 1. (a) System architecture diagram. (b) An example structure extracted from the mission model (represented in PDDL).

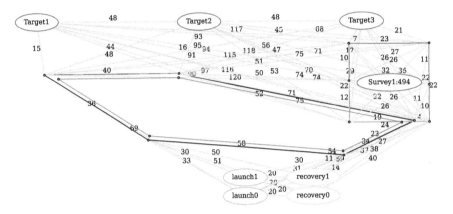

Fig. 1. (*continued*)

side. Three new components: *NewRegime*, *MIRIAM*, and XAIP (see Fig. 1a), have been specifically designed to provide support and effective assistance to a human *operator* in planning and observing UAV mission scenarios.

The system architecture consists of several parts. At the bottom layer are the assets themselves, i.e., the UAVs with their specific hardware configurations. Running on the assets is what is called the *Wetside* component from the Neptune software, which operates the asset using queues of actions, tasks, and behaviours. These queues are populated from above from what is called the *Topside* component of the Neptune software. The SeeTrack software provides a user interface to the *Topside* component, allowing *operators* to specify launch, recovery, survey, and target objectives, and other important structure for the mission, including exclusion zones. The *Topside* component subsequently allows *operators* to allocate assets to the mission, providing a visual rehearsal of what would happen if the mission was run as it was defined thus far. Using this feedback, *operators* can then adjust the mission plan until, eventually, *Topside* is used to upload the mission plan to the *Wetside* software on the assets.

NewRegime acts as a middleware component between the existing system and the new components. It provides (gRPC) interfaces to apply programmatically, similar functions *operators* would apply through the *Topside* user interface (e.g., adding or removing objectives, or allocating resources). Additionally,

NewRegime allows extra or external information relevant to the mission specification (e.g., locations of hazardous areas such as sandbanks) to be injected. This allows for (parts of) the mission specification process to be automated or supported by other systems or information sources.

On top of NewRegime, we developed a task-bot assistant based on the MIRIAM (Multimodal Intelligent interactIon for Autonomous systems) system [10]. MIRIAM guides operators in actively shaping the underlying task using planning-oriented interaction.

Finally, we have developed a component for supporting the operator in understanding plans and exploring alternative plans. This component was developed in order to decouple the new features away from the proprietary software, allowing us to exploit a concise representation of the planning problem, and cleanly separate it from the solver. We have defined a PDDL 2.1 [8] planning domain model that represents the necessary UAV actions: navigation, diving, climbing, surveying, and observing. The main mission elements are extracted through NewRegime from the mission specification (e.g., assets, exclusion zones, objectives, asset launch and recovery points, and transponder positions), and used to define a problem model (see Fig. 1b). The extracted models are used to support query interpretation, to construct answers to the operator queries, and to generate explanations of the mission plans. Underpinning our approach is the *Optic* planner [1], which is sensitive to alternative metric functions.

5 Natural and Visual Language Interface

As the operator and the machine have their own models of the scenario, it is important that the system can facilitate their communication. As a consequence, we have focused on developing approaches for mapping natural language queries onto the machine's problem representation and mission concepts, as well as providing visual and textual methods of communicating the results back to the operator. These approaches extend the MIRIAM system [10], allowing the operator to benefit from support during model specification and to both deepen their understanding of the trade-offs between different planning decisions and also to influence the eventual mission plan by adding additional (soft) constraints that were missing from the initial problem specification.

5.1 User Input and Mapping

MIRIAM's core is based on Keith Sterling's chatbot framework, programm-y[1]. This framework leverages the benefits of using Artificial Intelligence Markup Language (AIML) to quickly generate representations of task-specific indexable knowledge for chatbot question/answer and command and control (C2) interaction. The MIRIAM interface has two main components [10]. Firstly, an AIML-based Natural Language Processing (NLP) engine parses the user's input into

---

[1] A fully compliant AIML 2.1 chatbot framework written in Python3: https://github.com/keiffster/program-y.

a category, a basic unit of knowledge, formalising it as a semantic representation. In addition, each category designed is complemented by a series of reductions and recursion to add flexibility to the pattern-matching and normalisation-denormalisation process. Secondly, a set of extension libraries written in Python3 create a suitable reply by processing and retrieving the relevant information from the NewRegime and XAIP modules.

During execution, the NLP engine analyses the user input to categorise it, identifies the relevant parameters in the query, and populates the slots with this information. Then, the system matches each identified category with a function from the extensions library and processes the filled slots accordingly.

The extension library currently supports the execution of functions grouped into three topics: mission planning, mission monitoring, and plan contrastive explainability. Using MIRIAM, the operator can create and manipulate any objective, assign tasks to available assets, and monitor the assets' behaviours, properties, and progress after they have been launched. Possible natural language interactions include asking about the asset's current status, the mission and its current objectives, the asset's estimated arrival time at specific locations, and the mission's objectives completed. For example, given a user input *'Could you show me the current mission progress, please?'*, the NLP engine will first normalise and remove punctuation and, subsequently, perform a word-level search on the pattern tree. Once a pattern is detected, the AIML interpreter matches wild cards, invokes the Python extension and passes them through. Figure 2 shows an example of category definition and the AIML interpreter process.

While operators create the mission, MIRIAM provides them access to NewRegime shortcut functions, locations database, and previous mission information, such as hazardous areas, the target's location, and survey areas. Offshore wind farms are an example of underwater mission inspection, where relevant locations are known and stored in a database. In that scenario, an operator could upload the locations into NewRegime and access them using the extension libraries available on MIRIAM. Additionally, during the planning phase, MIRIAM provides the user with access to essential functions using NL to explore the planning model. This includes functions for each type of XAIP query, and additional functions to keep or drop constraints and to show the current plan.

5.2 Rendering System Responses

We developed a web-based user interface as the front end of the MIRIAM interface. This provides a mechanism for the user to understand and explore the planning task using different multimedia resources. In the context of XAIP, when the user asks query, the XAIP extension response is rendered by applying visual and textual representations as follows:

Textual Representation: Our approach to rendering plans in text follows [2] and relies on an annotated domain file to generate relevant sets of candidate rules for matching action descriptions. Each action is associated with tags detailing verbs and prepositions that might be used in describing the action. For example, for

```
<category>
  <pattern># MISSION PROGRESS ^</pattern>
  <template>
    <extension
    path="miriam2.extensions.core.mission.MissionExtension">
    MISSION_PROGRESS <star/>
    </extension>
  </template>
</category>
```

```
>> Normalising input from
   [Could you show me the current mission progress , please ?] to
   [COULD YOU SHOW ME THE CURRENT MISSION PROGRESS , PLEASE ?]
 - Removing punctuation...
 - Topic pattern detected = [MONITORING]
 - Matching [COULD YOU SHOW ME THE CURRENT MISSION PROGRESS PLEASE]
 - Matches...
 - 1:Match=word Node=ZEROORMORE [#] Matched=COULD YOU SHOW ME THE CURRENT
 - 2:Match=word Node=WORD [MISSION] Matched=MISSION
 - 3:Match=word Node=WORD [PROGRESS] Matched=PROGRESS
 - 4:Match=word Node=ZEROORMORE [#] [PROGRESS] Matched=PROGRESS
 - AIML Parser evaluating template resolved to [MISSION_PROGRESS]
 - Importing module [miriam2.extensions.core.mission.MissionExtension]
 - [[EXTENSION miriam2.extensions.core.mission.MissionExtension]]
   resolved to
   [Up to this point, 35 percent of the mission has been executed]
   [and the estimated time remaining is 45 minutes]
```

Fig. 2. World-level matching pattern process: a key task of the NLP engine.

the move action, we include 'move', 'navigate', 'go', and 'transit' as alternative verbs, and prepositions such as 'from [the] *from*', 'starting at [the] *from*', and 'from [the] *?from*'. We also support summarising rules, which allow certain chains of actions to be condensed where an effect summarisation exists (e.g., as in the case of sequences of transit actions).

Visual Representation: The plans are plotted over a top-down view, identifying the mission's key aspects, landmarks, areas, objectives, etc. This representation is based on the visual language that operators are familiar with, and should support effective communication of the plans.

6 Supporting Mission Specification in NewRegime

NewRegime sits between the task-bot interface and the mission system. During modelling, NewRegime can exploit its access to additional information, such as the topology of the seabed, weather and current information, so that while the operator builds the model through the task-bot, NewRegime proactively interrogates the partially specified mission in order to identify potential missing structure, which it can then propose that it adds. For example, in a mission where the assets are supposed to survey areas deep underwater, the location of dangerous shallow waters (e.g., sandbanks) will have a major effect on the mission specification. It could lead to the software suggesting the addition of an exclusion zone over the sandbank.

Exclusion zones come with their own set of rules as well, and NewRegime is capable of checking whether they are satisfied or violated. Assets, for example, are to avoid entering (or going too near to) an exclusion zone while attempting an objective, or while in transit from one objective to the next. Launch and recovery locations cannot be inside exclusion zones either, nor can a survey area be entirely encompassed inside an exclusion. If an exclusion zone partially overlaps a survey area, then only the part outside the zone will be surveyed. Likewise, the location of an acquire objective cannot be inside an exclusion zone.

As such, the addition of an exclusion zone, simply to ensure that assets do not go near to a dangerous part of the operation area, can have a profound effect on what the eventual specification of the plan. NewRegime can respond reactively or proactively to this use of extraneous information. It can respond reactively by alerting the operator to a violation of the rules, e.g., by explaining that part of a survey area is now covered by an exclusion zone, and will therefore not be surveyed during the mission. Or it can respond proactively by, e.g., moving the launch and recovery locations of the mission out of the newly added exclusion zone to new locations; locations selected based on operator supplied parameters or some other logic.

7 Supporting Plan Space Exploration and Understanding

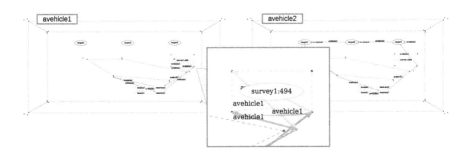

Fig. 3. Comparing two partial plans. For *avehicle2*, the plans are identical. For *avehicle1*, the plans differ in the specific survey activity (shown in the inset).

An important part of the human-machine interaction is in the planning of the mission. The standard method requires a hands-on approach, with the operator allocating assets to objectives, and creating the order in which they should be attempted. We propose using XAIP techniques to allow the operator to use natural language queries in order to understand the generated plans and explore the plan space. Our system can use the queries to generate explanations of the current plan, or generate alternative constrastive plans. In the latter case, through a series of operator queries, the system can allow the user to guide a joint exploration through the plan space.

We support contrastive explanations (following the approach in [11]), which allow the operator to explore and understand the space of possible plans by asking queries to the system. In [11] they consider queries including: Q1: *'Why is action A in the plan?'*, Q2: *'Why is action A not in the plan?'*, and Q3: *'Why is action A used before action B?'*. The approach relies on mapping these queries into constraints that can be added to the planning model. For example, in the case of a Q1 type query a constraint would be added that forced any plan to not use action A. A new plan will not include A and provides a contrastive example, which can then be compared with the original plan. As an example, the operator can ask a query, such as *'why is avehicle1 being used to survey area 1?'*. In this case, the system uses the query to generate a plan where the opposite is true. In the example above, a plan would be generated where *avehicle1* is not used to survey *area 1*. These new plans can be used to generate an explanation, such as *'If avehicle2 is used to survey area 1 then the plan becomes 10 s longer'*.

In [11], the focus was on properties of specific actions in a plan. However, in our context it is natural to ask queries ranging across other problem structures, including numeric functions. For example, in the context of UAVs maintaining communication can become important. This is particularly challenging in the context of missions in the proximity of windturbines, which can lead to interference in communications. As a consequence, we have extended the work in [11] to also support additional model structures, including on numerical variables.

Through chaining queries together, the operator can influence the next generated plan, and build an understanding of the plan space. For example, consider the situation where the operator prefers the plan with *avehicle2* doing the survey (denoted π_1) to the original plan (π_0). They then ask why the communication cost (currently at 22) is so high in the plan. Another constraint is added forcing this to be lowered and generates a new plan (π_2). As a consequence the system supports the operator in an exploration of the plan space (illustrated below).

$$\pi_0 \xrightarrow[+10]{\text{(survey avehicle2)}} \pi_1 \xrightarrow[+3]{(< (\text{comm\_cost}) \ 22)} \pi_2$$

7.1 Understanding the Difference Between Plans

To assist the operator in comparing the plan pairs (π_i, π_{i+1}), we have implemented a plan comparison visualisation, which splits the plan comparison into one separate visualisations per asset. For example, Fig. 3 presents the visualisation for *avehicle1* (left) and *avehicle2* (right). For a particular asset, ϕ_i, we can extract the actions in the plan that are relevant, denoted π^{ϕ_i}. For the original plan (π_0) and the new plan (π_1), and asset ϕ_i, we can identify the pair of asset specific partial plans: $\pi_0^{\phi_i}$ and $\pi_1^{\phi_i}$. We then use the plan to generate three line segments: the parts where the plans match (presented in blue); the part of the original plan $\pi_0^{\phi_i}$ that is not in the new plan $\pi_1^{\phi_i}$ (in green), and the part of the new plan $\pi_1^{\phi_i}$, which is not in the old plan (in red). To do this, we used the Levenshtein distance [13]: the distance between two-word sequences, which provides a measure of the edit difference between the sequences while also respecting order. In our case, we use unique words for each ground action and extract the best match between the two action sequences. In Fig. 3, the partial plans on the right are the same (all blue). On the left, the specific survey start/stop points have changed (shown in the inset).

8 Interaction Walkthrough

Our approach has been implemented and tested using simulated UAVs on an example mission situated off the east coast of Scotland. We are currently preparing for trials using real UAVs. In this section, we present a walkthrough of the operator support components to highlight the key elements of our approach.

```
User: First, create a new mission
MIRIAM: OK, it's called Scenario 1
User: OK, please move the view to the NnG offshore wind farm location.
MIRIAM: OK, the view location is 56.26 degrees north -2.34 degrees east.
User: Let's create a Target for pile_1, pile_2 and pile_3
       <Looking up pile locations in NnG db>
MIRIAM: The targets have been created successfully
User: Add an exclusion zone over the sandbank_1 300 meters south of the pile_3
       <Looking up sandbank_1 location in NnG db>
MIRIAM: I have added an exclusion zone south of the piles successfully
User: Please create a launch and recovery point
       <Looking up objectives' default locations in NnG db>
MIRIAM: I have created launch and recovery points outside the exclusion zone
User: can you move objective launch1 500 meters south from target pile_1?
MIRIAM: The launch1 objective has been moved successfully
...
User: Does the mission look OK?
MIRIAM: Do you want to put a boundary around the mission?
User: Yes, please.
MIRIAM: OK, I have created boundary exclusion zones around the mission.
User: Let's allocate assets vehicle1 and vehicle2 and then do a rehearsal.
MIRIAM: OK done! Here's my current plan.
       <Shows plan>
User: Why not move vehicle2 from survey1_2 to target3?
MIRIAM: Well, the plan becomes 30 seconds longer if we use that action.
       Do you want to see the comparison?
       <Loading comparison visualisation>
```

Fig. 4. An example interaction using the system. The operator specifies the mission and sets the view location over a wind farm. Next, targets are added for three pylons, and then hazardous areas and exclusion zones are added. Finally, the operator conducts a rehearsal and queries the generated plan using XAIP.

In Fig. 4, we present an example walkthrough, demonstrating how the interface can be used to support the user in specifying a mission by incorporating the query for known locations and appropriate default rules where possible. For example, after setting up a mission, the operator positions the simulator's view at the centre of the wind farm of interest. Next, the operator adds an inspecting task for three pylons and assigns a target to each one. The locations of the pylons have already been stored in a database. The operator then uses information from the database again to identify hazardous areas, such as sandbanks and proceeds to create an exclusion zone around them. After other default interactions, such as defining launch and recovery points and bounding the area, the operator decides to conduct a rehearsal, ending the modelling stage.

Once modelling is complete, the planning component pulls the relevant structure from the mission and makes a plan. The user can then use MIRIAM to examine the plan and ask for a visualisation, a comparison, or an explanation. In each case, MIRIAM identifies the appropriate type of query and requests the appropriate function. Finally, the appropriate parameters are extracted from user action descriptions (see description above).

In the example from Fig. 4, the user queries why a particular transit was not used (i.e., moving *avehicle2* from *survey1_2* to *target3*). The system first extracts an appropriate action header as a parameter for the Q2 query; in this case (*move avehicle2 survey1_2 target3*). It then adds a constraint to the planning model in order to force the planner to select this action as part of any valid plan. It then generates a new plan and uses the original and new plan as the basis of a comparison. The new plan, in this case, is longer. The user can view a comparison visualisation (e.g., similar to Fig. 3).

9 Conclusion and Future Work

In this paper, we presented ongoing work that aims to build a tool to support human operators to improve planning understanding and awareness throughout underwater autonomous vehicle missions. We presented an architecture which incorporates modelling assistance, plan generation, explanation, plan space exploration, and execution monitoring. We developed a user interface that extends the task-bot interface with visual content. Our system exploits existing information resources to provide modelling assistance, making suggestions to the operator to reduce the aspects of the model that must be specified. Our planning support component uses textual and visual explanation approaches and exploits explainable planning techniques in order to facilitate the operator's explanation and understanding of the plan space. We presented a walkthrough to demonstrate the main features of our system. We are planning a user study to test our system with human operators. The aim of this study is to evaluate the approach in terms of usability and trust. We are also in the process of developing LLM-enhanced plugin modules for natural language processing, to resolve ambiguities in classifying patterns, and to identify multiple commands in a single statement.

Acknowledgments. This work was partially supported by EPSRC Grant No. EP/V05676X/1 through the HUME Prosperity Partnership.

References

1. Benton, J., Coles, A., Coles, A.: Temporal planning with preferences and time-dependent continuous costs. In: Proceedings of the International Conference on Automated Planning and Scheduling, vol. 22, pp. 2–10 (2012)
2. Canal, G., Krivic, S., Luff, P., Coles, A.: Task plan verbalizations with causal justifications. In: ICAPS 2021 Workshop on Explainable AI Planning (XAIP) (2021)
3. Carreno, Y., Pairet, È., Petillot, Y., Petrick, R.P.: A decentralised strategy for heterogeneous AUV missions via goal distribution and temporal planning. In: Proceedings of the International Conference on Automated Planning and Scheduling, vol. 30, pp. 431–439 (2020)
4. Cashmore, M., Fox, M., Long, D., Magazzeni, D., Ridder, B.: Opportunistic planning in autonomous underwater missions. IEEE Trans. Autom. Sci. Eng. **15**(2), 519–530 (2017)
5. Crosby, M., Petrick, R., Rovida, F., Krueger, V.: Integrating mission and task planning in an industrial robotics framework. In: Proceedings of the International Conference on Automated Planning and Scheduling (2017)
6. De Pellegrin, E., Petrick, R.P.: Planning domain simulation: an interactive system for plan visualisation. In: Proceedings of the International Conference on Automated Planning and Scheduling (2024)
7. Ferri, G., et al.: Cooperative robotic networks for underwater surveillance: an overview. IET Radar Sonar Navig. **11**(12), 1740–1761 (2017)
8. Fox, M., Long, D.: PDDL2.1: an extension to PDDL for expressing temporal planning domains. J. Artif. Intell. Res. (JAIR) **20**, 61–124 (2003)
9. Hamilton, A., et al.: Adaptable underwater networks: the relation between autonomy and communications. Remote Sensing **12**(20) (2020)
10. Hastie, H., Garcia, F.J.C., Robb, D.A., Patron, P., Laskov, A.: Miriam: a multimodal chat-based interface for autonomous systems. In: Proceedings of the 19th ACM International Conference on Multimodal Interaction (2017)
11. Krarup, B., Krivic, S., Magazzeni, D., Long, D., Cashmore, M., Smith, D.: Contrastive explanations of plans through model restrictions. JAIR (2021)
12. Kumar, A., Vasileiou, S.L., Bancilhon, M., Ottley, A., Yeoh, W.: VizXP: a visualization framework for conveying explanations to users in model reconciliation problems. In: Proceedings of the International Conference on Automated Planning and Scheduling (2022)
13. Levenshtein, V.I.: Binary codes capable of correcting deletions, insertions and reversals. Cybern. Control Theory **10**, 707–710 (1966)
14. Magnaguagno, M.C., Pereira, R.F., Móre, M.D., Meneguzzi, F.: WEB PLANNER: a tool to develop classical planning domains and visualize heuristic state-space search. In: Proceedings of the Workshop on User Interfaces and Scheduling and Planning, UISP, pp. 32–38 (2017)
15. Mingyue Ma, L., Fong, T., Micire, M.J., Kim, Y.K., Feigh, K.: Human-robot teaming: concepts and components for design. In: Hutter, M., Siegwart, R. (eds.) Field and Service Robotics, pp. 649–663. Springer International Publishing, Cham (2018)
16. Porteous, J., Lindsay, A., Charles, F.: Communicating agent intentions for human-agent decision making under uncertainty. In: Proceedings of the International Conference on Autonomous Agents and Multiagent Systems (2023)
17. Rajan, K., Py, F., Barreiro, J.: Marine Robot Autonomy. Towards Deliberative Control in Marine Robotics. Springer, New York (2013)

18. SeeByte: Neptune technical whitepaper (2018). https://www.seebyte.com/media/1427/neptune-technical-whitepaper.pdf. Accessed Mar 2024
19. SeeByte: Seetrack v4 technical whitepaper (2018). https://www.seebyte.com/media/1136/seetrack-v4-technical-whitepaper.pdf. Accessed Mar 2024
20. Vaquero, T.S., Romero, V., Tonidandel, F., Silva, J.R.: itsimple 2.0: an integrated tool for designing planning domains. In: International Conference on Automated Planning and Scheduling, pp. 336–343 (2007)
21. Wickler, G., Chrpa, L., McCluskey, T.L.: KEWI - A knowledge engineering tool for modelling AI planning tasks. In: International Conference on Knowledge Engineering and Ontology Development, pp. 36–47 (2014)

Improved Computation Efficiency 2D Visual SLAM Based on Particle Filter With Distance Sliding Window

Zhixin Zhang[✉], Yichen Liang, and Alexandru Stancu

Department of Electrical and Electronic Engineering, The University of Manchester, Manchester M13 9PL, UK
`zhixin.zhang-2@postgrad.manchester.ac.uk`

Abstract. Simultaneous localization and mapping (SLAM) is one of the fundamental challenges for motion robot systems. Over time, SLAM methodologies have evolved from conventional filter-based approaches to contemporary optimization approaches. SLAM based on the particle filter has been widely used in mobile robot technology, especially in 2D ground robot SLAM. However, traditional implementation of particle filter visual SLAM algorithm typically only considers 2D position states, overlooking the orientation along the z-axis. Moreover, they entail substantial computational overhead as the state needs to be updated once there is a new visual observation received. To address these limitations, this paper presents a novel 2D visual SLAM framework that incorporates three freedoms and leverages Aruco Marker as the robust observation. A distance sliding window is introduced to avoid the computational workload resulting from the expansion of state dimensionality. Several simulation experimental results demonstrate that the proposed algorithm significantly enhances computational efficiency while maintaining accuracy and robustness in both localization and mapping tasks, through a reduced requirement for visual updates. The whole implementation is open source (https://github.com/Happy-ZZX/Puzzlebot).

Keywords: SLAM · Wheel Robots · ROS · Localization · Sensor Fusion

1 Introduction

Mobile robot systems have received widespread attention due to their application in more and more scenarios in recent years, and the corresponding technology is also gaining attention. SLAM (Simultaneous localization and mapping) technology is a technical solution to solve the problem of robot localization and mapping, and the goal is to complete the tasks of localization and mapping simultaneously in real time. SLAM technology was originally designed for localizing underwater submarines [1]. For mobile robotics, localization and mapping are two closely related tasks where localization results can improve mapping accuracy while mapping results can aid in relocation and loop closure.

Early-stage SLAM systems are usually based on the Extended Kalman Filter (EKF), like [2,3]. EKF-based SLAM method uses First-order Taylor linearization to transform the nonlinear system into a linearized Gaussian system, then uses the Kalman filter (KF) to maximize the system state posterior probability. For short-distance movements, this type of method works well. However, due to the Markov property of the Kalman Filter, the state error will keep increasing with time. At the same time, as the map size increases, the number of states that need to be maintained in the state vector also increases, which makes it difficult for the system to achieve real-time performance.

Another filter method used in SLAM is particle filter (PF). Unlike the former, PF uses many samples to simulate the state distribution and measures the posterior distribution probability by calculating weights in subsequent updates. This feature makes PF-SLAM can be implemented on a system that is not a linear Gaussian system. The most representative PF-based SLAM is the Fast-SLAM [4,5]. In their work, they use a Rao-Blackwellization particle filter (RBPF) [6] and each particle includes both robot pose and map information. Each iteration only filters the observed landmark points instead of all landmarks to guarantee efficiency. This method proved to be effective and was widely used in subsequent 2D Laser SLAM [7]. The accuracy of the PF-SLAM is related to the number of particles. The greater the number of particles, the better the algorithm performs and more computation is required. So balancing the particle number and algorithm accuracy often needs to be considered in these methods.

Whether SLAM methods are based on EKF or PF, they both belong to progressive SLAM methods, that is, each estimation is only based on the results of the previous period. This estimation method will not only accumulate previous errors into current estimates but also can not use more past information. So modern SLAM research usually depends on optimization to maximize information usage and perform better. For camera-only SLAM systems, ORB-SLAM [8] proposed a SLAM structure that includes loop closure and re-localization based on the ORB feature. Another popular SLAM research branch is Visual-Inertial (VI) SLAM, representative works including VINS-Mono [9] and OKVIS [10]. These two SLAM systems propose the concept of pre-integrating IMU information between keyframes. Then they optimise both the feature points and keyframe camera poses together.

Particle-filter-based SLAM methods typically involve sampling and updating every time new observation data is received, like [4,7]. However, in low-speed scenarios, frequent sampling may not yield performance improvements but instead consume computing power, potentially compromising accuracy. Furthermore, traditional 2D particle filter-based SLAM usually only considers the robot position x and y and overlooks the yaw angle. Inspired by the appeal papers and prior research [11], this paper proposed a particle filter SLAM system with 2D position and orientation along with z-axis, leveraging distance sliding window to reduce the computation loss and enhance robustness. Additionally, an online noise definition method based on Aruco marker [12] is defined as an indicator of observation uncertainty. For the experiments, use the two-wheel differential model as the motion model and the Aruco Marker as the observation model. To intuitively verify the effectiveness of the algorithm, some simulation experiments with simple scenarios are tested and results show that the proposed SLAM framework has improved efficiency with similar localization and mapping outcomes compared with classical Fast-SLAM.

2 Methodology

The proposed SLAM system is rooted in the Fast-SLAM [4] aided with a sliding distance window. Implementation occurs on a two-wheel robot equipped with a fish-eye camera, leveraging the Robot Operating System (ROS) [13]. Within this setup, the SLAM node subscribes to the camera and wheel encoder topic. Then publishes the robot pose and mapping results topic.

2.1 Problem Statement

For a ground mobile robot, the pose states s_k need to be estimated, usually including three elements: s_x, s_y, and s_θ. s_x and s_y represent the coordinates of a robot relative to a world frame, while the element s_θ represents the angle between the x-axis of the robot frame and the x-axis of the world frame. Then the state transition model can be defined as below:

$$s_k = h(s_{k-1}, u_k) + q_k \tag{1}$$

where $h(s_{k-1}, u_k)$ represent the function between the current robot state s_k and last state s_{k-1} by using the system control input u_k. And q_k is the motion model noise which is assumed to obey Normal distribution with zero mean.

$$q_k \sim \mathcal{N}(0, \Sigma_k) \tag{2}$$

This paper uses Aruco Marker to estimate the robot's pose and the observation model is defined as follows:

$$z_{i,k} = g(m_i, s_k) + r_{i,k} \tag{3}$$

where m_i is pose of the marker i. Different from the previous work, the marker pose in this paper considers the yaw angle of each marker in the world frame, which is defined as $m_i = [m_x, m_y, m_\theta]^\mathsf{T}$. The observation $z_{i,k}$ represent the distance and angle between the landmark m_i and the camera in timestep k. It is also defined as a dimension vector $z_{i,k} = [z_x, z_y, z_\theta]^\mathsf{T}$.

Like the motion model, observation also has a zero-mean Gaussian noise $r_{i,k}$, calculated by the re-projection error. According to the marker length, the 3D positions of the four corner points of the marker in the marker frame can be obtained. The 2D projection point of each corner point can be calculated based on the transformation matrix from the marker frame to the camera frame by using the camera projection model. This is a standard reprojection problem which is shown below:

$$s_i \begin{bmatrix} u_i \\ v_i \\ 1 \end{bmatrix} = K {}^c T_m \begin{bmatrix} X_i \\ Y_i \\ Z_i \\ 1 \end{bmatrix} \quad (4)$$

In this equation, s_i is the scale. u_i and v_i represent the pixel coordinate after Normalization. K is the camera intrinsic matrix and ${}^c T_m$ is the transformation matrix from marker frame to camera frame. $\begin{bmatrix} X_i & Y_i & Z_i \end{bmatrix}$ refers to the 3D coordinate of the marker corner point in the marker frame. Because all corner points lie on the $x - y$ plane of the marker frame, Z_i is 0 for all four corners. As for X_i and Y_i, their value should be half the marker length, as shown in Fig. 1. Then the error between the reprojected position of the 3D corner point and the corresponding observed image coordinate position is used to indicate the noise range of observation. A more detailed derivation of motion and observation models can be found in paper [11].

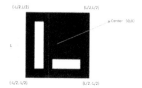

Fig. 1. Aruco Marker Example

SLAM problems can be seen as a problem of maximizing the posterior probability based on prior information. In this experiment, the aim is to estimate the system state based on information from the wheel encoder and camera, which is

$$Max(p(s_{0:t}, m_{1:M} | z_{1,t}, u_{1:t})) \quad (5)$$

where M is the number of Aruco markers. In the Rao-Blackwellization Particle Filter(RBPF) [6], the robot localization problem and the mapping problem are assumed to be independent. Then this SLAM problem can be decomposed into a robot pose localization problem and M landmark estimation problems conditioned on the robot pose estimate.

$$\begin{aligned} p(s_{0:t}, m_{1:M} | z_{1:t}, u_{1:t}) &= p(s_{0:t} | z_{1:t}, u_{1:t}) p(m_{1:M} | s_{0,t}, z_{1:t}) \\ &= p(s_{0:t} | z_{1:t}, u_{1:t}) \prod_{i=1}^{M} p(m_i | s_{0:t}, z_{1:t}) \end{aligned} \quad (6)$$

2.2 Algorithm Structure

The overview of this SLAM structure is shown in Fig. 2. The whole system starts with initialization, in which parameters need to be read, such as camera intrinsic and external parameters. Then n particles need to be created with the same weight 1 and pose 0. After that, the whole system will keep subscribing to sensor information and updating each particle until the robot stops.

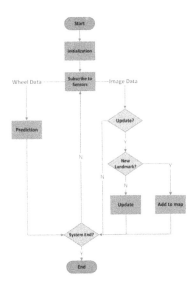

Fig. 2. The Proposed SLAM System Flow Chart

The core of PF-SLAM uses a large number of weighted particles to simulate the distribution of the robot's pose. So in the prediction step, each particle is updated by using the motion function with added noise.

$$s_k^n = h(s_{k-1}^n, u_k) + random(\sigma_k) \tag{7}$$

This noise σ_k is the motion model noise, which is assumed to have a Gaussian distribution as $\sigma_k \sim \mathcal{N}(0, \Sigma_k)$. The covariance matrix Σ of particle n is updated by using the Jacobian matrix of motion model H and wheel encoder noise q_k with the following equation:

$$\Sigma_k^n = H_k \Sigma_{k-1}^n H_k^T + Q_k \tag{8}$$

For the observation information, the first thing is to extract the marker information first. Then check if an update is needed, which is the main contribution of this structure. Using this strategy can not only reduce the update frequency but also avoid errors caused by low-speed motion. There are two conditions for judging whether an update is needed.

The first is there is a new marker has been observed. For instance, if two markers are observed, it will be updated if at least one of the markers is observed for the first time. Another situation is after the robot moves for a certain distance, it will be updated. Specifically, using wheel encoder information as a reference and only considering translation. After the update, the accumulated translation will be set to zero.

The weight of each particle will be updated depending on the observation. For the new observed markers, it needs to be added to the map. Each landmark in the map is represented as a 3D pose vector and a 3*3 covariance matrix. These two variables are initialized when the landmark is first observed and updated with each subsequent observation across the following function.

$$m_{i,k}^n = g^{-1}(s_k^n, z_k) \tag{9}$$

$$\Sigma_{i,k}^n = G^{-1}\Sigma_k^n(G^{-1})^T \tag{10}$$

where g is the observation model and its Jacobian matrix is denoted as G. And $m_{i,k}^n$ represents the marker i in the particle n at timestep k which is represented as a 3D vector including the translation of the x-y plane and the orientation of the z-axis. As $m_{i,k}^n$ has three related variables to the observation g, G should be a 3*3 matrix.

If the marker is not new and the update condition is satisfied, then a 3-dimensional EKF filter has been used to update its mean value and covariance. The currently saved value of the landmark is used as the predicted value of the Kalman filter and the latest observation is used as the observed value of the Kalman filter to update the pose and covariance information of the landmark. The measurement prediction value $\hat{z_{i,k}}$ is defined flowing:

$$\hat{z_{i,k}} = g(m_{i,k-1}^n, s_k) \tag{11}$$

To calculate the predicted covariance of this observation in the current time, the Jacobian matrix of the observation equation G needs to be recalculated.

$$G = g'(s_k^n, m_{i,k-1}^n) \tag{12}$$

By using this 3 * 3 Jaccabian matrix and the mark covariance in the last period, the observed covariance in the current period Q can be obtained.

$$Q = G\Sigma_{i,k-1}^n G^T + Q_k \tag{13}$$

The last part is to calculate the Kalman gain K of this observation and update the status of this landmark.

$$K = \Sigma_{i,k-1}^n G^T Q^{-1} \tag{14}$$

$$m_{i,k}^n = m_{i,k-1}^n + K(z_{i,k} - \hat{z_{i,k}}) \tag{15}$$

$$\Sigma_{i,k}^n = (I - KH)\Sigma_{i,k-1}^n \tag{16}$$

At this point, the map saved by each particle has been updated, but due to the difference in the robot pose recorded by each particle, the current observation value is different from the predicted observation value. If the currently observed value is used as a random variable and the predicted value is used as the mean, then their 1 relationship can be regarded as a Gaussian distribution of multi-variable quantities. The probability density function (PDF) of multi-variable Gaussian distribution is shown below.

$$f_X(x_1, x_2, ..., x_k) = \frac{exp(-\frac{1}{2}(x-\mu)\Sigma(x-\mu))}{\sqrt{(2\pi)^k |\Sigma|}} \quad (17)$$

where k represents the dimension of variable X and $|\Sigma|$ is the determinant of Σ. In this problem, observation has 3 dimensions and the covariance is the observation covariance Q. So based on this PDF, the possibility of each particle on this observation can be calculated. Theoretically, the closer the particle is to the real robot pose, the smaller the difference between the observed value and the predicted value, and the greater the calculated probability value. Based on this formula, the weight of each particle can be updated in the following way:

$$w_k^n = w_{k-1}^n * |2\pi Q|^{-\frac{1}{2}} exp(-\frac{1}{2}(z_{i,k}\hat{z_{i,k}})^T Q^{-1}(z_{i,k} - \hat{z_{i,k}})) \quad (18)$$

In this equation, w_k^n represents the weight of particle n in timestep k and its value range is $(0, 1)$. The particle with a larger weight is closer to the true value, and vice versa. For other unobserved landmarks, due to no added information being obtained, the pose vector and covariance matrix remain unchanged.

After updating the weight, each particle has a different weight and particles with greater weights have higher priority. In particle-related algorithms, the particles with the lower weight become lower after each observation, which reduces the diversity of random sampling and causes particle dissipation problems. To avoid this problem, it is often necessary to eliminate particles with low weights and retain particles with high weights. This step is called resampling. The basic idea is to sample N new particles from the old particles, depending on the probability of weight. The particle with a bigger weight will be chosen with greater probability and vice versa. In addition, each particle after resampling should have the same weight. The implementation method is as follows:

3 Experimental Result

The proposed algorithm is implemented with a simulation environment employing a two-wheel robot called Puzzlebot. For the result, both the localization and mapping results are evaluated by comparison with the true value. Since the simulation results of the traditional Fast-SLAM and the proposed SLAM are approximately equivalent (The difference is that the latter is more efficient, but the results may be different in real scenarios), encoder estimation results are utilized for comparison to highlight the improvements more distinctly. The

Algorithm 1. Resampling

Require: Current particle set P_c, the number of particles N
Ensure: New particle set P_n
1: Find the maximum weight W_{max} in P_c
2: $index = rand(0,1) * N$
3: **for** i=0 to N **do**
4:     $\beta + = rand(0,1) * 2 * W_{max}$
5:     **while** the weight of $P_c[i] < \beta$ **do**
6:         $\beta = \beta - P_c[i].weight$
7:         $index = (index + 1)\%N$
8:     **end while**
9:     $NewParticle = P_c[index]$
10:    $NewParticle.Weight = 1$
11:    Pull $NewParticle$ into P_n
12: **end for**
13: return P_n

true localization values are derived from the Gazebo joint-state data, while the mapping information is obtained from the settings file. The experimental results are presented below.

3.1 Experimental Setup

This paper focuses on simulation-based validation to assess the proposed SLAM system, utilizing a laptop equipped with an Intel i5 processor and NVIDIA 1650Ti graphic card. The simulation test environment is built in Gazebo (Ubuntu 18.04), a simulation tool compatible with ROS and enabling the replication of real physical world environments. By setting different parameters in the Gazebo, the robots can subjected to various forces, such as friction and gravity.

The test environment consists of four walls with a rectangular shape, each adorned with several Markers, as depicted in Fig. 3. The robot's initial position is located at the centre point of the environment, facing the positive direction of the y-axis of the robot frame. The entire environment is approximately eight meters long and six meters wide, satisfying the requirements for long-term automatic navigation by the robot.

For autonomous navigation, the robot deploys a multi-point tracking algorithm by using a PID controller. To verify the algorithm's feasibility in different scenes, two sequences with different trajectories were used for the test. They are shown in Fig. 4 respectively. The trajectory of the sequence 1 follows the path $(2,-2) \to (4,2) \to (0,0) \to (-2,2) \to (-4,-2) \to (0,0)$ while the sequence traces 2 is $(2,-2) \to (4,2) \to (0,2.5) \to (-2,2) \to (-4,-2) \to (0,-2) \to (2,-2)$. To enhance observation effectiveness, an observation criterion established only when the distance between the marker position and the robot is less than the threshold. In both test sequences, a phase exists where the landmark recedes to a distance, rendering it invalid. Given the stochastic nature of noise introduction,

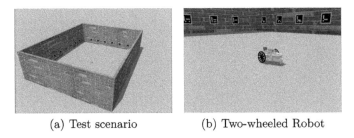

(a) Test scenario (b) Two-wheeled Robot

Fig. 3. Test Environment in Gazebo

it is necessary to ensure that the wheel encoder estimation and the PF-SLAM estimation calculate the same test sequence.

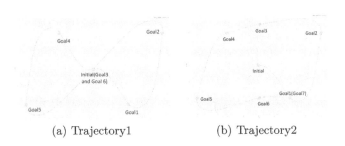

(a) Trajectory1 (b) Trajectory2

Fig. 4. Test trajectories

3.2 Results

The Absolute Pose Error (APE) [14] serves as a typical evaluation metric for V-SLAM and VI-SLAM to evaluate the performance of the SLAM algorithm. It reflects the accuracy of the SLAM algorithm by calculating the difference between the estimated pose and the true value. In this test, as the robot is a ground-move robot, the calculation of APE only considers the translation on the XY-axis and rotation on the Z-axis. At the same time, the average value, maximum value, and variance of APE are computed as metrics to evaluate the algorithm's performance. Additionally, to highlight the differences between the two algorithms, the trajectory results of both sequences and the corresponding true values are plotted within the same figure for direct comparison. The parameter settings in this experiment, including the size of the distance sliding window and particle number, need to be carefully tuned. After extensive testing, a distance sliding window size of 0.1 m and a particle number of 100 were determined to be optimal.

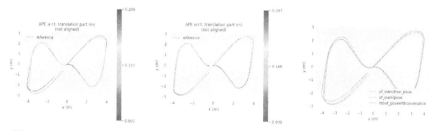

(a) Encoder Estimation (b) PF SLAM Estimation (c) Trajectory Comparison

Fig. 5. Trajectory Estimation Results of Sequence 1

The trajectory estimation results of encoder estimation and particle filter-SLAM for test sequence 1 are shown in Fig. 5. The first two figures illustrate the Absolute Pose Error (APE) color trajectory of the wheel encoder and PF-SLAM estimation, respectively. The colored line represents the estimated result, while the dotted line is the real trajectory of the robot. In the encoder estimation, the lack of observational corrections leads to the color shifting from blue to red, signifying a progressively increasing error. In the middle of the path, the errors decrease because the newly introduced noise offsets the previous noise, resulting in the cancellation of errors. The second figure illustrates that the estimation accuracy is notably improved after adding observation information. The overall estimated APE values remain within a stable range, with error growth only observed during periods of failed observations in the middle portion of the trajectory. In the third figure, the red and blue trajectories represent the trajectory estimation results of the encoder and PF-SLAM estimation, respectively. It is evident that the PF-SLAM result closely aligns with the true trajectory. Throughout sequence 1, the robot received approximately 6400 images, which would result in 6400 updates in the classical Fast-SLAM. However, due to the implementation of the distance sliding window, updates were reduced to only 142 here.

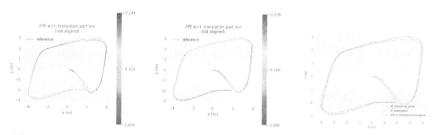

(a) Encoder Estimation (b) PF SLAM Estimation (c) Trajectory Comparison

Fig. 6. Trajectory Estimation Results of Sequence 2

The trajectory estimation results of sequence 2 are shown in Fig. 6. Diverging from the previous sequence, sequence 2 introduces large encoder noise at the beginning of the navigation, resulting in the right part of the encoder APE trajectory being red. However, it transitions to blue after the noise is offset in the medium part. Most of the PF-SLAM trajectory in the second picture is blue and there is no large area of red. Indeed, in the estimation results of PF-SLAM, despite the motion model exhibiting a significant error at the beginning, the presence of observations enables the algorithm to still yield relatively accurate results. In the third picture, noticeable differences can be observed between the reference trajectory and the wheel speed estimation results, particularly in the lower right corner and the upper left corner. However, the differences of PF-SLAM are comparatively smaller. This test sequence demonstrates that PF-SLAM can effectively estimate robot poses when the motion model exhibits significant errors at the beginning. Like sequence 1, the update frequency in sequence 2 is significantly reduced, occurring only 179 times despite receiving about 8232 images. The results, which include the mean value, max value, and

Table 1. APE Mean, Max, and Standard Deviation

	Sequence 1			Sequence 2		
	mean	max	std	mean	max	std
PF-SLAM	**0.1171**	**0.2967**	**0.0399**	**0.0982**	0.3380	**0.0341**
Encoder	0.1626	0.2992	0.0718	0.1442	**0.2439**	0.0527

standard deviation of absolute pose error (APE) for both test sequences, are presented in Table 1. Across these two test sequences, the proposed SLAM method has a smaller APE mean and standard deviation compared to the estimation results derived from the wheel encoder. However, the maximum value of the estimation error of PF-SLAM is bigger than that of the encoder. This disparity arises due to the nature of particle filtering, which requires samples to be sampled within a certain range randomly. Consequently, there are periods of no valid visuals occurring in each sequence, which means there is no visual update during this period. As a result, the results of PF-SLAM will be greater than those of encoder estimation.

4 Conclusion

In this paper, a novel particle filter-based SLAM system utilizing a distance sliding window is proposed. By using the Aruco marker as the observation model and the wheel encoder as the motion model, abandon the traditional visual update method and update the state only upon the detection of new markers or after a certain distance has been traversed. The results demonstrate that by fusing these two types of data within this SLAM system, more accurate localization and mapping can be achieved, along with improved computational efficiency. It is important to note that this SLAM system is tailored for environments densely populated with markers. Future work will focus on integrating features as visual observations to enhance the system's robustness and versatility.

References

1. Smith, R., Cheeseman, P.: On the representation and estimation of spatial uncertainty. Int. J. Robot. Res. **5**, 56–68 (1986)
2. Davison: Real-time simultaneous localisation and mapping with a single camera. In: Proceedings Ninth IEEE International Conference On Computer Vision, pp. 1403–1410 (2003)
3. Davison, A., Reid, I., Molton, N., Stasse, O.: MonoSLAM: real-time single camera SLAM. IEEE Trans. Pattern Anal. Mach. Intell. **29**, 1052–1067 (2007)
4. Montemerlo, M., Thrun, S., Koller, D., Wegbreit, B., Others: FastSLAM: a factored solution to the simultaneous localization and mapping problem. In: Eighteenth National Conference on Artificial Intelligence, pp. 593–598 (2002)
5. Montemerlo, M., Thrun, S., Koller, D., Wegbreit, B., Others: FastSLAM 2.0: an improved particle filtering algorithm for simultaneous localization and mapping that provably converges. IJCAI. **3**, 1151–1156 (2003)
6. Murphy, K., Russell, S.: Rao-Blackwellised particle filtering for dynamic Bayesian networks. In: Sequential Monte Carlo Methods in Practice, pp. 499–515 (2001)
7. Grisetti, G., Stachniss, C., Burgard, W.: Improved techniques for grid mapping with rao-blackwellized particle filters. IEEE Trans. Rob. **23**, 34–46 (2007)
8. Mur-Artal, R., Montiel, J., Tardos, J.: ORB-SLAM: a versatile and accurate monocular SLAM system. IEEE Trans. Rob. **31**, 1147–1163 (2015)
9. Qin, T., Li, P., Shen, S.: VINS-Mono: a robust and versatile monocular visual-inertial state estimator. IEEE Trans. Rob. **34**, 1004–1020 (2018)
10. Leutenegger, S., Lynen, S., Bosse, M., Siegwart, R., Furgale, P.: Keyframe-based visual–inertial odometry using nonlinear optimization. Int. J. Robot. Res. **34**, 314–334 (2015)
11. Liang, Y., Stancu, A., Zhang, Z.: Simultaneous localisation and mapping with fiducial markers based on extended Kalman filter. In: 2023 6th International Conference on Intelligent Robotics and Control Engineering (IRCE), pp. 154–162 (2023)
12. Fiala, M.: ARTag, a fiducial marker system using digital techniques. In: 2005 IEEE Computer Society Conference on Computer Vision and Pattern Recognition (CVPR'05), vol. 2, pp. 590–596 (2005)
13. Quigley, M., et al.: ROS: an open-source robot operating system. In: ICRA Workshop on Open Source Software, vol. 3, p. 5 (2009)
14. Grupp, M.: evo: Python package for the evaluation of odometry and SLAM (2017). https://github.com/MichaelGrupp/evo

The Benefits of Ordinal Regression Under Domain Shift

Andy Perrett[✉], James M. Brown, and Petra Bosilj

University of Lincoln, Brayford Pool, Lincoln LN6 7TS, UK
`aperrett@lincoln.ac.uk`

Abstract. Deep neural networks (DNNs) are trained under the assumption that the training data are independent and identically distributed. In the real world, autonomous systems typically receive out of distribution images causing a domain shift that hinders performance. One important consideration in the context of ordinal image data (i.e., their labels have an intrinsic order) is the choice of loss function and whether it takes the ordinality into account. In this paper, we examine the benefits of ordinal formulations over nominal classification using a human age estimation task. Experiments using DNNs with ordinal data suggest that performance on out of distribution data can be improved by over 240% if trained using ordinal regression methods as compared to classification.

Keywords: Ordinal Regression · Domain Shift · Domain Adaptation

1 Introduction

Deep neural networks (DNNs) in vision pipelines for robotic and autonomous systems assume that their *source* training images are independent and identically distributed (IID) to the *target* images they expect to encounter when deployed [19]. Typically, in real-world scenarios, this is not the case and the target imagery is out of distribution (OOD). This difference in distribution represents a *domain shift* between the source and target domains, which hampers their real-world performance. Domain adaptation (DA) methods are related to transfer learning [15] and seek to train DNNs by aligning the two distributions. This can be achieved via supervised learning for the source domain and adapting to the target using unsupervised domain alignment and pseudo-labeling [17]. In contexts where the data are ordinal, ordinal regression (OR) aims to capture the structured relationships between samples (i.e., their rank order) which is not utilised by nominal classification approaches. This work investigates the benefits of training DNNs using OR to take advantage of the underlying structure within both source and target distributions to aid DA. We conduct our investigation using well established ordinal datasets for age estimation, and argue that OR affords performance benefits in contexts where domain shift is apparent.

2 Related Work

Human face age estimation aims to detect the age of an individual from photographs. Previous work [5,7,11] has curated numerous datasets of human faces allowing age estimation tasks to be conducted as either a classification or OR task. The progression of age causes facial distortions that vary by age and from person to person [3] making it difficult for humans and computers to estimate a person's age [21]. These distortions violate the IID assumption and research shows that by normalising variation the performance of face recognition systems could be improved [11]. Age estimation was regarded [3] as a classification problem, with most algorithms designed as such, and solved by using shortest distance classifiers or neural networks [4]. The conclusion is that machines can estimate age as reliably as humans [10], but few studies have contrasted nominal and ordinal formulations in the presence of domain shift.

Treating age estimation as a multi-class problem does not convey the additional property of age being ordinal [4,13]. Previous work has sought to split the problem into feature extraction and OR [14], although with the success of DNNs, loss functions are designed to capture the ordinal ranking information through end-to-end training [3]. Many approaches are based on extended binary classification [12] for OR, due to multi-class formulations ignoring ordinality [4]. Various works [8,13,21] are opting to pre-train models using large datasets scraped from web images, such as IMDB-WIKI [18] to improve performance. Commonly used face datasets have biases and multiple issues [9,16] and efforts to improve them as baselines have seen multiple datasets re-processed and combined into one further enhancing performance [2], such as B3F and UGAD datasets [2,9]. The data itself, along with it's underlying structure is now seen as important [16], suggesting future research will focus more in this area.

3 Datasets

The Biometrically Filtered Famous Figure Dataset (B3FD) [2] is a combination of IMDB-WIKI [18] (523,051 images) and CACD 2000 [5] (163,446 images) containing 375,592 face images of size 256 × 256 pixels post-curation. Ages range from 1 to 101. In this study, the B3F dataset was used to pre-train all model backbones as per standard practice [8,13,21], with class labels set to age −1, (e.g. if face age is 12 then label is 11). The UTKFace [24] dataset comprises 24,108 images of faces in the age range 1 to 116. We pre-processed this data with *RetinaFace* [20] to detect facial landmarks of the eyes, nose and mouth, along with a bounding box of the face. Images were cropped to the bounding box and aligned using the landmarks to give $23,401$ images of widely varying aspect ratios. The AppaReal [1] dataset ($7,591$ images) was similarly pre-processed using RetinaFace to produce $7,565$ images with ages ranging from 1 to 100, with varying aspect ratios. RetinaFace was unable to detect all faces therefore some images were unused. Additionally, both the UTKFace and AppaReal datasets had all faces above 100 years old removed, along with any age group containing less than

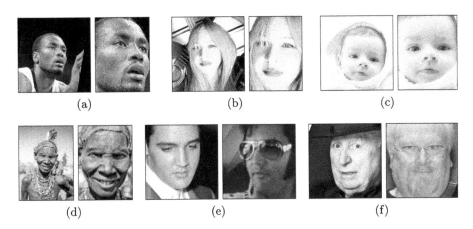

Fig. 1. Example images from the three datasets used in this paper. Examples of before and after applying RetinaFace: (a) AppaReal male, (b) AppaReal female, (c) UTKFace young, (d) UTKFace old. B3F examples (no RetinaFace): (e) same person younger and older, (f) different people, poses and scales.

5 samples. UTKFace and AppaReal were the main datasets used to report the results and their age ranges were *binned* into 10 uniform categories: 0−9, 10−19, ..., 90 − 99. The images from all datasets were resized to 224 × 224 pixels before training. Figure 1 shows example images from each dataset.

4 Methodology

Classification. A Convolutional Neural Network, ResNeXt-50 [23], was used for its state-of-the-art performance for image classification tasks. Dropout regularisation of 5% was employed to reduce overfitting and improve generalisation [22], and applied after each Rectified Linear Unit (ReLU) activation layer. The final linear output layer was changed to 100 output nodes for pre-training and 10 for the main experiments, with cross entropy loss employed:

$$\mathcal{L}_{\text{CE}} = -\sum_{i=1}^{C} \mathbb{1}(y = i) \log(p_i), \tag{1}$$

where C is the number of classes, y is the target class, $\mathbb{1}(\cdot)$ is an indicator function returning 1 if the expression in the parenthesis is true, and p_i is the logit probability of the output of the final linear layer corresponding to class i calculated by softmax, $p_i = \text{Softmax}(x_i)$.

Ordinal Regression. The same ResNeXt-50 backbone was used replacing the output layer for a three-layer fully connected neural network (300, 200 and 100/10 nodes, batch normalisation, 10% dropout, ReLU/sigmoid activation).

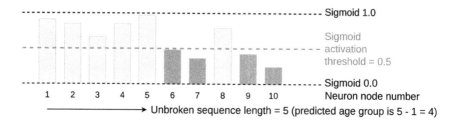

Fig. 2. An unbroken sequence of sigmoid activations predicting the age.

An OR prediction is calculated as the length of the longest unbroken sequence of sigmoid activations above a threshold of 0.5, starting from the first output neuron (Fig. 2). This imposed a single ordinal rank consistent output inspired by the CORAL method [3], from 0 to 99. The loss function employed was $\mathcal{L}_1$:

$$\mathcal{L}_1 = \frac{1}{C} \sum_{i=1}^{C} |s_i - \mathbb{1}(y \leq i)|, \qquad (2)$$

where C is the number of categories, and s_i the logit probability of each output node x_i, calculated using the sigmoid function $s_i = \text{Sigmoid}(x_i)$.

Metrics. Two performance metrics are reported. Accuracy treats all errors equally, and is defined as:

$$\text{Acc} = \frac{1}{N} \sum_{i=1}^{N} \mathbb{1}(\hat{y}_i = y_i), \qquad (3)$$

where N is the number of images and $\hat{y}$ the predicted label. Mean Absolute Error (MAE), unlike accuracy, has errors that are proportional to the differences between the predicted and true class, defined as:

$$\text{MAE} = \frac{1}{N} \sum_{i=1}^{N} |\hat{y}_i - y_i|. \qquad (4)$$

5 Experiments and Results

All the datasets were split into training, validation and tests sets with the ratio of 70:10:20. The backbone models were initialised with ImageNet weights [6] and trained on the B3F dataset, in both the OR and classification regimes for 25 epochs each with a learning rate (LR) of 5×10^{-3}. LR was reduced by a factor of 10 after 12 epochs for OR. The best models are selected based on the validation loss, and referred to as B3F-Class and B3F-OR. These two models were then used to initialise their respective architectures (classification and OR), and fine tuned using the UTKFace and AppaReal datasets. The augmentation policy for

Table 1. Results for classification and OR using initialised weights trained on each dataset giving both IID and OOD results. ↑ indicates high values are better, ↓ indicates lower values are better.

Model	UTKFace		AppaReal	
	Accuracy % (↑)	MAE (↓)	Accuracy % (↑)	MAE (↓)
UTK-Class	61.53	0.46	7.42	4.07
AppaReal-Class	6.11	4.04	55.45	0.55
UTK-OR	60.18	0.46	18.59	1.68
AppaReal-OR	19.59	1.68	58.10	0.48

all experiments consisted of random horizontal flipping at rate $p = 0.5$, random rectangular block erasing at rate $p = 0.7$ with sizes between 0.02 to 0.33 of the image, random rotation between $-5°$ and $+5°$, and normalisation to ImageNet. The final models were then trained for 50 epochs with the Adam optimizer and LRs of 5×10^{-5} and 5×10^{-6} for OR and classification respectively. All model selection was based on the lowest validation loss during training. We report the final performance of each model on both the UTKFace and AppaReal test sets to measure both the IID the OOD performances of the models.

Table 1 shows enhanced performance of OR for OOD with both datasets and metrics. Accuracy and MAE are improved by 11.96% and 2.39 respectively for AppaReal. A similar improvement is seen for UTKFace with 12.28% and 2.36. This represents over 240% gain in MAE and over 270% gain in accuracy for both AppaReal and UTKFace with OOD tests. This is in contrast with the difference between classification and OR performance reported on IDD data, which is not noticeable between different training regimes.

6 Conclusion

Initial results suggest utilising OR to take advantage of the underlying ordinally structured data could give DA a stronger start when learning from OOD datasets. Future work will test whether these results can be replicated in other domains, i.e. medical and ecology, and if this advantage leads to better overall DA performance. The advantage of using OR is not apparent for IID tests. This serves as additional evidence for the presence of domain shift within the OOD results, making OOD tests a better indicator of real world performance.

Acknowledgments. This work was supported by the Engineering and Physical Sciences Research Council [EP/S023917/1].

References

1. Agustsson, E., Timofte, R., Escalera, S., Baro, X., Guyon, I., Rothe, R.: Apparent and real age estimation in still images with deep residual regressors on appa-real database. In: 2017 12th IEEE International Conference on Automatic Face & Gesture Recognition (FG 2017), pp. 87–94. IEEE (2017)
2. Bešenić, K., Ahlberg, J., Pandžić, I.S.: Picking out the bad apples: unsupervised biometric data filtering for refined age estimation. Vis. Comput. 39, 1–19 (2022)
3. Cao, W., Mirjalili, V., Raschka, S.: Rank consistent ordinal regression for neural networks with application to age estimation. Pattern Recogn. Lett. **140**, 325–331 (2020)
4. Chang, K.Y., Chen, C.S., Hung, Y.P.: Ordinal hyperplanes ranker with cost sensitivities for age estimation. In: Proceedings of the CVPR, pp. 585–592. IEEE (2011)
5. Chen, B.C., Chen, C.S., Hsu, W.H.: Cross-age reference coding for age-invariant face recognition and retrieval. In: Proceedings of the ECCV, pp. 768–783. Springer (2014)
6. Deng, J., Dong, W., Socher, R., Li, L.J., Li, K., Fei-Fei, L.: ImageNet: a large-scale hierarchical image database. In: Proceedings of the CVPR, pp. 248–255. IEEE (2009)
7. Escalera, S., et al.: Chalearn looking at people and faces of the world: face analysis workshop and challenge 2016. In: Proceedings of the CVPR Workshop, pp. 1–8. IEEE (2016)
8. Georgopoulos, M., Panagakis, Y., Pantic, M.: Investigating bias in deep face analysis: the kanface dataset and empirical study. Image Vis. Comput. **102**, 103954 (2020)
9. Kong, C., Luo, Q., Chen, G.: A comparison study: the impact of age and gender distribution on age estimation. In: Proceedings of the 3rd ACM International Conference on Multimedia in Asia, pp. 1–5 (2021)
10. Lanitis, A., Draganova, C., Christodoulou, C.: Comparing different classifiers for automatic age estimation. IEEE Trans. Syst. Man Cybern. Part B (Cybernetics) **34**(1), 621–628 (2004)
11. Lanitis, A., Taylor, C.J., Cootes, T.F.: Toward automatic simulation of aging effects on face images. Trans. PAMI **24**(4), 442–455 (2002)
12. Li, L., Lin, H.T.: Ordinal regression by extended binary classification. In: Advances in Neural Information Processing Systems, vol. 19 (2006)
13. Li, W., Huang, X., Lu, J., Feng, J., Zhou, J.: Learning probabilistic ordinal embeddings for uncertainty-aware regression. In: Proceedings of the CVPR, pp. 13896–13905 (2021)
14. Niu, Z., Zhou, M., Wang, L., Gao, X., Hua, G.: Ordinal regression with multiple output CNN for age estimation. In: Proceedings of the CVPR (2016)
15. Pan, S.J., Yang, Q.: A survey on transfer learning. IEEE Trans. Knowl. Data Eng. **22**(10), 1345–1359 (2009)
16. Paplham, J., Franc, V.: A call to reflect on evaluation practices for age estimation: comparative analysis of the state-of-the-art and a unified benchmark. In: Proceedings of the IEEE/CVF Conference on Computer Vision and Pattern Recognition, pp. 1196–1205 (2024)
17. Rodriguez-Vazquez, J., Fernandez-Cortizas, M., Perez-Saura, D., Molina, M., Campoy, P.: Overcoming domain shift in neural networks for accurate plant counting in aerial images. Remote Sensing **15**(6), 1700 (2023)

18. Rothe, R., Timofte, R., Van Gool, L.: Deep expectation of real and apparent age from a single image without facial landmarks. IJCV **126**(2–4), 144–157 (2018)
19. Seemakurthy, K., Bosilj, P., Aptoula, E., Fox, C.: Domain generalised fully convolutional one stage detection. In: Proceedings of the ICRA, pp. 7002–7009. IEEE (2023)
20. Serengil, S.I., Ozpinar, A.: Hyperextended lightface: a facial attribute analysis framework. In: Proceedings of the ICEET, pp. 1–4. IEEE (2021)
21. Shin, N.H., Lee, S.H., Kim, C.S.: Moving window regression: a novel approach to ordinal regression. In: Proceedings of the CVPR, pp. 18760–18769 (2022)
22. Srivastava, N., Hinton, G., Krizhevsky, A., Sutskever, I., Salakhutdinov, R.: Dropout: a simple way to prevent neural networks from overfitting. J. Mach. Learn. Res. **15**(56), 1929–1958 (2014)
23. Xie, S., Girshick, R., Dollár, P., Tu, Z., He, K.: Aggregated residual transformations for deep neural networks. In: Proceedings of the IEEE Conference on Computer Vision and Pattern Recognition, pp. 1492–1500 (2017)
24. Zhang, Z., Song, Y., Qi, H.: Age progression/regression by conditional adversarial autoencoder. In: Proceedings of the CVPR, pp. 5810–5818 (2017)

Pretrained Visual Representations in Reinforcement Learning

Emlyn Williams[✉] and Athanasios Polydoros

University of Lincoln, Lincoln, UK
{emwilliams,apolydoros}@lincoln.ac.uk

Abstract. Visual reinforcement learning (RL) has made significant progress in recent years, but the choice of visual feature extractor remains a crucial design decision. This paper compares the performance of RL algorithms that train a convolutional neural network (CNN) from scratch with those that utilize pre-trained visual representations (PVRs). We evaluate the Dormant Ratio Minimization (DRM) algorithm, a state-of-the-art visual RL method, against three PVRs: ResNet18, DINOv2, and Visual Cortex (VC). We use the Metaworld Push-v2 and Drawer-Open-v2 tasks for our comparison. Our results show that the choice of training from scratch compared to using PVRs for maximising performance is task-dependent, but PVRs offer advantages in terms of reduced replay buffer size and faster training times. We also identify a strong correlation between the dormant ratio and model performance, highlighting the importance of exploration in visual RL. Our study provides insights into the trade-offs between training from scratch and using PVRs, informing the design of future visual RL algorithms.

Keywords: Reinforcement Learning · Computer Vision · Robotics

1 Introduction

Reinforcement learning (RL) has made tremendous progress in recent years, with the development of algorithms that enable robots to learn complex behaviors from high-dimensional sensory inputs. In visual RL, where robots learn from raw image observations, a key challenge lies in extracting relevant features from the visual data. Despite significant advances in computer vision over the last decade, current state of the art models in visual RL still use simple convolutional neural networks (CNNs) trained from scratch for different tasks. This approach has been successfully employed in algorithms such as DRQ-V2 [14] and DRM [13], which have solved some of the most challenging continuous control tasks in common RL benchmarks such as the humanoid run and dog tasks in the deepmind control suite [13].

Notwithstanding these successes, training a CNN from scratch does not leverage the knowledge captured by pre-trained models on large datasets. Recently, there has been growing interest in using pre-trained image encoders as feature extractors for Imitation Learning (IL) and RL [1,5]. These pre-trained models have been trained on massive datasets and have learned to extract rich and generalizable features that can be useful for a wide range of downstream tasks [1,9,16]. By leveraging pre-trained models, RL algorithms may learn more efficiently.

Applying RL to robotics poses unique challenges however, particularly in terms of sample efficiency and control frequency. Specifically, data collection is often time-consuming and expensive as it requires physical interactions with the environment. Moreover, robots typically operate at high frequencies, requiring control signals to be generated at rates of 10–100 Hz or more. When training a real robot from scratch, this demands RL algorithms that can learn quickly and make decisions rapidly.

Despite the potential benefits of using pre-trained image encoders in RL, there are limited comparisons of the two methods in surrounding literature. In their investigation of an "artificial visual cortex" for embodied AI, Majumdar et al. [8] compare pre-trained visual representations (PVRs) across multiple robotic manipulation and navigation tasks, but do not compare these models to a learning-from-scratch RL implementation. In their comparisons, they find that no single PVR is universally dominant across multiple domains. Parisi et al. [10] compare learning from scratch to using PVRs, however their study only considers learning from demonstrations using imitation learning.

In this paper, we compare visual RL algorithms that train a CNN from scratch with counterparts that utilize frozen PVRs. We investigate the performance of these algorithms on a simulated robotic pushing task, and analyze the trade-offs between training a CNN from scratch and using pre-trained feature extractors.

2 Related Work

2.1 Visual Reinforcement Learning

In domains such as robotics, choosing the right RL method to maximize learning efficiency is important for minimising the risk of robot degradation. In this context, off-policy methods such as SAC [4] and DDPG [7] are generally more suitable than on-policy methods because off-policy methods can learn from experiences gathered without following the same policy that is being learned. This allows them to leverage data collected from various sources such as demonstrations, simulations, or other robots. In contrast, on-policy methods such as policy gradient methods require data to be collected using the same policy that is being learned. This can be limiting in applications where data acquisition is expensive.

More recent advances in off-policy visual RL, such as the Data-Regularized Q (DRQ) [6] algorithm and its successor, DRQ-V2 [14], have improved the efficiency and effectiveness of off-policy learning in visual RL tasks by using random shift augmentations for regularisation. These methods have been shown to achieve state-of-the-art performance in a variety of simulated robotics tasks such as grasping and manipulation, from a relatively small amount of data.

In this paper, we use the Dormant Ratio Minimization (DRM) [13] visual RL algorithm as the baseline. DRM was chosen because it is the most sample efficient RL algorithm for visual control and has exhibited very good performance on continuous control tasks [13]. DRM builds from the observation that there is a connection between the reduction of an agent's dormant ratio and the agent's skill acquisition in visual control tasks [13]. A dormant neuron is a neuron that has become nearly inactive, displaying a minimal activation level relative to other neurons in a layer. The dormant ratio is the percentage of these dormant neurons in the entire network [13]. DRM uses the dormant ratio to guide exploration, and periodically perturbs the agent's weights when the dormant ratio is high.

2.2 Reinforcement Learning with Pre-trained Visual Representations

Yuan et al. [16] adapt the DRQ-v2 algorithm [14] to use an ImageNet pretrained ResNet18 as a PVR. The authors find that using features from early layers provides better performance for continuous control tasks when compared to use of the full network. Their approach outperforms DRQ-v2 on Deepmind control suite tasks with video backgrounds.

Xiao et al. [12] demonstrate that self-supervised pre-training from real world images is effective for learning motor control from pixels. They train a vision transformer (ViT) model by masked modelling of natural images. This visual encoder is then frozen and the RL policy is trained using the frozen encoder. Another finding of the paper is that "in the wild" images, such as those from YouTube or egocentric videos, lead to better visual representations for manipulation tasks when compared to ImageNet images.

2.3 Learning from Video Demonstrations

In recent years, there has been a growing interest in using RL as a method of learning from video demonstrations (LfVD) [11,17]. LfVD involves learning a policy from a set of video demonstrations, often collected from humans or other agents. This is motivated by the fact that learning policies from expert videos is a scalable way of learning a wide array of tasks without access to ground truth rewards or actions [3]. By leveraging the knowledge and expertise encoded in these demonstrations, LfVD can significantly reduce the amount of data required to learn a task [11,17]. However, LfVD also relies on the ability to extract relevant features from the video data. Sharing a common PVR for both LfVD and RL could enable more efficient learning.

Zakka et al. [17] use an ImageNet pretrained ResNet18 with a self-supervised Temporal Cycle Consistency loss to learn reward functions from video demonstrations. They demonstrate learning robot policies from videos of human demonstrations on the Metaworld Push environment. Whilst they use a PVR for learning the reward, they then use state based RL for learning from the videos.

Finding a suitable PVR for visual RL could allow both the visual state observation and reward to be given by the PVR when using reward functions learned from videos.

3 Methodology

3.1 Benchmarking

We use the Push-v2 task from the Metaworld benchmarking suite for our evaluation. Metaworld consists of 50 robotic manipulation tasks such as opening and closing doors and drawers and pressing buttons [15]. The chosen task consists of pushing a small red puck towards a green sphere, with the initial puck and sphere positions randomised at the beginning of each episode. An episode is considered a success when the puck is within 5cm of the target at the final timestep. The reward function rewards minimizing the distance from the gripper to the puck, from the puck to the target site, and gives a bonus when the object is grasped by the robot. A maximum reward of 20 is given when the object is within 5cm of the target site.

From the Metaworld suite, this task was chosen as it involves manipulating a small object which could be contained entirely within one patch of a ViT. It was also selected due to it being one of the hardest tasks to solve in the Metaworld suite according to the original paper [15]. One explanation for the task's difficulty could be that the puck can fall on its side and roll away from the target position if not carefully placed, resulting in a failed episode even if the object temporarily reaches the target zone.

For further evaluation we use the Drawer-Open-v2 task. The task consists of opening a drawer, with the drawer position randomised at the beginning of each episode. The reward function rewards minimizing the distance of the gripper to the handle, and the distance of the handle to a target position. An episode is considered a success when the handle is less than 3 cm from a target position and the gripper is less than 3cm from the handle. This task was chosen due to it involving a larger and more natural looking object (Fig. 1).

All experiments use an image resolution of 112×112 and maintain a consistent third-person camera position unless otherwise noted. The default DRM hyperparameters [13] are used for all experiments.

3.2 Pretrained Visual Representations

We compare the baseline DRM algorithm to three PVRs: ResNet18, DINOv2, and Visual Cortex (VC). The PVR's weights are frozen throughout the RL training.

ResNet18. An ImageNet pretrained ResNet18 was chosen due it being a widely used and well established CNN that has demonstrated its efficacy in a variety of computer vision tasks such as classification, object detection and segmentation.

(a) Push-v2 (b) Drawer-Open-v2

Fig. 1. Metaworld Tasks

Other reasons for its use are its relatively small size and its high performance in visual RL, shown in "Pre-Trained Image Encoder for Generalizable Visual Reinforcement Learning" (PIEG) [16].

We adopt a similar approach to PIEG, where the images are fed through the first two layers of a ResNet18 CNN. To mitigate the high dimensionality of the output, we flatten the CNN output and pass it through a trainable fully connected layer with a reduced output size. Consistent with DRM's CNN encoder, we omit pooling layers after the CNN output to preserve spatial information. However, this design choice comes with a limitation: higher resolutions would result in an explosion of parameters in the fully connected layer, making it computationally prohibitive. The output of this layer is then fed into the actor and critic networks of DrM.

DINOv2. DINOv2 is a self-supervised vision backbone based on a ViT model. It is trained using a student teacher approach at both a whole image level and at a patch level. A 1 billion parameter model is trained on a curated dataset of 142 million images. This large model is then distilled down into smaller models. The backbone gives state of the art results for self-supervised networks in downstream tasks such as image classification, semantic segmentation and depth estimation. In this paper we utilise the 86 million parameter ViT-B model. DINOv2 was chosen as a PVR for comparison as it is one of the best performing self-supervised ViT based models for downstream tasks noted above.

Darcet et al. [1] find that artefacts in transformer feature maps correspond to high norm patch tokens appearing in low information background areas of images. The authors hypothesize that large, sufficiently trained models recognize these redundant tokens and use them as places to store, process and retrieve global information. By adding additional tokens (register tokens) to the transformer input to store and process global information, these high norm patch patch tokens disappear.

To use DINOv2 as a PVR, we pass the image through the model and feed the CLS token directly into the actor and critic networks. We also compare this to concatenating the register tokens to the CLS token and passing these concatenated tokens into the actor and critic networks.

Visual Cortex. Visual cortex (VC) [8] is a ViT based PVR model trained using Masked Auto-Encoding (MAE) on egocentric data, with the aim of being used for embodied intelligence. VC was chosen for this comparison because it was designed for use in embodied AI. It was also found to be the best performing PVR on average in comparisons performed in [8]. As with DINOv2, we feed the CLS token directly into the actor and critic networks.

3.3 Visual RL Algorithm

As we focus on the performance of the vision systems, we do not include robot proprioception information in the state space. To compare between the baseline and the pretrained backbones, the CNN based encoder is replaced with the pretrained backbone.

DRM stores observations as images in the replay buffer. At each update step a batch of these observations is augmented and passed through the image encoder, actor and critic networks. When using the ViT based PVRs, the image observations are passed through the frozen vision backbone at each environment step. The CLS token, which is a learnable vector that serves as a representation of the entire image [2], is given by the backbone and is stored in the replay buffer. This reduces the memory footprint of the replay buffer by over 90%, since the CLS token is a 768 dimensional float 32 vector, compared to the $3 \times 112 \times 112$ 8 bit integer image. Passing one image through the backbone at each time step also avoids the need to pass mini batches of 256 through the backbone at each update step.

Another advantage of storing the CLS tokens in the replay buffer is that the size of the replay buffer does not increase with higher resolutions, since only the CLS token is stored. The original DRM [13] paper and PIEG [16] use image observations of size 84×84 with a replay buffer size of 1M. Uncompressed, this results in a replay buffer size of 21.17 GB. At a resolution of 112×112, this increases to 37.63 GB, and at 224×224 (the resolution commonly used by ViT models), this increases to 150.53 GB. These large replay buffers pose challenges to the training of CNNs from scratch for visual RL.

Since no pooling layers are used in DRM's image encoder, the number of parameters in the actor and critic network increases with the input resolution. This limits the maximum resolution that can be used while still being able to run at a frequency suitable for robotic control. Table 1 shows how the number of parameters differs at different resolutions.

Table 1. Trainable Parameters

Model	112 × 112	224 × 224
DRM	15978281	57373481
PIEG	107340233	415621577
DINOv2	4768713	4768713
DINOV2 REG	6151113	6151113
VC	4768713	4768713

3.4 Performance Metrics

We present graphs of the episode reward and success rate of the evaluation episodes. The episode reward is the total reward obtained by the agent for an entire episode, and the success rate is the proportion of evaluation episodes where the agent successfully completed the task. All results are plotted as the mean performance over 5 seeds with a standard deviation shading of ± 0.5.

4 Experiments and Results

4.1 Data Augmentation

DRM applies a padding and random shift augmentation to all images before they are passed through the encoder to regularize the input data [13,14]. We first test whether this augmentation is beneficial when using PVRs which have been trained on large datasets with augmentations [9]. We compare the results of the DINOv2 CLS token on the Push-v2 task with and without the random shift augmentation.

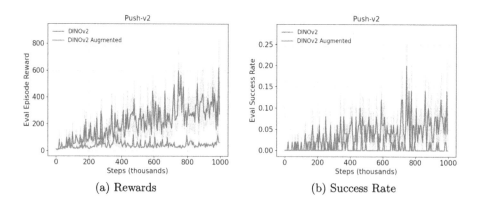

(a) Rewards (b) Success Rate

Fig. 2. Comparison of using plain images to using random shift augmentations.

As shown in Fig. 2, the DINOv2 PVR performed better when the random shift augmentations used in the baseline DRM algorithm were not applied.

Therefore, we do not apply these augmentations to images passed into the ViT based PVRs that are included in this comparative study.

4.2 PVR Performance

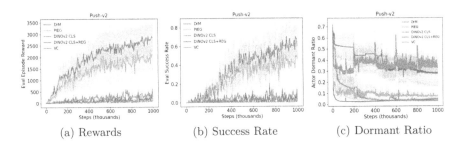

Fig. 3. Mean performance on Push-v2 Metaworld task.

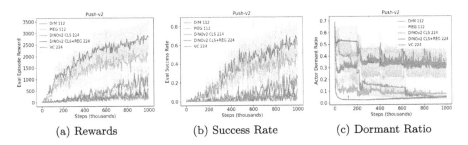

Fig. 4. Mean performance on Push-v2 Metaworld task with increased resolutions for ViT PVRs.

Figure 3 presents the performance of the different models on the Push-v2 task. The baseline DRM algorithm achieves the best performance, closely followed by the PIEG-based ResNet model. In contrast, the ViT-based models exhibit significantly poorer performance, highlighting a substantial gap between CNN-based and ViT-based models.

Figure 3 also reveals a strong correlation between the dormant ratio and model performance. The top-performing models, including the DRM and PIEG-based ResNet, demonstrate a drastic reduction in the dormant ratio throughout training. In contrast, the poorly performing models, including the ViT-based models, maintain a high dormant ratio throughout training.

To explore the potential benefits of using ViT-based models at higher resolutions, we additionally tested these models at a resolution of 224×224. Whilst

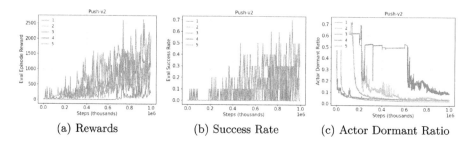

Fig. 5. Per Seed Performance of DINOv2 CLS token at 224 × 224 resolution

the CNN-based models are unable to operate at this resolution due to the computational constraints of the large replay buffer size, the ViT-based models can process images at this resolution with only a marginal decrease in control frequency.

As shown in Fig. 4, the performance of the ViT models improves slightly at the higher resolution, with the DINOv2 CLS token achieving the best performance. However, even at this higher resolution, the ViT models are still significantly outperformed by the CNN-based models at the lower resolution.

Figure 5 illustrates the high variance in performance across different seeds for the DINOv2 CLS model at the 224 × 224 resolution. Notably, the worst performing seeds are those that fail to reduce their dormant ratio throughout training. This trend is consistent across the ViT-based models, with the VC model exhibiting only one convergent seed out of five, and the DINOv2 with register tokens showing three convergent seeds out of five.

To investigate whether these results are task specific, we compare the ViT models at 224 × 224 to the CNN models at 112 × 112 on a drawer opening task. The results of this experiment are shown in Fig. 6. We find that the DINOv2 PVR outperforms both DRM and PIEG. This could be explained by the drawer providing more visual features due to its size and shape when compared to the rendered cylinder and sphere in the pushing task. The large difference between the rewards and success rate are due to some seeds ending stuck in a local minima of grasping the handle without opening the door, indicating the need for further exploration.

The videos of the evaluation episodes show that the robot sometimes learned different strategies to complete the task. Some agents would try to hit the puck towards the target with the outside of the gripper, whilst others would use the gripper to grasp the puck and place it near the target. These strategies are shown in Fig. 7.

5 Discussion

Our comparison reveals that the choice of training from scratch or using a PVR for maximising performance in visual RL is task dependant. With the DRM

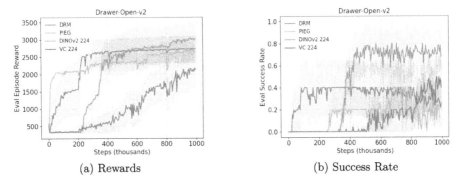

Fig. 6. Mean performance on Drawer-Open-v2 Metaworld task with increased resolutions for ViT PVRs.

Fig. 7. Examples of different learned approaches. Top: robot hits puck towards target. Bottom: Robot grabs object and moves it to target.

algorithm performing best on the pushing task and the DINOv2 PVR performing best on the drawer opening task, we hypothesize that PVRs perform better when the objects being manipulated are larger and more distinguishable. We also identify several advantages of using PVRs, including a significant reduction in replay buffer size and operating at a higher control frequency.

We find that on the push-v2 task the PIEG PVR outperforms other PVRs, and the DINOv2 CLS token is the top-performing ViT-based PVR. Interestingly, our results indicate that the DINOv2 ViT outperforms the Visual Cortex (VC) model, which was specifically designed for embodied AI. Whilst Darcet et al. [1] demonstrate the effectiveness of register tokens in training ViTs to avoid storing image-level information in patch tokens, we find that including these register tokens in the state space does not improve visual RL performance when compared to sole use of the CLS token.

Furthermore, our analysis highlights the importance of the dormant ratio in visual RL. We found that a lower dormant ratio is a strong indicator of successful learning, and the best-performing models demonstrate a drastic reduction in the dormant ratio throughout training. In contrast, poorly performing models maintained a high dormant ratio, suggesting that the robot was unable to effectively explore and learn from the environment. The PVR based approaches resulted in a higher number of failed seeds where the dormant ratio did not decrease throughout training. This finding has significant implications for incorporating PVRs into visual RL algorithms, as it suggests that incorporating alternative mechanisms to reduce the dormant ratio may be essential for achieving successful learning outcomes.

Using PVRs also reduces the number of trainable parameters, which can lead to faster training times and lower computational costs. This is crucial for tasks such as robotic manipulation and navigation in the real world, where data collection is costly and time-consuming. This trade-off between performance and complexity is important in visual RL, where large models can be prohibitive for real time control. While PVRs may not yet always match the performance of training from scratch, they offer a promising approach for balancing performance and complexity.

6 Conclusion

Our study provides a comprehensive analysis of the effectiveness of pre-trained PVRs in visual RL. Our results demonstrate that the optimal performance of PVRs or from scratch learning is task-dependent and highlight the importance of evaluating visual RL algorithms across multiple tasks and domains. We identify several advantages of using PVRs, including reduced replay buffer size and computational costs. We also find that the general purpose DINOv2 PVR outperforms the Visual Cortex PVR that was specifically designed for use in embodied AI. Our findings suggest that mechanisms to reduce the dormant ratio may be essential to achieve successful learning outcomes when incorporating PVRs into visual RL algorithms.

Future work will explore alternative RL algorithms, regularization approaches for PVRs, analysis across further tasks, and fine-tuning PVRs during training to provide a more comprehensive understanding of the strengths and limitations of PVRs in visual RL.

Acknowledgments. This work was supported by the Engineering and Physical Sciences Research Council [EP/S023917/1] and Dogtooth Robotics.

References

1. Darcet, T., Oquab, M., Mairal, J., Bojanowski, P.: Vision transformers need registers. In: The Twelfth International Conference on Learning Representations (2023)
2. Dosovitskiy, A., et al.: An image is worth 16×16 words: transformers for image recognition at scale. In: International Conference on Learning Representations (2020)

3. Escontrela, A., et al.: Video prediction models as rewards for reinforcement learning. Adv. Neural Inf. Process. Syst. **36** (2024)
4. Haarnoja, T., et al.: Soft actor-critic algorithms and applications. arXiv preprint arXiv:1812.05905 (2018)
5. Karamcheti, S., et al.: Language-driven representation learning for robotics. In: Proceedings of Robotics: Science and Systems. Daegu, Republic of Korea (2023). https://doi.org/10.15607/RSS.2023.XIX.032
6. Kostrikov, I., Yarats, D., Fergus, R.: Image augmentation is all you need: regularizing deep reinforcement learning from pixels. arXiv preprint arXiv:2004.13649 (2020)
7. Lillicrap, T.P., et al.: Continuous control with deep reinforcement learning. arXiv preprint arXiv:1509.02971 (2015)
8. Majumdar, A., et al.: Where are we in the search for an artificial visual cortex for embodied intelligence? Adv. Neural Inf. Process. Syst. **36** (2024)
9. Oquab, M., et al.: Dinov2: learning robust visual features without supervision. Trans. Mach. Learn. Res. (2023)
10. Parisi, S., Rajeswaran, A., Purushwalkam, S., Gupta, A.: The unsurprising effectiveness of pre-trained vision models for control. In: International Conference on Machine Learning, pp. 17359–17371. PMLR (2022)
11. Schmeckpeper, K., Rybkin, O., Daniilidis, K., Levine, S., Finn, C.: Reinforcement learning with videos: combining offline observations with interaction. In: Conference on Robot Learning, pp. 339–354. PMLR (2021)
12. Xiao, T., Radosavovic, I., Darrell, T., Malik, J.: Masked visual pre-training for motor control. arXiv preprint arXiv:2203.06173 (2022)
13. Xu, G., et al.: Drm: mastering visual reinforcement learning through dormant ratio minimization. In: The Twelfth International Conference on Learning Representations (2023)
14. Yarats, D., Fergus, R., Lazaric, A., Pinto, L.: Mastering visual continuous control: improved data-augmented reinforcement learning. arXiv preprint arXiv:2107.09645 (2021)
15. Yu, T., Quillen, D., He, Z., Julian, R., Hausman, K., Finn, C., Levine, S.: Metaworld: a benchmark and evaluation for multi-task and meta reinforcement learning. In: Conference on Robot Learning, pp. 1094–1100. PMLR (2020)
16. Yuan, Z., et al.: Pre-trained image encoder for generalizable visual reinforcement learning. Adv. Neural. Inf. Process. Syst. **35**, 13022–13037 (2022)
17. Zakka, K., Zeng, A., Florence, P., Tompson, J., Bohg, J., Dwibedi, D.: Xirl: cross-embodiment inverse reinforcement learning. In: Conference on Robot Learning, pp. 537–546. PMLR (2022)

IntelliMove: Enhancing Robotic Planning with Semantic Mapping

Fama Ngom[1,3]([✉]), Huaxi Yulin Zhang[2], Lei Zhang[1], Karen Godary-Dejean[3], and Marianne Huchard[3]

[1] S.A.S. euroDAO, 161 rue Ada, 34095 Montpellier cedex 5, France
{fama.ngom,lei.zhang}@eurodao.com
[2] Université de Picardie Jules Verne, 48 Rue d'Ostende, 02100 Saint Quentin, France
yulin.zhang@u-picardie.fr
[3] LIRMM, CNRS, Université de Montpellier, Montpellier, France
{karen.godary-dejean,marianne.huchard}@lirmm.fr

Abstract. Semantic navigation enables robots to understand their environments beyond basic geometry, allowing them to reason about objects, their functions, and their interrelationships. In semantic robotic navigation, creating accurate and semantically enriched maps is fundamental. Planning based on semantic maps not only enhances the robot's planning efficiency and computational speed but also makes the planning more meaningful, supporting a broader range of semantic tasks.

In this paper, we introduce two core modules of IntelliMove: IntelliMap, a generic hierarchical semantic topometric map framework developed through an analysis of current technologies strengths and weaknesses, and Semantic Planning, which utilizes the semantic maps from IntelliMap. We showcase use cases that highlight IntelliMove's adaptability and effectiveness. Through experiments in simulated environments, we further demonstrate IntelliMove's capability in semantic navigation.

Keywords: Semantic navigation · Semantic mapping · Semantic planning

1 Introduction

Recent advancements in artificial intelligence have significantly shifted the focus towards augmenting the autonomy of robots by integrating high-level information into robotic systems. This shift has particularly influenced the domain of mobile robotics, where embedding semantic data into navigational functions has catalyzed the development of semantic navigation. Semantic navigation is defined as a system that incorporates semantic information into environmental representations, leveraging this data during robot localization and navigation processes to enrich understanding and functionality [1–3].

This research was funded by the EUROCLUSTER DREAM Project under the subproject *IntelliMove* and by France BPI Innovation under Grant No. *0208161/00*.

Semantic mapping plays a pivotal role in robotics and autonomous systems, enhancing environmental models with high-level abstract elements that effectively bridge the gap between human understanding and robotic interpretation. The development of hierarchical semantic maps, integrating metric, topological, and semantic layers, is seen as a key advancement. The literature on hierarchical representations for semantic navigation, including hierarchical topometric models [4-6] and 3D scene graphs [7-9], reveals a trade-off between operational efficiency and semantic richness. Thus, a primary challenge remains: finding a balance between achieving computational efficiency and incorporating rich semantic details into map representations for effective semantic navigation.

The IntelliMove framework introduced in this paper comprises two main components: IntelliMap and Semantic Planning. IntelliMap is a hierarchical semantic topometric map framework that combines metric-semantic and topological-semantic data within a multi-layered structure to balance computational efficiency with environmental comprehension. Semantic Planning uses these enriched maps to enable robots to develop and execute contextually relevant navigation and task strategies, adapting to dynamic environmental changes. This planning process includes traditional graph-based pathfinding integrated with advanced semantic analysis, enhancing path generation from the start to designated goals, which could be specific objects or rooms, thus improving flexibility and efficiency.

The paper is organized as follows: Sect. 2 reviews related work, Sect. 3 describes the IntelliMap framework and its implementation, Sect. 4 explains the semantic planning algorithm, Sect. 5 details experiments and evaluations, and Sect. 6 concludes with future work directions.

2 Related Work

Semantic Mapping. Within the present research context, two primary categories of semantic topometric maps have emerged as prominent.

1) *Hierarchical Topometric Map.* Predominantly based on hierarchical topometric maps [4-6], this mapping approach utilizes inference methods on metric maps but lacks object-centric semantic details. It only recognizes basic categories such as rooms, corridors, and doors, showing limitations in early research works due to the lack of scmantic scene understandings.
2) *3D Scene Graph.* To improve upon these maps, the 3D scene graph technique [7,8,10,11] provides an all-encompassing representation of the environment. It integrates a 3D semantic mesh layer, object layer, complex topological map, room layer, and building layer into the mapping process, requiring significant computational resources [12,13].

Each of the mentioned map types has its own limitations, highlighting the need for a map type that balances efficiency, detail, and semantic understanding. In this paper, we proposes IntelliMap, a hierarchical semantic topometric mapping framework that seamlessly integrates metric-semantic and topological-semantic representations in a balanced manner.

Semantic Planning. Recent research has proposed methodologies for generating path planning that integrate semantic information of the environment. For instance, [14] introduces S-Nav, a semantic-geometric planner that enhances the performance of geometric planning through a hierarchical architecture. It exploits semantic information concerning room boundaries and doorways but does not assign semantic labels to rooms based on specific functions such as kitchens or offices. Similarly, [15] presents the SMaNa navigation stack, which comprises an online 2.5D semantic navigation graph builder and a weighted A* pathfinder. While [16] utilizes a semantic map consisting of three layers: a costmap, a conventional exploration grid, and a binary grid that records observations of distinct semantic categories. These works [17–19] generate paths to the designated semantic object goals based on object semantic map. However, they lack the ability to consider a higher-level semantic abstraction of the room environment, which would be beneficial for improved planning and decision-making. Finally, [20] conducts indoor path planning by utilizing a methodology that employs three layers for a 2D floor map, integrating predefined semantic encoding for objects, obstacles, dynamic entities, and room properties

As mentioned in the previous section, hierarchical semantic mapping integrated with metric maps is crucial not only for semantic planning but also for maintaining the precision of navigation. The works mentioned often focus on a single semantic layer of the environment, with or without a metric map. This limitation restricts their ability to achieve deeper semantic planning.

3 IntelliMap: Semantic Mapping

This section presents our framework, IntelliMap, a hierarchical semantic topometric mapping framework and one of its implementation.

3.1 Framework

The IntelliMap framework in Fig. 1 provides a robust approach to creating semantic maps for autonomous robotic navigation and decision-making across various settings. It starts with data collection via multiple sensors and utilizes a metric SLAM system to produce a geometrically accurate map. An Object Mapping component then integrates this data to identify and classify objects, organizing them spatially within the metric map to form a metric-semantic map or developing a separate topological semantic layer for specific needs. The process culminates in spatial organization and relational mapping, achieving a comprehensive topological semantic map.

Input Data: Sensor Data. The IntelliMap framework primarily utilizes camera inputs to capture essential visual information. Additional sensors like LiDAR and IMUs can be integrated for enhanced perception and accuracy.

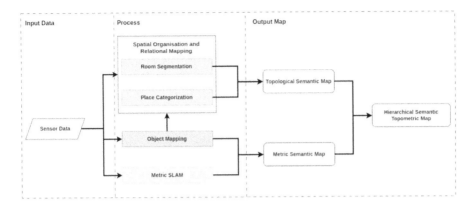

Fig. 1. IntelliMap framework for semantic robot navigation. This figure illustrates the core components of the IntelliMap framework, designed to generate hierarchical semantic topometric maps for robot navigation in diverse environments.

Map Construction Process

- **Metric SLAM.** In our IntelliMap framework, metric Simultaneous Localization and Mapping (SLAM) module serves a dual purpose. First, it constructs a foundational metric map that captures the physical layout and geometry of the environment. Second, it continuously determines the robot's pose (position and orientation) within the constructed metric map.
- **Object Mapping.** The Object Mapping module, integral to the SLAM system, captures and localizes objects using the robot's pose data. It transitions raw data into semantic insights, enabling interaction and navigation within a semantically rich environment. This module not only recognizes and locates distinct objects within visual data but also assigns precise positions relative to the metric map. It plays a crucial role in spatial organization by categorizing places based on the objects' presence and distribution. Object information can be integrated into the metric semantic map, where objects are directly incorporated into the metric map, or in a separate object map layer.
- **Room Segmentation.** The room segmentation algorithm can identify discernible environmental features that differentiate one room from another, such as physical boundaries (walls, floor-level variations), or object clusters. Approaches to this segmentation process are detailed on earlier studies.
- **Place Categorization.** Without place categorization, the map might simply show rectangular rooms labeled as "Room 1", "Room 2", and so on. Thanks to place categorization, these rooms are identified as "Office", "Living Room", each carrying a specific meaning and context. Various approaches to achieve place categorization include ontology-based or LLM-based.

Output Maps. The previous two steps generate two separate maps: a metric semantic map and a semantic topological map. While each map can be used for semantic robot navigation, our goal is to create a unified map that leverages both.

- **Metric Semantic Map:** Combines metric SLAM data with semantic object mapping to detail both the environment's geometry and its object labels.
- **Topological Semantic Map:** Arises from Spatial Organisation and Relational Mapping, illustrating room connectivity and their functional relationships, useful for abstracted navigation and planning.
- **Hierarchical Semantic Topometric Map:** Integrates topological and metric information in a layered, hierarchical format, enhancing environmental representation. Its adaptable structure supports customization for specific applications, merging precise spatial details with a broader contextual understanding of space and object interactions.

3.2 A Three-Layered Semantic Map Based on Costmap

This section introduces a three-layered semantic map integrating the IntelliMap framework to improve semantic robot navigation and spatial understanding in varied environments. The layers include Metric, Object, and Room categories. Figure 2 displays the IntelliMap output in a simulated office setting, utilizing the uHumans2 dataset [13]. This photo-realistic Unity simulation, enriched with diverse objects and spaces like desks, chairs, and corridors, serves as a preliminary evaluation of the framework's adaptability and effectiveness in complex settings. The three-layered structure significantly enhances navigation in indoor environments, with detailed results.

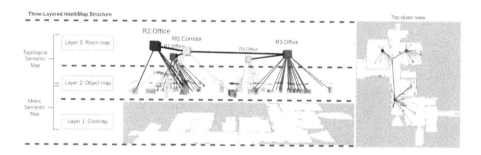

Fig. 2. Three-Layered IntelliMap in a simulated office environment

Our proposed IntelliMap framework, tailored for semantic robot navigation in diverse indoor environments utilizes data from various modules to construct a three-layered map (Fig. 2).

- **Metric Layer.** The Metric Map, generated through the SLAM process, encapsulates various spatial representations of the environment, such as costmaps, octomaps, or grid maps.

– **Object Layer.** Derived from the Object Mapping module, this layer builds upon the Metric Map by identifying and incorporating objects within the environment. It features identifiable objects, each mapped with respect to its location and orientation, overlaying the Metric Map with detailed object information.
– **Room Layer.** The highest abstraction layer emerges from the place categorization and room segmentation modules. By providing a topological overview of space categorization and room segmentation, this layer facilitates advanced decision-making and planning.

The metric layer and the object layer collectively form a metric semantic map, but are separated into distinct layers. The semantic topological map, serving as the third layer, merges with the previous layers to create the three-layered semantic map.

In the three-layered semantic map, semantic information is organized within a graph. The object layer represents objects as nodes, while the room layer assigns nodes to specific rooms. Graph edges denote physical connections between rooms, and connections between object and room layers indicate the presence of specific objects within their respective rooms.

4 Semantic Planning Based on IntelliMap

In this section, we present a semantic planning based on the IntelliMap introduced previously. By leveraging semantic mapping, robots achieve a deeper comprehension of their environments, enriched with semantic significance. This advanced paradigm in robotic navigation transcends traditional methods by incorporating extensive semantic information into navigation systems. Our approach to semantic planning not only utilizes fine-grained semantic details, such as objects, but also incorporates coarse-grained semantic information, such as rooms. This dual-level granularity enhances the robot's ability to navigate and interact within complex environments effectively.

4.1 Overview

Semantic Planning Definition. In this paper, we discuss semantic path planning, which combines graph-theoretical constructs with real-world navigation by leveraging IntelliMap's semantic graph. Nodes in this graph represent elements like rooms and objects, while edges, can be weighted by distance and energy, also encode semantic relationships between these elements. This integration not only enhances the robot's understanding of its environment but also optimizes navigation paths contextually and spatially. Consequently, the robot's movements and tasks align more effectively with the environment's semantic structure, improving task performance efficiency.

IntelliPlanner output: Semantic path from the bookcase in office room 1 to the desk in office room 3

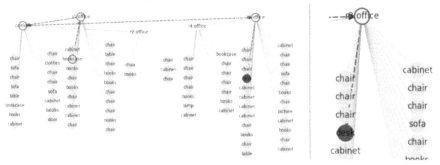

Fig. 3. Example of a red semantic path generated in an office environment, navigating from the bookcase in office room 1 to a desk in office room 3. (Color figure online)

Example. An example of semantic planning in an office environment is illustrated in Fig. 3, showing a path from a bookcase in office room 1 to a desk in office room 3. This example highlights our semantic planning capabilities for navigating complex semantic contexts. For example, when a robot is tasked with moving from an office to a conference room, our algorithm generates a path that includes all necessary intermediate rooms or nodes. This strategic navigation plan optimizes the route based on semantic relevance and travel costs, thus enhancing planning efficiency in large-scale environments compared to traditional metric maps.

4.2 Algorithm of Planning

The semantic planning process, detailed in Algorithm 1, involves assessing the type of goal, integrating semantic data, and executing optimal path calculations to ensure that the robot's navigation is both effective and adaptively responsive to the environment's semantic attributes.

Path Input. Our semantic planning algorithm introduces a sophisticated approach that interprets and interacts with the environment, utilizing the following structured inputs to facilitate intelligent pathfinding:

- Semantic_map: Represents the navigational space, organized into nodes that denote important locations such as rooms or objects. Edges connect rooms and delineate feasible paths, facilitating navigation within the environment. These edges also capture the semantic relationships between rooms and objects, with each edge can be weighted by factors such as distance, time, etc.
- Start Node: Denotes the initial position of the robot within the semantic map, serving as the start point of the navigation task.
- Target Goal: Specifies the target endpoint for the navigation, which could be either a specific object or a room.

Algorithm 1. Enhanced Semantic Planning with IntelliMove

1: **Inputs:** *semantic_map, start, goal*
2: **Output:** *best_path*
3: **procedure** INTELLIMOVESEMANTICPLANNING(*semantic_map, start, goal*)
4:     *best_path* ← empty path
5:     *goal_state* ← FINDGOALSTATE(*semantic_map, goal*)
6:     **if** *goal_state* is empty **then**
7:         **Discovery Mode:**
8:         *goal_LLM* ← GOALLLMRESPONSE(*semantic_map, goal*)
9:         *best_path* ← DIJKSTRA(*semantic_map, start, goal_LLM*)
10:         **return** *best_path*
11:     **else if** length of *goal_state* = 1 **then**
12:         **Targeted Navigation Mode:**
13:         *best_path* ← DIJKSTRA(*semantic_map, start, goal*)
14:         **return** *best_path*
15:     **else**
16:         **Multi-target Exploration Mode:**
17:         Initialize *min_path_length* ← ∞
18:         **for** each *node* in *goal_state* **do**
19:             *path* ← DIJKSTRA(*semantic_map, start, node*)
20:             **if** *path* is not None and *path.length*() < *min_path_length* **then**
21:                 *best_path* ← *path*
22:                 *min_path_length* ← *path.length*()
23:             **end if**
24:         **end for**
25:         **return** *best_path*
26:     **end if**
27: **end procedure**

Path Output. In the semantic planning, the output path plays an important role as the tangible result of the semantic planning process. This path represents the calculated route that the robot will follow to reach its designated goal. Below, the nature and significance of the output path are detailed, emphasizing its role in practical robot navigation.

The path is an ordered list of nodes starting from the initial position (start node) and ending at the target destination (goal node). Each node corresponds to a significant location within the environment, such as a specific room or object, defined in the semantic map. The chosen path is the most efficient route calculated based on the weighted edges of the graph. Each segment of the path is contextually relevant, reflecting the robot's understanding of the environment's semantic layout. This includes navigating through areas that are semantically tagged with specific functionalities or roles.

Process. The semantic planning process is meticulously outlined in the algorithm 1, showing how it determines the most efficient route from the start node to the destination. This process involves assessing the type of goal, integrating semantic data, and executing optimal path calculations. By doing so, the algorithm ensures that the robot's navigation is both effective and adaptively responsive to the environment's semantic attributes. The algorithm orchestrates sophisticated semantic planning by leveraging a structured approach based on the nature of the navigation goal.

Operational phases of the semantic planning algorithm are:

- **Goal Evaluation and Path Planning**: The algorithm distinguishes between three types of navigation mode: Discovery Mode, Targeted Navigation Mode and Multi-target Exploration Mode.
  - **Discovery Mode**: When the destination object is not predefined in the map, the algorithm extracts room attributes to understand the contextual setup of the environment. It then employs Large Language Model(LLM) to identify the most probable locations for the object, based on the semantic attributes of existing rooms within the semantic map. Once these probable locations are identified, Dijkstra's algorithm is applied to calculate the most efficient routes to these rooms, ensuring that the navigation is directed towards the most likely locations as predicted by the Large Language Model (LLM). In our implementation, we employed the ChatGPT 3.5 large language model.
  - **Targeted Navigation Mode**: The target point is singular. For navigation to a known destination, relevant nodes are identified. Subsequently, Dijkstra's algorithm calculates the shortest paths from the start node to these identified targets, optimizing the route based on the weights assigned to the edges in the graph.
  - **Multi-target Exploration Mode**: If there are multiple target points in the map, the robot will choose to explore the nearest target point.
- **Path Execution**: the algorithm is designed to manage navigation efficiently by automatically selecting the optimal route. It evaluates and prioritizes paths based on their feasibility and proximity, thereby directing the robot to the most accessible location effectively.

5 Experimentation

This section presents the findings of our comprehensive evaluation of our proposed IntelliMap framework and Semantic Planning in office environments.

5.1 Evaluation

Semantic Information Coverage. Table 1 provides a comparative analysis of the structure of IntelliMove's two components: IntelliMap and Semantic Planning, in comparison to other studies. This analysis evaluates various features, including metric map representation, object and room representation, and object-based and room-based semantic planning. IntelliMove demonstrates the most extensive coverage across these features.

Table 1. Comparative analysis of layer configurations and semantic planning in IntelliMove and existing methods

Method	Metric Map	Semantic Map		Semantic Planning	
		Object	Room	Object Based	Room Based
IntelliMove	Costmap	Yes	Yes	Yes	Yes
Zimmerman et al., 2023 [21]	Costmap	Yes	No	No	No
Hughes et al., 2022 [10]	Mesh	Yes	Yes	No	No
He et al., 2021 [6]	Pointcloud	No	Yes	No	No
Hiller et al., 2019 [5]	Costmap	No	Yes	No	No
Gemignani et al., 2016 [22]	Costmap	Yes	Yes	No	No
Paul et al., 2023 [14]	Room contours and SDF	Yes	Yes	No	Yes
Devendra et al., 2020 [17]	Costmap	Yes	No	Yes	No

Computational Efficiency. The experiments were conducted on a Dell Precision 7550 workstation equipped with 64.0 GiB of memory, an Intel® Core™ i9-10885H CPU @ 2.40 GHz × 16 processor, and an NVIDIA GeForce RTX 3000 Mobile/Max-Q graphics card.

We evaluated the semantic planning algorithm's performance focusing on its runtime and success rate. Runtime is defined as the duration from algorithm initiation to result generation, while the success rate measures the likelihood of accurately generating a path. In 50 tests within a simulated office environment, the average success time for existing targets was 7 ms, with a maximum runtime of 10 milliseconds. The algorithm achieved a 99% success rate for known targets and 55% for goal discovery scenarios.

5.2 Discussion

Based on the experimental tests conducted, we can conclude that our algorithm provides a framework for semantic robot navigation, characterized by its efficiency, adaptability, and deep integration of semantic context:

– **Efficiency and Precision**: It incorporates Dijkstra's algorithm to ensure accurate and optimal pathfinding, continuously optimized through real-time data adjustments.
– **Adaptability**: Designed to handle both specific navigational targets and exploratory goals, the semantic planning seamlessly adapts to a wide range of operational scenarios, enhancing its functionality in complex environments.
– **Semantic Awareness**: The system's use of AI for semantic data analysis guarantees navigation strategies that are deeply aware of the context, significantly improving robot interactions within their operational environment.

Our semantic planning approach to semantic robot navigation merges traditional graph-based pathfinding techniques with advanced semantic analysis, enabling efficient and context-aware navigation. This dual capability ensures that robots equipped with this algorithm are not only capable of navigating through intricate settings but can also adjust their paths in response to changing conditions and sophisticated mission requirements.

6 Conclusion

In this paper, we present the IntelliMove framework, which comprises two fundamental components: IntelliMap, a hierarchical semantic topometric mapping framework, and Semantic Planning. IntelliMap is engineered to generate nuanced and adaptable maps tailored for semantic robot navigation. By harnessing existing methodologies like hierarchical topometric maps and 3D scene graph while mitigating their drawbacks (e.g., high computational overhead, limited scalability), IntelliMap offers a modular, flexible, and resilient solution.

Building upon the IntelliMap framework, semantic planning merges traditional pathfinding techniques with advanced semantic analysis to provide not just efficient but also context-aware navigation. This dual functionality ensures navigation strategies that are highly adaptable and precise, enabling robots to traverse intricate environments with a level of understanding akin to human perception. The successful deployment of IntelliMap and our semantic planning algorithm within the IntelliMove project underscores their efficacy. This paper illustrates how these frameworks facilitate swifter and more efficient navigation by integrating semantic insights at various stages of the process.

References

1. Kostavelis, I., Gasteratos, A.: Semantic mapping for mobile robotics tasks: a survey. Rob. Auton. Syst. **66**, 86–103 (2015)
2. Wang, L., et al.: Visual semantic navigation based on deep learning for indoor mobile robots. Complexity **2018** (2018)
3. Barber, R., Crespo, J., Gómez, C., Hernámdez, A.C., Galli, M.: Mobile robot navigation in indoor environments: geometric, topological, and semantic navigation. In: Applications of Mobile Robots. IntechOpen (2018)
4. Kuipers, B.: The spatial semantic hierarchy. Artif. Intell. **119**(1), 191–233 (2000)
5. Hiller, M., Qiu, C., Particke, F., Hofmann, C., Thielecke, J.: Learning topometric semantic maps from occupancy grids. In: IEEE/RSJ International Conference on Intelligent Robots and Systems (IROS) 2019, pp. 4190–4197 (2019)
6. He, Z., Sun, H., Hou, J., Ha, Y., Schwertfeger, S.: Hierarchical topometric representation of 3d robotic maps. Auton. Robots **45**(5), 755–771 (2021)
7. Armeni, I., et al.: 3d scene graph: a structure for unified semantics, 3d space, and camera. In: Proceedings of the IEEE/CVF International Conference on Computer Vision, pp. 5663–5672. IEEE, Seoul (2019)
8. Kim, U.H., Park, J.M., Song, T.J., Kim, J.H.: 3-D scene graph: a sparse and semantic representation of physical environments for intelligent agents. IEEE Trans. Cybern. **50**(12), 4921–4933 (2019)
9. Gu, Q., et al.: Conceptgraphs: open-vocabulary 3d scene graphs for perception and planning. arXiv preprint arXiv:2309.16650 (2023)
10. Hughes, N., Chang, Y., Carlone, L.: Hydra: A real-time spatial perception system for 3d scene graph construction and optimization. In: Robotics, Science and Systems XVIII (2022)
11. Wu, S.-C., Wald, J., Tateno, K., Navab, N., Tombari, F.: Scenegraphfusion: incremental 3d scene graph prediction from rgb-d sequences. In: Proceedings of the IEEE/CVF Conference on Computer Vision and Pattern Recognition, pp. 7515–7525 (2021)

12. Rosinol, A., Gupta, A., Abate, M., Shi, J., Carlone, L.: 3D dynamic scene graphs: actionable spatial perception with places, objects, and humans. In: Toussaint, M., Bicchi, A., Hermans, T. (eds.) Robotics: Science and Systems XVI, Oregon, USA (2020)
13. Rosinol, A., et al.: Kimera: from SLAM to spatial perception with 3D dynamic scene graphs. Int. J. Rob. Res. **40**(12–14), 1510–1546 (2021)
14. Kremer, P., Bavle, H., Sanchez-Lopez, J.L., Voos, H.: S-nav: semantic-geometric planning for mobile robots. arXiv preprint arXiv:2307.01613 (2023)
15. Serdel, Q., Marzat, J., Moras, J.: Smana: semantic mapping and navigation architecture for autonomous robots. In: Gini, G., Nijmeijer, H., Filev, D.P. (eds.) Proceedings of the 20th International Conference on Informatics in Control, Automation and Robotics, ICINCO 2023, Rome, Italy, 13–15 November 2023, vol. 1, pp. 453–464. SCITEPRESS (2023)
16. Achat, S., Marzat, J., Moras, J.: Path planning incorporating semantic information for autonomous robot navigation. In: 19th International Conference on Informatics in Control, Automation and Robotics (ICINCO) (2022)
17. Chaplot, D.S., Gandhi, D.P., Gupta, A., Salakhutdinov, R.R.: Object goal navigation using goal-oriented semantic exploration. Adv. Neural. Inf. Process. Syst. **33**, 4247–4258 (2020)
18. Fukushima, R., Ota, K., Kanezaki, A., Sasaki, Y., Yoshiyasu, Y.: Object memory transformer for object goal navigation. In: 2022 International Conference on Robotics and Automation (ICRA), pp. 11 288–11 294. IEEE (2022)
19. Majumdar, A., Aggarwal, G., Devnani, B., Hoffman, J., Batra, D.: Zson: zero-shot object-goal navigation using multimodal goal embeddings. Adv. Neural Inf. Process. Syst. **35**, 32340–32352 (2022)
20. Sun, N., Yang, E., Corney, J., Chen, Y.: Semantic path planning for indoor navigation and household tasks. In: Althoefer, K., Konstantinova, J., Zhang, K. (eds.) TAROS 2019. LNCS (LNAI), vol. 11650, pp. 191–201. Springer, Cham (2019). https://doi.org/10.1007/978-3-030-25332-5_17
21. Zimmerman, N., Sodano, M., Marks, E., Behley, J., Stachniss, C.: Constructing Metric-semantic maps using floor plan priors for long-term indoor localization. In: IEEE/RSJ International Conference on Intelligent Robots and Systems (IROS) (2023)
22. Gemignani, G., Capobianco, R., Bastianelli, E., Bloisi, D.D., Iocchi, L., Nardi, D.: Living with robots: interactive environmental knowledge acquisition. Robot. Auton. Syst. **78**, 1–16 (2016)

Localisation-Aware Fine-Tuning for Realistic PointGoal Navigation

Fraser McGhan[1], Ze Ji[1], and Raphael Grech[2]

[1] Cardiff University, Cardiff, UK
{mcghanf,jiz1}@cardiff.ac.uk
[2] Spirent Communications, Paignton, UK
raphael.grech@spirent.com

Abstract. Prior research has demonstrated the effectiveness of end-to-end reinforcement learning for PointGoal navigation tasks within indoor environments. Given 2.5 billion frames of experience, a navigation policy can be trained to achieve a success rate of 0.94 when deployed in unseen environments. However, a limitation of this approach is its reliance on perfect localisation, which is unrealistic for real-world deployment scenarios where localisation must be estimated, inevitably introducing errors. In this paper, we present a study on the effectiveness of integrating a traditional vision-based SLAM algorithm with a reinforcement learning-based PointGoal navigation policy. Through our experimentation, we demonstrate how fine-tuning a pre-trained navigation policy on realistic localisation estimates can increase the success rate by 14% (0.71 → 0.85) and SPL by 15% (0.66 → 0.81) when compared to deploying policies in a zero-shot manner.

Keywords: Reinforcement Learning · PointGoal Navigation · Sim-to-Real Transfer

1 Introduction

Navigation in indoor environments is a fundamental challenge for autonomous agents, with applications ranging from household assistance to industrial automation. Recent advancements in reinforcement learning (RL) have shown promising results in addressing PointGoal navigation tasks [1], where agents must navigate to a specific target location within a given environment. End-to-end RL approaches, which learn navigation policies directly from sensory inputs, have achieved high success rates in simulated environments, leveraging large-scale datasets and distributed training algorithms [12].

However, the transition from simulation to real-world deployment presents significant challenges, particularly concerning the robustness of navigation policies. A limitation of existing approaches is their reliance on perfect localisation, an assumption rarely met in practice where localisation must be estimated. This discrepancy between simulation and reality underscores the need to evaluate the robustness of pre-trained navigation policies to localisation inaccuracies and explore strategies for improving their performance in real-world settings.

We address this gap by investigating the effectiveness of fine-tuning pre-trained navigation policies using realistic localisation estimates from ORB-SLAM2. Our study aims to provide practical insights into enhancing navigation capabilities for autonomous agents and robots operating in real-world environments. By evaluating the impact of localisation errors and proposing effective mitigation strategies, we contribute to the advancement of navigation algorithms that are robust and reliable in indoor settings.

The remainder of the paper is structured as follows: Sect. 2 introduces previous works related to PointGoal navigation, Sect. 3 details the methodology proposed in this work, Sect. 4 describes the experiments performed and the results we obtained, and Sect. 5 summarises the contributions of this work.

2 Related Work

2.1 PointGoal Navigation Task

The objective of the PointGoal navigation task [1] is for an autonomous agent to navigate from its current location to a specified target location within a given indoor environment. Depending on the environment specification, the agent may utilise RGB, depth, semantic segmentation, or even a top-down egocentric map to perceive its surroundings. The target location is typically specified by polar coordinates representing the current distance and angle to the goal. This measurement is updated at every step in the environment and therefore requires access to the true position and heading of the agent. The success of the agent is defined by the agent stopping within the success radius of the goal location and not exceeding the maximum allowed number of steps.

2.2 Reinforcement Learning for PointGoal Navigation

Savva et al. proposed a reinforcement learning agent for PointGoal navigation in their work that introduced the Habitat simulator [11]. The authors trained an end-to-end agent that uses depth information and perfect localisation for 75 million steps on the Gibson dataset and achieved a success rate of 0.89 on unseen scenes.

Chaplot et al. presented a modular and hierarchical approach to learn policies for exploring indoor environments [2]. The authors combined both classical and learning-based methods by using analytical path planners with a learned mapping module, and global and local reinforcement learning-based policies. The designed approach uses multiple input modalities for each module, including depth images, a predicted egocentric top-down map, and notably ground-truth agent pose. The proposed model was trained for 10 million steps on the Gibson dataset and achieved a success rate of 0.95 when transferred to the PointGoal task.

Wijmans et al. proposed Decentralized Distributed Proximal Policy Optimization (DD-PPO), a method for distributed reinforcement learning that is able to run multiple simulation environments in parallel across numerous GPUs [12]. The authors leverage this scalability to train an agent for 2.5 billion environment steps in under 3 days of wall-clock time with 64 GPUs. The trained agent achieves a success rate of 0.997 on the Gibson validation split utilising RGB and depth input. However, the authors also require access to the agent's ground-truth pose.

Datta et al. were one of the first to propose an approach to the PointGoal navigation task that does not rely on ground-truth localisation [3]. The authors first use supervised learning to train a CNN-based visual odometry model that estimates egomotion using consecutive depth images. They then use the trained visual odometry model to train an LSTM-based navigation policy. The navigation policy uses the pre-trained policy from [12] and is trained for a further 60M steps, achieving a success rate of 0.54 when evaluated without ground-truth localisation.

Zhao et al. train an ensemble of visual odometry models using top-down egocentric map prediction as an auxiliary loss [14]. They train their ensemble on a dataset of 1 million image pairs using supervised learning. During evaluation, they directly replace ground-truth localisation with estimates from their visual odometry model. They use the pre-trained DD-PPO navigation policy from [12] and fine-tune with their visual odometry estimates for a further 14.7 million frames, achieving a success rate of 0.717 on the Gibson validation split.

Partsey et al. follow the same principle as [3,14] whereby they train a visual odometry model. However, their approach scales dataset and model size, and applies data augmentation techniques during model training [10]. With a larger egomotion dataset (5M image pairs) and a deeper visual odometry model architecture (ResNet-50), their approach achieves a success rate of 0.96 on the Gibson validation split. However, their approach required significant computation, with training requiring approximately 5000 GPU hours in total. Moreover, the authors highlight that their visual odometry model is also somewhat dataset-specific as it did not transfer well to similar indoor datasets.

There has also been significant work related to the more general problem of combining traditional localisation methods with reinforcement learning-based navigation. Notably, Lin et al. demonstrated that an RL-based policy can be trained to avoid obstacles while also avoiding feature-poor parts of the environment that may degrade localisation performance [7,8].

Although many of the works mentioned above have demonstrated good performance on the PointGoal navigation task, most assume access to ground-truth localisation and those that do not require significant computation to train an additional visual odometry model which is dataset-specific and does not transfer well to different environment datasets. Therefore, our work aims to improve the performance of a pre-trained PointGoal policy in realistic scenarios by fine-tuning on localisation estimates from ORB-SLAM2, a traditional feature-based visual SLAM algorithm [9].

2.3 Simulation Platforms

There are many open-source simulators capable of simulating a robot in an indoor environment, including iGibson [13], Habitat [11], and Gazebo [6]. We utilise Habitat for all of our experiments as it provides pre-trained navigation agents and implements a simple interface for training agents using DD-PPO.

The Habitat simulator [11] is a platform designed for conducting research in embodied AI, particularly in the domain of navigation and interaction in realistic 3D environments. It leverages the Gibson dataset of indoor spaces to provide high-fidelity simulations for training and evaluating AI agents. The Gibson [13] dataset consists of 3D mesh models created from scans of real indoor environments such as homes, offices, and schools. Habitat provides agents with a range of onboard sensors such as RGB, depth, semantic segmentation, and ground-truth pose. Habitat also supports a variety of navigation tasks, including the PointGoal navigation task, where agents are tasked with navigating to a specified target location within an environment using only an RGB-D camera. This task serves as a benchmark for evaluating the navigation capabilities of embodied AI agents.

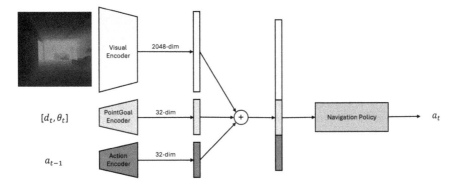

Fig. 1. An overview of the agent's architecture consisting of a pre-trained visual encoder that provides a compact representation of the egocentric depth image, a PointGoal encoder that learns a representation of agent's PointGoal estimate, an action encoder that learns a representation of the agent's last action, and a navigation policy that learns which actions to execute.

3 Methodology

3.1 Agent Architecture

The agent operates with a discrete action space consisting of four actions: stop, move forward (0.25 m), turn left (10°), and turn right (10°). Our agent (Fig. 1) uses the same architecture as [12], which consists of four parts - a visual encoder, a PointGoal encoder, an action encoder, and a navigation policy. The visual encoder is a ResNet-50 [4], which takes a 256 × 256 depth image and encodes it as a 2048-dimensional vector. The agent uses only depth information, as previous works showed that using RGB-D images degrades the performance of agents on the PointGoal navigation task [12]. The PointGoal encoder first converts the PointGoal polar coordinates [d, θ] to a magnitude and unit vector [d, $\cos(\theta)$, $\sin(\theta)$]. This is then encoded into 32-dimensional vector using a fully-connected layer. The action encoder computes a 32-dimensional embedding of the previous action. The navigation policy is a two-layer LSTM [5] with a 512-dimensional hidden state, which takes the depth image encoding, PointGoal encoding, and previous action embedding as input. The output of the LSTM is fed to a fully connected layer, which returns a 4-dimensional softmax distribution over the set of actions.

3.2 Fine-Tuning Procedure

We use the pre-trained weights provided by [12] as our baseline agent. We take this pre-trained agent and train it for a further 10 million frames from the Gibson training split. During the fine-tuning stage, the weights of the visual encoder are frozen and only the PointGoal encoder, action encoder, and the navigation policy are updated. We follow the same training procedure as [12], using DD-PPO to train with 12 workers on 12 Nvidia V100 GPUs.

The reward given to the agent at each step is formulated as follows:

$$R_t = R_s - R_d$$

where $R_s = 10$ is a terminal success reward given when the agent stops within the success radius (0.2 m) and R_d is the change in geodesic distance to the goal between consecutive steps. The geodesic distance is provided by the simulator and corresponds to the length of the shortest path from the agent's current position to the goal position.

The agent's discrete action space can lead to relatively large rotations and translations between frames, resulting in ORB-SLAM2 having poor tracking performance as few features can be matched between frames. Therefore, when training and evaluating agents that use ORB-SLAM2 for localisation, we render RGB-D images at a frequency of 10 Hz to prevent ORB-SLAM2 tracking failure. We implement a modified training environment that skips every 10 frames. This allows the navigation policy to run at 1 Hz while allowing ORB-SLAM2 to run at 10 Hz in the background.

3.3 Simulator and Datasets

We use the Habitat simulator and the Gibson 4+ dataset for all experiments. The training split consists of over 1 million navigation episodes across 72 reconstructed indoor scenes that were judged to have a quality score of at least 4 out of 5 [11]. The validation split contains 994 navigation episodes across 14 scenes that are not present in the training split.

4 Experiments

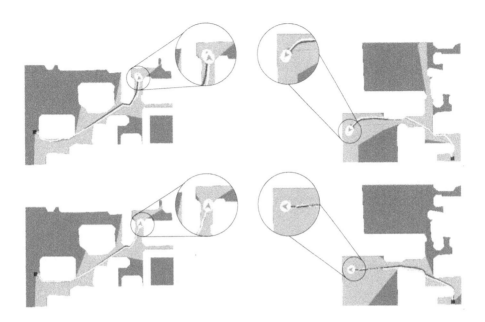

Fig. 2. Top: Examples of the zero-shot agent failing to stop within the success radius. **Bottom:** The fine-tuned agent successfully completes the episodes. The green, blue, and yellow lines indicate the shortest path between the start and goal location, the true path taken by the agent, and the path estimated by ORB-SLAM2, respectively.

4.1 Environment Configuration

During fine-tuning and evaluation, navigation episodes are sampled from the training and validation splits, respectively. At the start of an episode, an agent is spawned at the starting position and orientation and must navigate to target coordinates specified relative to the agent's start location.

In our experiments, agents use only an onboard RGB-D camera with a resolution of 256×256 pixels and a horizontal field-of-view of $90°$ to navigate. As mentioned previously, the navigation policy only uses depth information to compute actions. However, ORB-SLAM2 localisation makes use of both RGB and depth for computing pose estimates. Using our modified environment implementation, pose estimates from ORB-SLAM2 are then used to compute estimated PointGoal coordinates (distance and angle to the goal) at a rate of 10 Hz. The current depth image and current PointGoal estimate are passed to the navigation policy at a rate of 1 Hz.

A navigation episode terminates either when the agent calls the stop action or when 500 environment steps have elapsed.

4.2 Baseline Agents

We compare against two baseline agents, which we refer to as oracle and zero-shot. The oracle agent is the pre-trained policy that uses perfect localisation from the simulator during evaluation. The zero-shot agent is the same pre-trained policy, but it uses estimated localisation from ORB-SLAM2 during evaluation and does not perform any fine-tuning. We also compare our results to those reported by [3,10], which are currently the two best performing approaches to PointGoal navigation that do not rely on ground-truth localisation.

4.3 Evaluation Metrics

A navigation episode is deemed successful if the agent calls the stop action within the success radius of the target location. Success weighted by Path Length (SPL) is used to quantify the efficiency of the agent's path:

$$SPL = S\frac{l}{\max(l,p)}$$

where S is the success value, l is the length of the shortest path, and p is the length of the path taken by the agent. Therefore, a value of $SPL = 1$ indicates that the agent followed the shortest possible path between the start and goal position.

We also measure the distance from the agent to the goal at the end of an episode.

4.4 Evaluation on Unseen Environments

We evaluated the performance of the baseline and fine-tuned agents in simulation on the Gibson 4+ validation split. Figure 2 shows examples of failure cases of the zero-shot agent. Accumulated errors in ORB-SLAM2 estimates result in an erroneous PointGoal measurement, which causes the agent to call the stop action outside the success radius of the goal. However, this failure mode is not as common in the fine-tuned agent, suggesting that the navigation policy has learnt to adapt to the errors in the PointGoal estimates during fine-tuning. This hypothesis is supported by the results displayed in Fig. 3; there is a significant drop in performance when the pre-trained policy is first introduced to localisation estimates from ORB-SLAM2. However, the policy is able to adapt to this and performance begins to increase over the course of fine-tuning.

Table 1 shows the performance of the fine-tuned ORB-SLAM2 agent compared to the oracle and zero-shot baselines. Our approach of fine-tuning the pre-trained policy on localisation estimates from ORB-SLAM2 offers an improvement in success rate of 14% (0.71 → 0.85), a 15% increase in SPL (0.66 → 0.81), and a reduction in the final distance to the goal of 0.54 m when compared to zero-shot deployment.

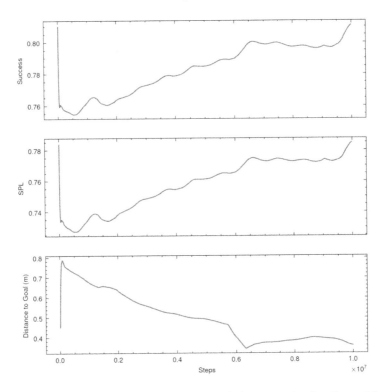

Fig. 3. Evaluation metrics during fine-tuning of the pre-trained policy with ORB-SLAM2 localisation.

Table 1. Performance of agents on the Gibson 4+ validation split.

	Success ↑	SPL ↑	Distance to Goal (m) ↓
Oracle	0.944	0.851	0.479
Zero-Shot	0.710	0.661	0.996
Fine-Tuned	0.853	0.809	0.456
Partsey et al. [10]	0.960	0.766	0.201
Zhao et al. [14]	0.717	0.525	0.802

We also notice that our zero-shot ORB-SLAM2 agent achieves a similar success rate to that of Zhao et al. [14] (0.710 vs. 0.717) despite the authors fine-tuning their navigation policy with estimates from their visual odometry model for a further 14.7 million steps. These results also suggest that fine-tuning the navigation policy with realistic localisation estimates may allow the agent to take more efficient paths as our fine-tuned agent outperforms the agent from [10] in terms of SPL, which does not perform any fine-tuning of the navigation policy.

5 Conclusion

In this paper, we have presented a study on the effectiveness of integrating a traditional vision-based SLAM algorithm with a reinforcement learning-based PointGoal navigation policy.

Through our experimentation, we have demonstrated how fine-tuning a pre-trained navigation policy on realistic localisation estimates can provide a significant increase in success rate and SPL when compared to deploying policies in a zero-shot manner. We have also shown that using an open-source traditional SLAM algorithm to provide localisation is competitive with learning-based visual odometry methods, while also being dataset-agnostic and not requiring a large amount of additional computation for training.

Overall, our findings contribute to the growing body of research in embodied AI and autonomous navigation. By addressing key challenges in PointGoal navigation and proposing effective solutions, we move closer to developing AI systems capable of navigating in realistic indoor environments.

In future work, we plan to explore additional techniques for improving navigation performance, such as incorporating semantic information and leveraging additional policy rewards. We also plan to evaluate our fine-tuned and zero-shot agents using real-world experiments to investigate the sim-to-real performance of our proposed approach.

References

1. Anderson, P., et al.: On evaluation of embodied navigation agents. arXiv preprint arXiv:1807.06757 (2018)
2. Chaplot, D.S., Gandhi, D., Gupta, S., Gupta, A., Salakhutdinov, R.: Learning to explore using active neural slam. arXiv preprint arXiv:2004.05155 (2020)
3. Datta, S., Maksymets, O., Hoffman, J., Lee, S., Batra, D., Parikh, D.: Integrating egocentric localization for more realistic point-goal navigation agents. In: Kober, J., Ramos, F., Tomlin, C. (eds.) Proceedings of the 2020 Conference on Robot Learning. Proceedings of Machine Learning Research, vol. 155, pp. 313–328. PMLR (2021). https://proceedings.mlr.press/v155/datta21a.html
4. He, K., Zhang, X., Ren, S., Sun, J.: Deep residual learning for image recognition. In: Proceedings of the IEEE Conference on Computer Vision and Pattern Recognition, pp. 770–778 (2016)
5. Hochreiter, S., Schmidhuber, J.: Long short-term memory. Neural Comput. 9(8), 1735–1780 (1997)
6. Koenig, N., Howard, A.: Design and use paradigms for gazebo, an open-source multi-robot simulator. In: 2004 IEEE/RSJ International Conference on Intelligent Robots and Systems (IROS) (IEEE Cat. No. 04CH37566), vol. 3, pp. 2149–2154. IEEE (2004)
7. Lin, F., Ji, Z., Wei, C., Grech, R.: Localisation-safe reinforcement learning for mapless navigation. In: 2022 IEEE International Conference on Robotics and Biomimetics (ROBIO), pp. 1327–1334. IEEE (2022)
8. Lin, F., Ji, Z., Wei, C., Niu, H.: Reinforcement learning-based mapless navigation with fail-safe localisation. In: Fox, C., Gao, J., Ghalamzan Esfahani, A., Saaj,

M., Hanheide, M., Parsons, S. (eds.) TAROS 2021. LNCS (LNAI), vol. 13054, pp. 100–111. Springer, Cham (2021). https://doi.org/10.1007/978-3-030-89177-0_10
9. Mur-Artal, R., Tardós, J.D.: Orb-slam2: an open-source slam system for monocular, stereo, and rgb-d cameras. IEEE Trans. Rob. **33**(5), 1255–1262 (2017)
10. Partsey, R., Wijmans, E., Yokoyama, N., Dobosevych, O., Batra, D., Maksymets, O.: Is mapping necessary for realistic pointgoal navigation? In: Proceedings of the IEEE/CVF Conference on Computer Vision and Pattern Recognition, pp. 17232–17241 (2022)
11. Savva, M., et al: Habitat: a platform for embodied ai research. In: Proceedings of the IEEE/CVF International Conference on Computer Vision, pp. 9339–9347 (2019)
12. Wijmans, E., et al.: Dd-ppo: learning near-perfect pointgoal navigators from 2.5 billion frames. arXiv preprint arXiv:1911.00357 (2019)
13. Xia, F., Zamir, A.R., He, Z., Sax, A., Malik, J., Savarese, S.: Gibson env: real-world perception for embodied agents. In: Proceedings of the IEEE Conference on Computer Vision and Pattern Recognition, pp. 9068–9079 (2018)
14. Zhao, X., Agrawal, H., Batra, D., Schwing, A.G.: The surprising effectiveness of visual odometry techniques for embodied pointgoal navigation. In: Proceedings of the IEEE/CVF International Conference on Computer Vision, pp. 16127–16136 (2021)

Towards Revisiting Visual Place Recognition for Joining Submaps in Multimap SLAM*

Markus Weißflog[1](✉), Stefan Schubert[1], Peter Protzel[1], and Peer Neubert[2]

[1] Process Automation, Chemnitz University of Technology, Chemnitz, Germany
markus.weissflog@etit.tu-chemnitz.de
[2] Intelligent Autonomous Systems, University of Koblenz, Koblenz, Germany

Abstract. Visual SLAM is a key technology for many autonomous systems. However, tracking loss can lead to the creation of disjoint submaps in multimap SLAM systems like ORB-SLAM3. Because of that, these systems employ submap merging strategies. As we show, these strategies are not always successful. In this paper, we investigate the impact of using modern VPR approaches for submap merging in visual SLAM. We argue that classical evaluation metrics are not sufficient to estimate the impact of a modern VPR component on the overall system. We show that naively replacing the VPR component does not leverage its full potential without requiring substantial interference to the original system. Because of that, we present a post-processing pipeline along with a set of metrics that allow us to estimate the impact of modern VPR components. We evaluate our approach on the NCLT and Newer College datasets using ORB-SLAM3 with the NetVLAD and HDC-DELF descriptors. Additionally, we present a simple approach for combining VPR with temporal consistency for map merging. We show that the map merging performance of ORB-SLAM3 can be improved. Building on these results, researchers in VPR can assess the potential of their approaches for SLAM systems.

Keywords: Visual SLAM · Visual Place Recognition · Mulimap SLAM

1 Introduction

Visual Simultaneous Localization And Mapping (visual SLAM) is a key technology for autonomous systems. It has applications in areas like robotics, autonomous driving, and augmented reality. In SLAM, an agent aims to create a map of its environment and tries to localize itself in this map at the same

---

* The work of Stefan Schubert was supported in part by the German Federal Ministry for Economic Affairs and Climate Action.

© The Author(s), under exclusive license to Springer Nature Switzerland AG 2025
M. N. Huda et al. (Eds.): TAROS 2024, LNAI 15041, pp. 94–106, 2025.
https://doi.org/10.1007/978-3-031-72059-8_9

Fig. 1. (a) Ground truth trajectory. (b) The output of ORB-SLAM3 is a set of disjoint submaps, with no known transformations between them. (c) The transformations are estimated using VPR (red dotted lines) and temporal consistency (blue dotted lines). (d) The final joined map. (Color figure online)

time. The resulting map can be used for surveying, navigation, obstacle avoidance, and other tasks.

However, the agent cannot always accurately estimate its current position. One of the reasons for this is tracking loss, which prevents the agent from estimating its pose in relation to the global map. Tracking loss can occur due to fast motion, textureless regions, occlusions, or various other factors. In multimap visual SLAM systems, the agent has no choice but to create new maps, resulting in a set of disjoint submaps whose relative pose to each other is unknown, as shown in Fig. 1. This is problematic for mapping applications, for example, as it is not possible to create a continuous map.

For this reason, modern visual SLAM systems like ORB-SLAM3 [4] use submap merging strategies. As we will show in Sect. 3.1, these strategies are not always successful, especially on challenging datasets. A promising approach to counteract this problem is Visual Place Recognition (VPR). VPR is the task of recognizing a previously visited place based on images. This can be used to merge submaps. We will show in Sect. 3.2 that there are newer VPR approaches that outperform ORB-SLAM3's bag-of-visual-words-based (BOW) approach.

However, as we show in Sect. 3.3, by just switching out the VPR component in isolation, ORB-SLAM3 cannot make use of the improved VPR performance. A complete integration into the overall system is required, which would involve considerable implementation effort and changes to the algorithm. Before these changes are made, it would be sensible to estimate the impact of using modern VPR for submap merging on the overall system. This estimation is the goal of our work. For this, we introduce a simplified experimental setup in Sect. 4. We evaluate this setup in Sect. 5. Finally, we discuss the results in Sect. 6 and conclude in Sect. 7. We start by discussing the related work in Sect. 2. Our main contributions are:

- We present a pipeline along with a set of metrics to evaluate the performance of VPR for submap merging in visual SLAM. Our pipeline does not require reparametrization or other modifications to the SLAM system.
- We evaluate our approach on two challenging datasets using a state-of-the-art SLAM system with modern VPR approaches.

– We present a simple approach for map merging that combines the advantages of modern VPR with temporal consistency.

2 Related Work

Visual Simultaneous Localization and Mapping. A general introduction to SLAM is given in [19]. This paper focuses on monocular visual SLAM, which uses a single camera as the only sensor. [16] provides a survey of visual SLAM methods. Modern SLAM algorithms include maplab 2.0 [6], VINS [13], Basalt [21] and ORB-SLAM3 [4].

Visual Place Recognition. Yin et al. [22] survey the general problem of place recognition, including VPR. Schubert et al. [17] provide a tutorial on VPR where they introduce the general problem, challenges, and evaluation metrics. Masone and Caputo [10] provide a comprehensive survey on the role of deep learning in VPR. A central part of VPR are holistic image descriptors. They convert the pixel data of an image into a vector representation, enabling comparison between two images to determine if they show the same location. Holistic descriptors can either be directly computed from the whole image [1], or they are an aggregation of local features [9,11].

Similar to our work, Khaliq et al. [8] compare the loop closing component of ORB-SLAM3 with a deep-learning-based VPR method. VPR-Bench [23] benchmarks different VPR metrics. However, both publications measure performance only on the VPR tasks and do not consider the underlying SLAM system.

2.1 ORB-SLAM3

We use ORB-SLAM3 [4] for our experiments, as it is considered a highly performant algorithm [2,20]. ORB-SLAM3 is a landmark-based visual-inertial SLAM approach, which can also handle stereo and RGB-D cameras. In essence, ORB-SLAM3 consists of four main threads, which run in parallel:

1. The *tracking thread* aims to localize the current frame in the submap using ORB features [15]. If localization fails, ORB-SLAM3 saves the current submap and starts a new one.
2. The *local mapping thread* inserts the current keyframe and its new landmarks into the map and optimizes a local window.
3. The *loop and map merging thread* searches the database for similar frames to the current keyframe. Upon finding a matching frame (which must satisfy certain checks, see Sect. 3.3), a loop or a merge operation is triggered. If the current keyframe and the matching frame are part of the same submap, a loop closure is performed. If they are part of different submaps, the submaps are merged.
4. In parallel, full bundle adjustment is performed.

For tracking, ORB-SLAM3 uses ORB features [15]. These features are also aggregated and used as holistic descriptors in the mapping and map merging thread. The descriptors are computed using the BOW [18] implementation DBoW2 [7]. In our experiments, we used ORB-SLAM3 in the monocular configuration and only changed the camera parameters according to the datasets.

3 Analysis of the Problem

3.1 Loops and Submap Merges in ORB-SLAM3

In this section, we analyze the submap merging performance of ORB-SLAM3 without post-processing. The results are shown in Table 1. The top part of the table shows the results for the NCLT [5] and Newer College [14] datasets, which will be introduced in Sect. 4.4. Noteworthy is the large number of unmerged submaps. The loop and map merging thread is unable to recover the relative transformations between the submaps after tracking loss, which significantly limits mapping capabilities.

We performed the same analysis on the EuRoC [3] dataset, which is shown in the bottom part of the table. This dataset consists of eleven sequences captured using a drone under different conditions and motion patterns. We evaluated all sequences, but only show the best-performing (MH_01) and worst-performing (V2_02) sequences in Table 1. On this dataset, ORB-SLAM3 has almost no tracking losses, which might be due to the short duration of the sequences.[1] This makes this dataset unsuitable for evaluating submap merging performance. Nevertheless, the number of loops and merges found never surpasses one.

3.2 Performance of Modern Holistic Descriptors

In this section, we analyze the performance of the BOW-based descriptors used in ORB-SLAM3 in isolation. We use the widely adapted metrics precision and recall [17] for evaluation. Good performance is indicated by high values for both metrics. Our evaluation setup is similar to [17]. As can be seen in Fig. 2, ORB-BOW

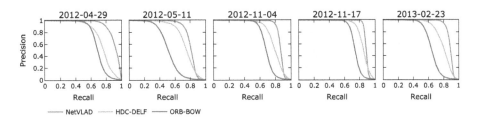

Fig. 2. Precision-recall curves for the analyzed descriptors on the NCLT dataset. Better performance is indicated by curves that are near the top right corner of the plots.

---

[1] The longest sequence lasts for 182 s [3].

Table 1. ORB-SLAM3 performance on the analyzed datasets. *Tracked Map* states the proportion of frames that were tracked successfully. *# Maps* states the number of unmerged submaps after the dataset was processed. *# Loops/ Merges* states the number of successful loop closures or submap merges.

Dataset		Tracked Map	# Maps	# Loops/ Merges
NCLT	2012-04-29	91.7 %	23	0
	2012-05-11	97.4 %	39	0
	2012-11-17	97.9 %	21	0
	2012-11-04	95.9 %	24	0
	2013-02-23	99.3 %	18	2
Newer College		80.7 %	33	1
EuRoC	MH_01	99.9 %	1	0
	⋮			
	V2_03	91.5 %	2	1

is outperformed by the more modern descriptors HDC-DELF and NetVLAD, which will be presented in more detail in Sect. 4.4. This evaluation hints at the fact that switching out the holistic descriptors in ORB-SLAM3 could lead to better performance in map merging.

However, the performance of the descriptors in isolation does not necessarily translate to better performance in the overall SLAM system. For example, matches within a submap contribute to improving the accuracy by providing loop closures; however, they cannot help with joining submaps. Matches between submaps, on the other hand, are crucial for map merging. The difference between these types of matches is not considered by precision and recall. In Sect. 4, we will present an evaluation metric that takes this difference into account.

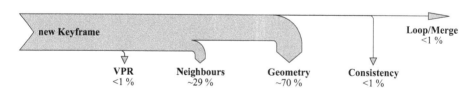

Fig. 3. Sankey diagram of the checks performed by ORB-SLAM3. The width of the arrows without the head corresponds to the number of frames that take this path. Downward pointing arrows mean that the loop/ merge is aborted.

3.3 Naive Approach: Only Replacing the VPR Descriptor

ORB-SLAM3 performs multiple checks before a loop closure or submap merge. These checks are visualized in Fig. 3. The most important for this work are:

1. *VPR*: ORB-BOW descriptors are utilized to find a matching place within the database of keyframes. The check fails if none are found.
2. *Neighbours*: The current keyframe and the potential match cannot be neighbours in ORB-SLAM3's covisibility graph.
3. *Geometry*: RANSAC and optimization are used on the ORB features of the current frame and the loop/ merge candidate to estimate the transformation between the two frames. The check is passed if there are enough inlying ORB keypoints.
4. *Consistency*: The checks have to be passed three times in a row to perform a loop closure/ map merge.

After a frame passes all checks, the loop closure/ map merge is performed. As can be seen in Fig. 3, most loops/ merges are aborted due to the strict check 3. We have experimented with different holistic descriptors for the VPR-check, which has only a marginal effect on Fig. 3 and no effect on check 3. To benefit from the full potential of modern VPR, the SLAM algorithm would need reparametrization and algorithmic adjustments to its loop/ merge pipeline. HDC-DELF, for example, works on local DELF [12] features, which could also be used during tracking and check 3. Such deep changes, which could potentially have a significant impact on system performance and involve high implementation efforts, should be tested in a simplified setup before implementation. In the remaining sections, we propose such a setup and evaluate the potential of these changes.

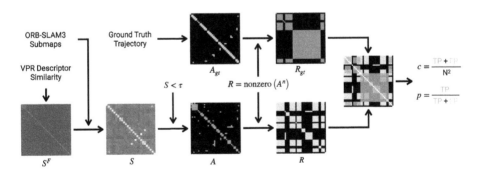

Fig. 4. The pipeline shows how precision and coverage are calculated using the example of rule VPR. The VPR descriptors, the ORB-SLAM trajectory, and the ground truth are input to the pipeline. S^F and S are continuous matrices, A, A_{gt}, R, and R_{gt} are boolean.

4 Approach

This section describes our evaluation pipeline and the metrics used. An exemplary pipeline is shown in Fig. 4. The pipeline starts by predicting the distance between submaps. We use the word distance to refer to the inverse of the similarity. Section 4.1 describes how VPR and the timestamps of a SLAM trajectory are used to predict the submap adjacency matrix A. Section 4.2 describes how this matrix is used to compute the metrics precision and coverage, and how the ground truth is obtained. Section 4.3 summarizes the assumptions we make for our pipeline. Finally, Sect. 4.4 describes the descriptors and the datasets used in our experiments.

Our pipeline is a post-processing step for a multimap SLAM system, which outputs a trajectory $\mathcal{T}$ containing N poses. For this work, we define a trajectory as $\mathcal{T} = \{(t_i, \mathbf{p}_i, R_i, I_i, j_i)\}_i^N$, where t_i is the timestamp of frame i, $\mathbf{p}_i$ and R_i are the position and orientation relative to the submap, I_i is the camera image, and j_i refers to the corresponding submap of frame i, with $j \in [1, M]$ and $M \ll N$.

4.1 Predicting Submap Adjacency

This section describes how the submap adjacency matrix A is computed, which is required by our post-processing pipeline. We use the word adjacency to describe that the transformation between two frames or two submaps is known. A is a boolean, square, symmetric matrix of size $M \times M$, where M is the number of submaps. If the transformation between submap i and submap j is known, the submaps are adjacent, and A contains a one at positions (i, j) and (j, i). If no post-processing is performed, A is the identity matrix because no transformations between submaps are known. We analyze three different rules for estimating A:

Rule Time. One natural approach to predict the adjacency matrix is to use the timestamps. We define the temporal distance between two submaps as the difference between the timestamp of the last keyframe of submap i and the timestamp of the first keyframe of submap j. By pairwise comparing all submaps like this, we can create a time distance matrix S_{time}.

Rule VPR. Another way to predict the adjacency is to use the visual information of the images I. For that, we compute a holistic descriptor for each image using methods from VPR. We define the visual distance between two images as the distance between their descriptors. Based on this, we can calculate the frame distance matrix S_{VPR}^F, which contains at position (i, j) the descriptor distance of the images I_i and I_j. This matrix can be converted into a submap distance matrix S_{VPR}. At positions (i, j) and (j, i), the matrix S_{VPR} contains the smallest visual distance between any two images of submap i and submap j. The computation of S_{VPR} is visualized in the left part of Fig. 4.

Combining Time and VPR. We test the combination of the two rules using a simple, exemplary algorithm, which can be expanded as required. Using the temporal distance matrix S_{time} and the visual distance matrix S_{VPR}, we define two maps as adjacent, if

- their visual distance is below a strict threshold τ_{VPR},
- or their temporal distance is below a strict threshold τ_{time},
- or their temporal distance is below a more relaxed threshold $f_{\text{time}} \cdot \tau_{\text{time}}$ and their visual distance is below a relaxed threshold $f_{VPR} \cdot \tau_{VPR}$.

f is the factor by which the threshold is relaxed. In the later evaluation, we will show exemplary results for the following two combinations of thresholds[2]: *Rule Comb. 1* uses $\tau_{\text{time}} = 2\,s$, $f_{\text{time}} = 10$ and $f_{VPR} = 2$. *Rule Comb. 2* uses $\tau_{\text{time}} = 0.5\,s$, $f_{\text{time}} = 10$ and $f_{VPR} = 4$. We leave τ_{VPR} as a free parameter.

To convert the distance matrices S into adjacency matrices A, a threshold τ (or τ_{VPR} in the case of the combined rules) is chosen and applied to S. All distances below the threshold are set to one, and distances above are set to zero.

Ground Truth. To obtain the ground truth, we define submaps as adjacent if they have two frames whose Euclidean distance (of the ground truth trajectory) and angle of rotation are smaller than two predefined thresholds $\epsilon_{\text{dist}} = 10\,m$ and $\epsilon_{\text{rot}} = 20°$. Similar definitions are typically used for VPR [23]. Thus, the ground truth matrix of adjacent submaps A_{gt} is a boolean matrix of shape $M \times M$.

4.2 Evaluation Pipeline

From Adjacency to Reachability. The adjacency matrix A can be further processed into a reachability matrix R. Two submaps are reachable if the transformation between them can be obtained directly or indirectly.[3] R is boolean, symmetric, and of size $M \times M$. It is computed as follows:

$$R = \text{nonzero}(A^n) \tag{1}$$

The function nonzero($\cdot$) sets all non-zero entries of a matrix to one. n is the number of times the matrix A is multiplied with itself until R converges. This step is visualized in the central portion of Fig. 4.

Computing Coverage and Precision. At first, a weight vector $\mathbf{w} \in \mathbb{N}^M$ is extracted from T, whose elements w_j are defined as the number of frames in submap j. A weight matrix $W \in \mathbb{N}^{M \times M}$ is computed using $W = \mathbf{w}\mathbf{w}^T$.

---

[2] These values were chosen by hand to demonstrate the feasibility of the approach. No parameter optimization was performed.

[3] Indirectly refers to the relationship where, if the transformation between the submaps (i, j) and the transformation between the submaps (j, k) are known, the transformation (i, k) can be obtained.

Comparing the ground truth matrix R_{gt} with the predicted reachability matrix R yields the true positive (TP), false positive (FP), true negative (TN), and false negative (FN) matches. From that, coverage and precision can be computed, which is visualized in the right part of Fig. 4. *Coverage* describes how much the reachability matrix is filled:

$$c = \frac{TP + FP}{N^2} = \frac{\sum (R \odot W)}{\sum (W)} \qquad (2)$$

The symbol $\odot$ refers to the Hadamard product, and the symbol $\sum(\cdot)$ refers to the sum over all matrix elements. *Precision* describes how many of the found matches are correct:

$$p = \frac{TP}{TP + FP} = \frac{\sum (R \odot R_{gt} \odot W)}{\sum (R \odot W)} \qquad (3)$$

Note that we exclude the main diagonal of all matrices, as all submaps are always reachable from themselves, and thus precision and coverage would be skewed. Note also that the adjacency matrix, the estimated reachability matrix, and thus coverage and precision depend all on the threshold value τ (or τ_{VPR} for the combined rules). By varying this threshold, pairs of precision and coverage values are obtained, forming the precision-coverage curve, which is used for evaluation in the following experiments. The area under the curve (AUC) can be obtained by integration.

4.3 Assumptions and Limitations

In this work, we investigate the isolated problem of finding possible matches between submaps of a SLAM system. Because of that, we make several assumptions about the system and the data we work with: (1) We assume that the SLAM system estimates accurate poses within the submaps. (2) We assume that the SLAM system can detect tracking loss reliably. (3) We assume that the transformations between matching frames can be estimated.

4.4 Experimental Setup

Datasets. We used the NCLT and Newer College datasets. *NCLT* [5] consists of multiple sequences that were collected onboard a robot covering a large outdoor area of a college campus. We selected five representative sequences that cover different times of the day and different seasons. They have a duration of 40 to 80 min. The *Newer College* dataset [14] consists of a single sequence collected using a handheld sensor setup. It covers a large trajectory over a courtyard and a park at New College, Oxford, and has a duration of approximately 44 min. We subsample it at 2 Hz. As we focus on monocular SLAM, we only employ the front camera for NCLT and the left camera for Newer College.

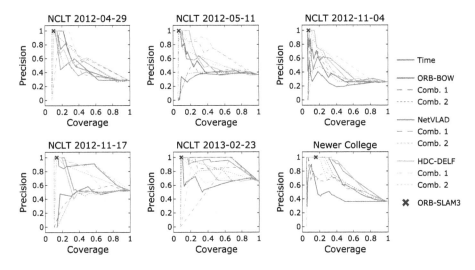

Fig. 5. Precision-coverage curves for the analyzed datasets.

Holistic Descriptors. We refer to the local BOW-based [7,18] aggregation of ORB features [15] used by ORB-SLAM3 as *ORB-BOW*. *HDC-DELF* is a deep learning-based descriptor that uses Hyperdimensional Computing to aggregate DELF features [12] to a holistic descriptor. *NetVLAD* [1] is a deep learning-based holistic descriptor as well. In our experiments, we used the implementations and parameters as proposed by the original authors.

5 Results

Our results can be seen in Fig. 5 and Table 2. In Fig. 5, the curves are not smooth because a small change in the threshold value τ could lead to a merge of two large submaps, causing coverage and precision to change abruptly.[4] On the left

Table 2. Area under precision-coverage curve. The best-performing rule is highlighted in **bold**, the second-best underlined, and the third-best in *italics*. Each row is summarized by the mean and worst-case performance across all datasets.

Rule	Newer College	NCLT 12-04-29	NCLT 12-05-11	NCLT 12-11-04	NCLT 12-11-17	NCLT 13-02-23	Mean	Worst Case
Time	0.418	0.461	0.454	0.334	*0.709*	**0.860**	0.539	0.334
ORB-BOW	0.637	0.453	0.337	0.240	0.425	0.594	0.448	0.240
Comb. 1 + ORB-BOW	0.637	*0.500*	0.347	0.294	0.383	0.680	0.474	0.294
Comb. 2 + ORB-BOW	0.559	**0.552**	0.370	0.330	0.426	0.785	0.504	0.330
NetVLAD	**0.713**	0.364	0.464	0.400	0.548	0.822	0.552	0.364
Comb. 1 + NetVLAD	0.614	0.475	0.516	0.359	0.668	0.760	*0.565*	0.359
Comb. 2 + NetVLAD	0.566	0.403	0.623	0.330	**0.733**	0.559	0.536	0.330
HDC-DELF	0.712	0.443	0.499	0.420	0.495	0.715	0.547	0.420
Comb. 1 + HDC-DELF	*0.706*	0.491	*0.528*	**0.437**	0.721	0.674	0.593	**0.437**
Comb. 2 + HDC-DELF	0.620	0.526	**0.641**	*0.410*	0.644	*0.785*	**0.604**	*0.410*

---

[4] In all experiments, we iterate over all possible values of τ to ensure that the smoothest possible curves are obtained.

of the coverage axis, few highly confident matches are found. On the right side, the threshold value is relaxed to the extent that the matrix R is completely filled with ones. In this case, the precision reflects the proportion of positive matches in the ground truth. Thus, all curves converge to this point. ORB-BOW performs poorly overall, except on the Newer College dataset. As discussed in Sect. 3.3, ORB-SLAM3's conservative approach to submap merging misses many potential merges, but the merges found are likely correct, reflected by the purple marker in Fig. 5 showing low coverage but high precision. ORB-SLAM3 is kept as originally proposed in [4], so there are no parameters that could be varied to get a curve instead of a single point. The Newer College dataset has large areas of failed tracking (as shown in Table 1). This causes time-based rules to perform worse than pure VPR-based rules on this dataset. NCLT 2013-02-23 shows the opposite case, where tracking, and thus the time-based rules, work well.

6 Discussion

Figure 5 and Table 2 show a large variation in the results, which could be due to the different characteristics of the datasets. It can be seen that the ORB-BOW strategy is outperformed in almost all cases.

These results indicate that a modern VPR component offers the potential to improve a SLAM system. HDC-DELF has a very good mean and worst-case performance, especially in combination with the temporal consistency rule, making it a promising choice for future research. This VPR approach could also improve other parts of the SLAM pipeline: the local DELF features, for example, could be used for tracking and transformation estimation. As this work provides an estimation, future research is needed to show the extent to which this potential is realized in a full SLAM pipeline. We propose that modern VPR methods for map merging as part of a SLAM pipeline should maximize coverage.

7 Conclusion

Multimap visual SLAM systems like ORB-SLAM3 suffer from tracking loss, which leads to the creation of disjoint submaps without relative pose information between them. VPR is a potential approach for merging these submaps. However, integrating a modern VPR component into a SLAM system requires considerable modifications to the system. In this work, we have presented a pipeline to estimate the performance of an improved VPR component for submap merging in visual SLAM. We have evaluated our approach using ORB-SLAM3. Additionally, we have presented a submap merging approach that combines VPR with temporal consistency. Our pipeline does not require reparametrization or changes to the SLAM system. Our results show that the map merging performance of ORB-SLAM3 can be improved by using modern VPR approaches. As this work only provides an estimation of possible improvements, future work includes fully integrating the new VPR components into the overall SLAM system to exploit their potential.

References

1. Arandjelovic, R., Gronat, P., Torii, A., Pajdla, T., Sivic, J.: NetVLAD: CNN architecture for weakly supervised place recognition. In: Proceedings of the Conference on Computer Vision and Pattern Recognition (CVPR) (2017). https://doi.org/10.1109/TPAMI.2017.2711011
2. Bujanca, M., Shi, X., Spear, M., Zhao, P., Lennox, B., Luján, M.: Robust SLAM systems: are we there yet? In: International Conference on Intelligent Robots and Systems (IROS) (2021). https://doi.org/10.1109/IROS51168.2021.9636814
3. Burri, M., et al.: The EuRoC micro aerial vehicle datasets. Int. J. Rob. Res. **35**(10), 1157–1163 (2016). https://doi.org/10.1177/0278364915620033
4. Campos, C., Elvira, R., Rodríguez, J.J.G., M. Montiel, J.M., D. Tardós, J.: ORB-SLAM3: an accurate open-source library for visual, visual–inertial, and multimap slam. IEEE Trans. Rob. **37**(6), 1874–1890 (2021). https://doi.org/10.1109/TRO.2021.3075644
5. Carlevaris-Bianco, N., Ushani, A.K., Eustice, R.M.: University of Michigan North campus long-term vision and lidar dataset. Int. J. Rob. Res. **35**(9), 1023–1035 (2016). https://doi.org/10.1177/0278364915614638
6. Cramariuc, A., et al.: maplab 2.0 - a modular and multi-modal mapping framework. IEEE Rob. Autom. Lett. **8**(2), 520–527 (2022). https://doi.org/10.1109/LRA.2022.3227865
7. Gálvez-López, D., Tardós, J.D.: Bags of binary words for fast place recognition in image sequences. IEEE Trans. Rob. **28**(5) (2012). https://doi.org/10.1109/TRO.2012.2197158
8. Khaliq, S., Anjum, M.L., Hussain, W., Khattak, M.U., Rasool, M.: Why ORB-SLAM is missing commonly occurring loop closures? Auton. Rob. **47**(8), 1519–1535 (2023). https://doi.org/10.1007/s10514-023-10149-x
9. Lowry, S., et al.: visual place recognition: a survey. IEEE Trans. Rob. **32**(1) (2016). https://doi.org/10.1109/TRO.2015.2496823
10. Masone, C., Caputo, B.: A survey on deep visual place recognition. IEEE Access **9**, 19516–19547 (2021). https://doi.org/10.1109/ACCESS.2021.3054937
11. Neubert, P., Schubert, S., Schlegel, K., Protzel, P.: Vector semantic representations as descriptors for visual place recognition. In: Proceedings of Robotics: Science and Systems (2021). https://doi.org/10.15607/RSS.2021.XVII.083
12. Noh, H., Araujo, A., Sim, J., Weyand, T., Han, B.: Large-scale image retrieval with attentive deep local features. In: Proceedings of the International Conference on Computer Vision (ICCV)
13. Qin, T., Pan, J., Cao, S., Shen, S.: A general optimization-based framework for local odometry estimation with multiple sensors (2019)
14. Ramezani, M., Wang, Y., Camurri, M., Wisth, D., Mattamala, M., Fallon, M.: The newer college dataset: handheld LiDAR, inertial and vision with ground truth. In: International Conference on Intelligent Robots and Systems (IROS). IEEE (2020). https://doi.org/10.1109/iros45743.2020.9340849
15. Rublee, E., Rabaud, V., Konolige, K., Bradski, G.: ORB: an efficient alternative to SIFT or SURF. In: International Conference on Computer Vision. IEEE (2011). https://doi.org/10.1109/iccv.2011.6126544
16. Sahili, A.R., et al.: A survey of Visual SLAM methods. IEEE Access **11** (2023). https://doi.org/10.1109/ACCESS.2023.3341489
17. Schubert, S., Neubert, P., Garg, S., Milford, M., Fischer, T.: Visual place recognition: a tutorial (2023). https://doi.org/10.1109/mra.2023.3310859

18. Sivic, J., Zisserman, A.: Video google: a text retrieval approach to object matching in videos. In: International Conference on Computer Vision (ICCV) (2003). https://doi.org/10.1109/ICCV.2003.1238663
19. Thrun, S., Burgard, W., Fox, D.: Probabilistic Robotics. Intelligent Robotics and Autonomous Agents, MIT Press, Cambridge (2006)
20. Tourani, A., Bavle, H., Sanchez-Lopez, J.L., Voos, H.: Visual SLAM: what are the current trends and what to expect? Sensors **22**(23), 9297 (2022). https://doi.org/10.3390/s22239297
21. Usenko, V., Demmel, N., Schubert, D., Stückler, J., Cremers, D.: Visual-inertial mapping with non-linear factor recovery. IEEE Rob. Autom. Lett. **5**(2), 422–429 (2019). https://doi.org/10.1109/LRA.2019.2961227
22. Yin, P., et al.: General place recognition survey: towards the real-world autonomy age (2022). https://doi.org/10.48550/ARXIV.2209.04497
23. Zaffar, M., et al.: VPR-Bench: an open-source visual place recognition evaluation framework with quantifiable viewpoint and appearance change. Int. J. Comput. Vision **129**(7), 2136–2174 (2021). https://doi.org/10.1007/s11263-021-01469-5

Consulting an Oracle; Repurposing Robots for the Circular Economy

Helen McGloin[1](✉), Matthew Studley[2], Richard Mawle[2], and Alan Winfield[2]

[1] FARSCOPE CDT, University of Bristol and University of West England, Bristol Robotics Laboratory, Bristol, UK
h.mcgloin@bristol.ac.uk
[2] College of Arts, Technology and Environment, University of West England, Bristol, UK
{Matthew2.Studley,Richard2.Mawle,Alan.Winfield}@uwe.ac.uk

Abstract. This paper reports a new process which can be used to repurpose a robot at the end of its primary life. Repurposing a robot enables it to continue delivering value beyond the point at which it might have been otherwise recycled or scrapped. The repurposing process was created using the future thinking forecasting methodology known as a Delphi Study. This paper shows there are some similarities between a process for repurposing, to processes for developing new systems and remanufacturing systems. However this new process contains unique elements, highlighting the value and potential of challenging accepted norms for the management of robot systems when they reach the end of their useful life and could otherwise be considered e-waste.

Keywords: Repurposing · Circular-economy · Delphi · Robotics · Sustainability · Environment · UNSDG · Recycling · Reuse · e-waste

1 Introduction

In [1] the authors of this paper introduced the concept of repurposing a robot to reduce electronic waste and increase the working life of robotic systems. Repurposing was defined as "providing new utility to an existing robotic system to give the system a new role which is independent of the robot's original utility" [1]. Here, the *utility* is comprised of the robot's *skill* or tasks it can complete, and the *application* in which the robot functions [1].

Repurposing a robot is a step which should be taken when a robot can no longer be reused in the same or similar utility, and before any attempts are made to dispose of, or recycle the system. This maintains the product at its highest value for as long as possible [2]. In doing this the robotics industry will move further towards a circular business model and away from a linear economy system which traditionally encourages the purchase, use and disposal of products

[3]. A circular business model should result in both economic and environmental benefits for the industry; "reversing trends towards waste accumulation and resource scarcity" [3], working towards the UN Sustainable Development Goal 12 (Ensure sustainable consumption and production patterns) [4]. Repurposing further delays the need for recycling, which cannot be relied on in a circular economy as a singular solution to e-waste as global recycling rates are consistently low - an average of only 17% of electronic waste is correctly recycled globally [5].

The concept of repurposing a robot is a relatively new and untested idea. While the authors found anecdotal evidence of repurposing examples in the robotics industry, there too few enough examples to carry out a quantitative analysis of how robots have been repurposed to date. Additionally, with so many permutations of robots in use currently, designing a useful quantitative experiment in the timescales available was not possible. Therefore an inductive method of enquiry was selected for this research which aimed to build a credible framework for the process of repurposing robots.

Utilising the Delphi study methodology presented in Sect. 2, this research produced a process for repurposing robots which is presented in this paper (Sect. 3). Analysis of this process, described in Sect. 4, found similarities to processes used both in the development of new systems and in the remanufacturing of old systems. Additionally, new process steps were identified which were unique to the repurposing process, reflecting the challenges and opportunities of implementing circular economy principles.

2 Methodology

2.1 Research Aim

This research aimed to investigate the process of repurposing robots at the end of their primary life. The repurposing process was assessed through forecasting methods which used the experience of a panel of experts to reach a consensus on how a robot could be viably repurposed from one utility to another, where viability included both economic and environmental factors. The scenario presented to participants to develop the repurposing process is shown in Fig. 1.

2.2 Qualitative Research Methodology

Social research philosophies include the science of forecasting - an "estimate of a value or condition in a future time period" [6]. Forecasting includes statistical methodology (such as operational research methods and multiple regression techniques) and qualitative methods (such as nominal groups, Delphi studies and focus groups) [6]. These qualitative methods may also be referred to as Futures Thinking [7]. A qualitative method of forecasting enquiry was selected for this study, requiring participants outside of the research group to apply their expertise to predict the process by which robots could be repurposed in the future. This method of hypothesis investigation works on the basis "that history will repeat itself, and it will continue to do so" [6].

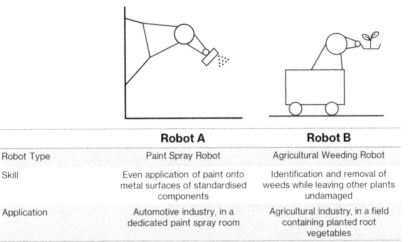

Fig. 1. The description of Robot A and Robot B presented to participants within the Delphi study questionnaires

The Delphi Study technique was compared to other qualitative future thinking methods and selected for its advantages over techniques including; focus groups, nominal groups and vignette studies. Specifically, a Delphi Study can be carried out online with participants accessing questionnaires asynchronously, which makes the study cheaper to run and more accessible for participants [8]. It also has a higher chance of reaching a consensus among participants compared to other qualitative forecasting methods [8] and participants remain anonymous to each other, thus avoiding the effects of confrontation of one or more dominant individuals from group settings seen in other qualitative methods [9]. Participants only remained semi-anonymous to the researchers as panellists were contacted directly with reminders to complete questionnaires.

Like other qualitative methods, a Delphi study does not aim to provide a hypothesis true to a "cross-sectional representation" of the population [10], but instead aims to build a hypothesis through methods of consensus on a given topic between independent participants [11].

2.3 Overview of Study Design

This study was split into three phases; Phase 0 - Participant Recruitment, Phase 1 - Brainstorming, and Phase 2 - Consensus Voting. Phases 1 and 2 are summarised in Fig. 2. All questionnaires used in the study were designed in Qualtrics and distributed via email. As part of the development process, questionnaires 1 to 3 were piloted ahead of distribution. This allowed for feedback to improve participant input.

2.4 Phase 0: Participant Recruitment

In phase 0 the research team recruited suitable experts to join the participant panel. Literature on the ideal number of participants is varied but generally suggests that between 10–18 participants are needed [7,9]. Participants were required to be 'informed individuals' [12] with current knowledge of robotics. An understanding of how robots are designed and/or made was listed as a desirable attribute. Participants self-selected to join the study and so were also likely to have an interest in this topic [12].

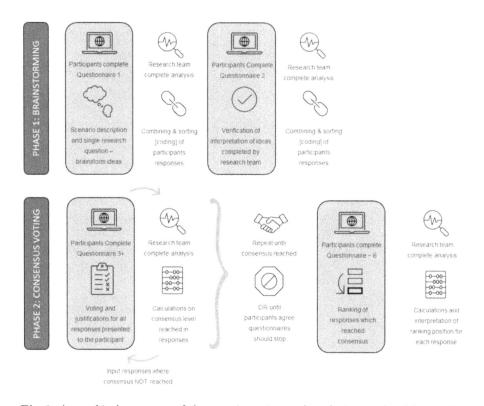

Fig. 2. A graphical summary of the questionnaires and analysis completed by participants and researchers respectively for the Delphi Study in repurposing robots

As recommended in [7], a diversity of experts was sought to increase the breadth of opinions. For this, no minimum number of years working in robotics was required for participants; instead, adjacent but relevant experience was considered by the researchers before approving a panellist. Additionally, the recruitment of participants targeted academia, industry and consulting to provide a greater range of experience.

This study recruited 16 participants to take part in the Delphi study. 25% of study participants were female which is comparable to 16% of the robotics workforce which is female [13]. The age demographic for all participants was split 19% aged 18–25 years, 38% aged 26–35 years, 25% aged 36–45 years and 19% aged 46+ years. The expertise of the participants was demonstrated through their years of experience in robotics (or equivalent fields of work) which is shown in Table 1. This information is shared to demonstrate the credibility of the experts recruited for this study [8].

Table 1. The expertise and employment information of study participants

Robotics Experience	Job Type	Employer
0 years*	Design Engineer	Frazer-Nash Consultancy
1–5 years	_ **	Oshen
1–5 years	Research Engineer	Manufacturing Technology Centre
1–5 years	Research Engineer	Manufacturing Technology Centre
1–5 years	Business Manager	Fairfield Control Systems Ltd.
1–5 years	_ **	Business using robots***
1–5 years	Research Engineer	Cranfield University
6–10 years	Senior Lecturer	University of Bristol
6–10 years	Research Associate	University of Bristol
6–10 years	Senior Specialist Technician	University of Bristol
11–15 years	_ **	Jacobs
6–10 years	Senior Developer	Jacobs
16+ years	Manufacturing Director	Business using robots***
16+ years	Senior Specialist	Business using robots***
16+ years	Principal Specialist	Jacobs
16+ years	Manufacturing Engineer	Rolls-Royce Plc

*had alternative relevant non-robotics experience, ** declined to share Job Title in publication, ***declined to share Employer name in publication

2.5 Phase 1: Brainstorming

Questionnaire 1. presented the research scenario of two robots - Robot A and Robot B (Fig. 1). Robots A and B were presented in a graphical and generalised form to help participants think "outside the current constraints" and "imagine futures with different assumptions" [7]. In this questionnaire, participants were asked to brainstorm: "What are the process steps that would need to be taken in order to viably repurpose Robot A into Robot B? In this example, viability includes economic and environmental factors". From the output of this questionnaire, researchers created a consolidated list of brainstormed ideas through the process of coding; the line-by-line analysis of written text raw data to identify repeating concepts [14,15]. This consolidated list was referred to as the "combined steps".

Questionnaire 2. presented the combined steps list to validate the analysis completed by the research team after Questionnaire 1. [9] recommends this validation step to assess the interpretation process completed by researchers during the coding process. Participants were presented their original idea and the combined steps their idea had been interpreted into to verify. All participants were also asked to review the combined steps list and add in any missing steps.

2.6 Phase 2: Consensus Voting

Questionnaire 3. asked the participants to review each of the combined steps and categorise them via voting as either: a *core process step* essential to the repurposing process; an *additional process step* supplementary to the core process; or a step which was *not required* in the repurposing process. Participants were given the option to justify their voting decision in a free-text box. An instruction manipulation check was included which looked like other questions, but which had specific instructions which participants had to follow. Research shows that concentration checks improve data quality and counter issues in online surveys where participants provide useless data [16]. Voting data from this questionnaire was used to calculate a consensus level between participants. Literature suggests several figures for an appropriate consensus level, from 50% to 80% [12,17]. For this study, a consensus level of 75% was selected.

Questionnaire 4–5. followed a similar format to Questionnaire 3, except participants were only required to vote on the steps which had not met the consensus threshold in the previous questionnaire.

Questionnaire 6. included a final round of voting for steps which had not reached consensus. Participants were also presented with a list of all core process steps and were required to provide an indicative ordering for the steps in order to suggest an acceptable process order.

3 Results

3.1 Brainstorming Phase

Participants individually brainstormed a total of 164 ideas in Questionnaire 1. These were coded by the research team into 54 'combined steps'. In Questionnaire 2 participants confirmed that 92% of the responses to the brainstorm had been correctly coded by the research team. Additionally, half of the participants reviewed the combined steps list and felt one or more ideas were missing from the process, which were then submitted as part of Questionnaire 2. The analysis of these changes resulted in a combined list of 70 steps to use in Phase 2.

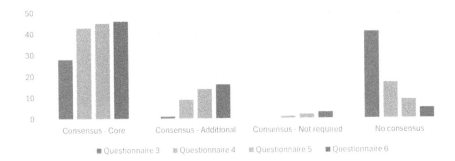

Fig. 3. Cumulative voting results of consensus for the combined steps

3.2 Voting Consensus

In phase 2, consensus was reached by the participants on 65 of the 70 proposed steps. The cumulative effect of voting for consensus can be seen in Fig. 3. In this paper, only those steps which were voted as core steps were used in the results for the finalised repurposing process (Sect. 4.1). Future work will include the analysis of steps voted as additional to the core process.

3.3 Finalised Repurposing Process

Figure 4 shows the finalised results of this study to produce a process for repurposing a robot from one utility to a new utility for which it was not originally designed. Writing shown in Fig. 4 in purple was added by the research team to aid clarification to the process. In particular, the concept of 'Gate checks' in step 1f was continued by the research team into later steps of the finalised process. Additionally the process steps were summarised and grouped by the research team into eight key stages needed in the repurposing process. Several steps which were voted as core process steps during the study were specific to the scenario given to participants in Fig. 1 as part of the research question. These steps were removed from the finalised process (Fig. 4) to generalise the results of this study so that they would be applicable across robotic systems. The list of the steps which were voted as core by the panel, but which were scenario specific (and therefore removed from Fig. 4) are shown in Fig. 5.

4 Discussion

4.1 Finalised Repurposing Process

The repurposing process created during this study can be compared to known processes for the build of new robotic systems, and compared to definitions of a remanufacturing process. The remanufacturing process proposed by [18] and summarised in Fig. 6 has five key steps which convert a used system at the end

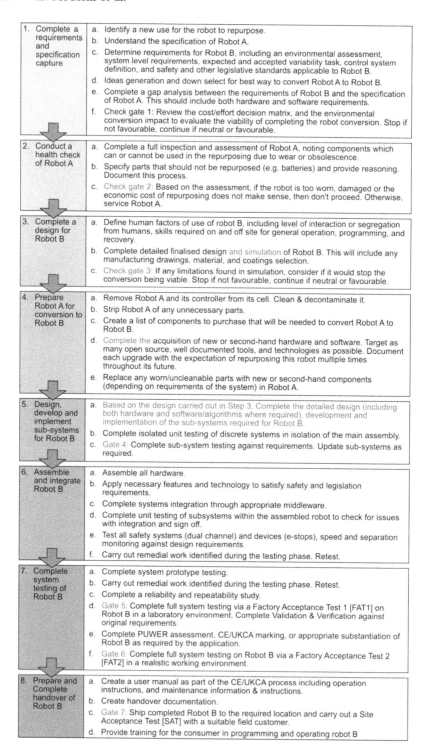

Fig. 4. The finalised repurposing processes, created in this Delphi Study by the expert panel through brainstorming and consensus voting

4b. Strip Robot A of any unnecessary parts. This may include:
  i. Removing Robot A from its static base,
  ii. Removing the end effector from Robot A.

5a: Dependant on the specification of the robot, identified in step 3, Complete a detailed design for the systems required for Robot B.

For the conversion of the paint spray robot to a weed removal robot, this should include:
  i. Design, develop and apply a vision system that allows the robot to identify weeds.
  ii. Devise and implement a suitable power supply and charging method capable of powering robot (and mobile base) for its anticipated work period.
  iii. Design and apply an All-Terrain mobile base to the robot arm. Consider a power supply (& batteries) able to cope with expected terrain and of suitable geometry & centre-of-gravity to effectively manage robot's kinematic ability whilst remaining safe and stable.
  iv. Integrate a locomotion system to the robot to enable it to be mobile.
  v. Design, develop and apply to the robotic system a gripper and associated algorithm suitable for picking and gripping weed. The system must be able to reach to ground level, and grip weeds identified during the requirements phase.
  vi. Design or select an appropriate robot function application pack. Integrate the software.
  vii. Design or select an appropriate robot loom pack. Integrate the software.
  viii. Design, modify and integrate a navigation system that will allow the robot to successfully path plan and navigate around its environment.
  ix. Design and apply a program capable of identifying the weed root type (depth of root) and an appropriate method for removing weeds.
  x. Integrate required sensors into the robot arm and mobile base.
  xi. Design and integrate an overarching Control System/Operating Software to tie in the Robot arm, base, and vision system into a cohesive machine.

6a. Assemble all hardware. This may include:
  i. Weatherproof the robot, including dust and water resistance.

Fig. 5. Steps voted by consensus as core to the repurposing process by the panel, which were also assessed as scenario specific to this study by the research team

of its primary life and recondition it to a "good-as-new" condition. Integrated throughout the described process is a quality assurance element.

The five steps presented in the remanufacturing process are echoed within stages 2 (Conduct a Health check of Robot A) and 4 (Prepare Robot A for conversion to Robot B) of the finalised repurposing process. The order in which these happen, however, is different. [18] starts with a full product disassembly, and must be completed prior to cleaning, inspection and sorting.

In the repurposing process the panel opted position the elements of inspection and sorting ahead of product disassembly. This reflects that systems and sub-systems may be kept whole in the repurposing process. [2] describes that a core principle of the circular economy is to keep products in the status which retains their highest embedded value - recognising the time and material effort in creating a product. Therefore, by similar argument, retaining as many systems and subsystems in their original form and not fully disassembling the entire robotic product renders the repurposing process more environmentally and economically viable. This also means that the repurposing process is working to meet the requirements identified in step 1, rather than making the entire system "good-as-new", negating the need for full disassembly.

Aspects of quality assurance can also be seen throughout the repurposing process. Inspections, testing and decision gates were placed into the core process

Fig. 6. Remanufacturing process described in [18]

steps by the panel, and highlighted by the research team in the form of gates. [18] notes that testing should not remain as a final step but instead should be integrated within the whole process. This ethos is reflected in the steps voted as core to the repurposing process.

Similarly, some specific forms of testing have been listed in the process in Fig. 4, alongside safety requirements during the design and build phases. These process steps selected by the participants work towards meeting the requirements of ISO 10218 Part 1 and 2 standards which include safety requirements and protective measures, and verification and validation processes [19,20]. These standards apply only to industrial robots, but the safety principles from this standard can be applied to other robots [19]. In this study, participants were presented with a scenario where an industrial robot was to be repurposed into an agricultural robot - which does not meet definitions of an industrial robot in the context of this standard. It is likely that old and second-hand robots will move from industrial to non-industrial settings within the context of repurposing in the future; it is recognised, therefore, that industrial standards for robots in the design, verification and validation will contribute to the repurposing of non-industrial robotic systems.

The proposed process for repurposing differs from the published standards in robotics in the inclusion of specific environmental considerations in the design process. Step 1c (Fig. 4) highlights the need for environmental assessments as part of the requirements capture for Robot B. British Standards are currently drafting a "Guide to the Sustainable Design and Application of Robotic Systems" [21]. This new standard may demonstrate additional links between the repurposing process developed in this study, and the wider principles of environmentally ethical design of robotic systems. Additionally, it verifies the gap in current processes and standards for methods to tackle the issue of robots at the end of their primary life in a useful and waste conscious manner - which this repurposing process can address.

4.2 Evaluation of the Delphi Study Methodology

Participants in this study were able to reach a consensus for the majority (93%) of ideas brainstormed in Phase 1 of this study. While engagement remained high, despite the study taking place over a period of four months, some participants found it difficult to imagine this study's particular future. In the facilitation of future theory studies, participants may find it "difficult to step outside of the current constraints to imagine alternative futures with different assumptions" [7]. The use of commentary boxes and concentration checks in the questionnaires gave individuals options to raise issues, while ensuring responses submitted remained useful to the context of the study. Participants were only removed from the study if they missed two questionnaires in a row, or if they did not meet the concentration checks presented to them in the questionnaires. This resulted in 16 participants completing Questionnaires 1 and 2; then for the remaining questionnaires (3 to 6) participant numbers were 15, 13, 14 and 12 respectively. This engagement level required active management of participants by the research team; chasing up missing responses in order to maintain the minimum participant levels outlined in Sect. 2.4.

5 Conclusion

This study took the expertise of a diverse panel of participants in the robotics industry to create a process which can be used for repurposing a robot from one utility to another. The repurposing process, created using a Delphi study methodology, utilised participants' knowledge of current industrial standards, while providing participants with the opportunity to identify gaps and new elements within the repurposing process which reflected environmental considerations needed for the sector to move towards a more circular economy. The repurposing process has similarities to the documented remanufacturing process, but goes beyond the intent to disassemble and reassemble in a 'like-new' status, and instead produces a method for providing robots with alternative uses outside of their original design intent.

Acknowledgements. The work of HM was supported by the EPSRC through FARSCOPE CDT, grant no. EP/S021795/1. Thank you to Shabaj Ahmed, Daniel Gosden, and Ella Maule for piloting the questionnaires used in this study.

References

1. McGloin, H., Studley, M., Mawle, R., Winfield, A.: Introducing the concept of repurposing robots; to increase their useful life, reduce waste, and improve sustainability in the robotics industry. In: International Conference on Robot Ethics and Standards (2023)
2. The technical cycle of the butterfly diagram. The Ellen MacArthur Foundation (2022). www.ellenmacarthurfoundation.org/articles/the-technical-cycle-of-the-butterfly-diagram. Accessed 20 Mar 2024
3. Elzinga, R., Reike, D., Negro, S.O., Boon, W.P.C.: Consumer acceptance of circular business models. J. Clean. Prod. **254** (2020)
4. The 17 Goals. United Nations. https://sdgs.un.org/goals. Accessed 20 Mar 2024

5. Forti, V., Baldé, C.P., Kuehr, R., Bel, G.: The Global E-waste Monitor 2020: Quantities, flows and the circular economy potential. United Nations University (UNU)/United Nations Institute for Training and Research (UNITAR) (2020). www.invest-data.com/eWebEditor/uploadfile/2020071100243938206180.pdf. Accessed 29 Mar 2022
6. Jupp, V. (ed.): The Sage Dictionary of Social Research Methods, 1st edn. SAGE Publications Ltd. (2006)
7. Groves, P., et al.:The Futures Toolkit. UK Government Office for Science (2023)
8. Robson, C., McCartan, K.: Real World Research, 4th edn. Wiley, Hoboken (2015)
9. Okoli, C., Pawlowski, S.D.: The Delphi method as a research tool: an example, design considerations and applications. Inf. Manag. **42**, 15–29 (2004). https://doi.org/10.1016/j.im.2003.11.002
10. Del Grande, C., Kaczorowski, J.: Rating versus ranking in a Delphi survey: a randomized controlled trial. In: Trials, vol. 24. BioMed Central Ltd. (2023)
11. Erffmeyer, R. C.: Decision-making formats: a comparison on an evaluative task of interacting groups, consensus groups, the nominal group technique, and the delphi technique (1981)
12. Hasson, F., Keeney, S., McKenna, H.: Research guidelines for the Delphi survey technique. J. Adv. Nurs. **32**, 1008–1015 (2000). https://doi.org/10.1046/j.1365-2648.2000.t01-1-01567.x
13. Wessling, B.: Celebrating the first International Women in Robotics Day. The Robot Report (2023). www.therobotreport.com/celebrating-the-first-international-women-in-robotics-day. Accessed 04 Apr 2024
14. Charmaz, K.: Constructing Grounded Theory: A Practical Guide Through Qualitative Analysis. Sage Publications Ltd. (2006)
15. Khaksar, S.M.S., Khosla, R., Chu, M.T.: Socially assistive robots in service innovation context to improve aged-care quality: a grounded theory approach. In: Proceedings of the 2015 7th IEEE International Conference on Cybernetics and Intelligent Systems, CIS 2015 and Robotics, Automation and Mechatronics, RAM 2015. Institute of Electrical and Electronics Engineers Inc. (2015). https://doi.org/10.1109/ICCIS.2015.7274614
16. Aust, F., Diedenhofen, B., Ullrich, S., Musch, J.: Seriousness checks are useful to improve data validity in online research. Behav. Res. Methods **45**, 527–535 (2013)
17. Parker, G., Kastner, M., Born, K,. Berta, W.: Development of an implementation process model: a delphi study. In: BMC Health Services Research, vol. 21. BioMed Central Ltd. (2021)
18. Steinhilper, R.: Remanufacturing: the ultimate form of recycling. Fraunhofer IRB Verlag (1998)
19. BS EN ISO 10218-1-2011: Robots and robotic devices - safety requirements for industrial robots. Part 1: Robots. In: BSI Standards Publication. The British Standards Institution (2011)
20. BS EN ISO 10218-2-2011: Robots and robotic devices - safety requirements for industrial robots. Part 2: Robot systems and integration. In: BSI Standards Publication. The British Standards Institution (2011)
21. BS XXXX Guide to the Sustainable Design and Application of Robotic Systems. The British Standards Institution. https://standardsdevelopment.bsigroup.com/projects/9021-05214#/section. Accessed 26 Mar 2024

"Incomplete Without Tech": Emotional Responses and the Psychology of AI Reliance

Mriganka Biswas[✉] and John Murray[✉]

University of Sunderland, Sunderland, UK
mriganka.biswas@sunderland.ac.uk

Abstract. The increasing integration of Artificial Intelligence (AI) technologies into daily life raises questions about our evolving relationship with technology. While quantitative studies have explored usage patterns and potential addiction-like behaviours, these models often fail to capture the depth of the psychological connection some individuals form with their devices. This mixed-methods study investigates the phenomenon of feeling "Incomplete Without Tech," exploring how reliance, trust, and emotional responses differ based on self-reported technical savviness. Quantitative analysis revealed significant differences in reliance on AI-powered recommendations and predictions for the tech-savvy group, but not for general AI usage or habit formation. Qualitative analysis of open-ended responses unpacked themes of convenience, task-specific trust, and concerns over autonomy and accuracy. Crucially, narratives surrounding tech absence revealed strong negative emotions, including a sense of 'incompleteness' across tech-savviness levels. This suggests reliance on AI might extend beyond functional utility and fulfill psychological needs traditionally met by other means. Findings challenge the adequacy of addiction-based models and highlight the need for new frameworks to understand how AI might shape our sense of self, potentially influencing well-being and resilience in an increasingly tech-saturated world.

Keywords: AI reliance · tech dependence · Incomplete Without Tech

1 Introduction

The omnipresence of technology in contemporary life is undeniable. From smartphones to social media platforms, our daily actions and interactions are intertwined with digital tools. While research into technology dependence often centers around usage patterns and addiction models (Griffiths, 1996; Kuss & Griffiths, 2011), emerging evidence suggests the need for a nuanced understanding focused on the emotional and psychological dimensions of our relationship with technology.

This study explores a specific but powerful theme – the feeling of being "incomplete" in the absence of technology. This subjective state points to a potential reliance that extends beyond the simple need to complete tasks. Existing literature on technology dependence often focuses on functional impairments or compulsive use (Billieux, 2012). The current study's mixed methods approach found an interesting pattern in the qualitative responses where the some of the participants mentioned they could potentially be feeling 'incomplete' in the absence of AI technologies. The feeling of incompleteness can stem from a sense of lacking something essential to one's sense of self or purpose. Psychologically, this aligns with self-determination theory, which posits that unmet needs for autonomy, competence, and relatedness can result in a feeling of deficiency (Ryan & Deci, 2000). This sense of incompleteness might manifest as a lack of fulfillment, disconnection, or a pervasive feeling that something is missing in one's life. By delving into such "Incomplete" response, this work aims to illuminate the possible psychological drivers underpinning individuals' deeply felt connections to their devices.

Initial quantitative and qualitative analyses reveal a statistically significant association between participants' self-described experiences without technology and their emotional intensity scores. Responses such as "frustration" and "uneasy" correlated with lower emotional intensity, while the "Incomplete" responses group demonstrated significantly higher intensity rankings. This aligns with a growing concern that for some individuals, technology might fulfill needs not strictly defined by functionality but rather by its impact on self-perception and social identity (Turkle, 2011). Drawing upon object relations theory (Winnicott, 1953) and the potential extension of Maslow's hierarchy (Koltko-Rivera, 2006), this study hypothesizes that tech absence induces feelings beyond practical inconvenience. For those who described themselves as "incomplete," technology potentially acts as a transitional object or even an extension of self. Disruption of this bond might evoke emotional responses mirroring separation anxiety or a threat to their sense of self. Through a mixed-methods approach, this paper investigates the "Incomplete Without Tech" phenomenon which reveals valuable implications for understanding healthy versus potentially harmful tech relationships in the context of individual well-being and societal adaptation in increasingly technology-saturated environments.

2 Background

The ubiquity of technology in modern life has fundamentally reshaped our relationship with the world around us. From instant communication to seamless access to information and entertainment, digital tools have become an extension of our daily routines and social interactions. While the convenience and efficiency benefits of technology are undeniable, a growing body of research suggests a critical need to delve deeper into the psychological implications of this ever-present digital presence.

Historically, investigations into technology dependence have often focused on compulsive behaviours and usage patterns, drawing heavily on addiction models (Griffiths, 1996; Kuss & Griffiths, 2011). These frameworks provide valuable insights, but they might not capture the full spectrum of human-technology interaction. A nuanced understanding requires acknowledging the potential for psychological reliance beyond mere usage frequency. This study takes a step in this direction by exploring a unique phenomenon: the feeling of being "incomplete" in the absence of technology. This subjective experience goes beyond simple frustration or inconvenience; it hints at a deeper psychological attachment. Participants expressing this feeling during the initial survey demonstrated a statistically significant increase in emotional intensity compared to those describing less intense emotions during tech absence.

To fully grasp the potential significance of the "Incomplete" response, we must examine established psychological theories that illuminate human motivation and the nature of our relationships with objects. One potential framework is Object Relations Theory (ORT), developed by Donald Winnicott (1953). This theory posits that humans rely on transitional objects, often physical items, to bridge the gap between self and the external world (Winnicott, 1953). These objects provide a sense of comfort and security, particularly in the early stages of development. Drawing a parallel to Winnicott's notion, technology, for some individuals, could fulfill a similar function.

Furthermore, Maslow's hierarchy of needs (Maslow, 1943) offers a valuable lens through which to understand the potential drivers of technology dependence. Maslow proposes a hierarchy of human needs, progressing from basic physiological needs to more complex ones like esteem and self-actualization. While Maslow's original framework doesn't explicitly consider technology, some researchers (Vedechkina and Borgonovi, 2021; Atske, 2021) suggest that technology might play a role in fulfilling needs beyond the purely functional. For example, social media platforms can contribute to a sense of belonging and social connection, potentially addressing esteem needs. Similarly, access to information and educational resources might facilitate self-actualization aspirations. In this context, the "Incomplete" feeling could signify a reliance on technology to fulfill needs that extend beyond basic functionality, potentially impacting the way individuals perceive themselves and interact with the world. While the current study offers a starting point, existing literature on the psychological dimensions of technology dependence offers a foundation to build upon. Turkle's (2011) work on "The Alone Together" phenomenon highlights society's paradoxical state of hyperconnectivity alongside increasing feelings of isolation. This suggests a potential reliance on technology to address social needs, even if the fulfillment itself is incomplete. Similarly, Przybylski et al. (2013) explore the concept of "Facebook Depression," suggesting that for some users, social media use can negatively impact self-esteem and well-being. Understanding these complexities is crucial for a comprehensive analysis of the "Incomplete Without Tech" phenomenon. The current study aims to delve deeper beyond existing quantitative findings by employing a mixed-methods approach. Qualitative exploration will shed light on the underlying narratives and experiences surrounding the "Incomplete" feeling, potentially revealing how technology functions to address psychological needs, shape identity, and influence emotional responses.

While previous work on technology dependence has provided substantial insights, certain limitations must be acknowledged. Firstly, a significant focus on quantifiable aspects, such as screen time or frequency of compulsive behaviours, neglects the potential depth of the psychological relationship between individuals and their devices. The intensity of an emotional response to tech absence offers a qualitative dimension often overlooked in traditional models (Przybylski et al., 2013). Moreover, existing addiction-based models might not fully encapsulate the complex nuances of tech reliance for certain individuals (Billieux, 2012). While these models provide a valuable framework to assess problematic behaviours related to excessive online gaming or social media use (Kuss & Griffiths, 2011), they may fall short in explaining a dependence driven primarily by psychological needs beyond the compulsion to use. However, by acknowledging the limitations of these models, particularly their focus on quantifiable aspects, we can explore the emotional dimensions of tech reliance through the concept of feeling "incomplete" in its absence.

The landscape of technology dependence is further complicated by the increasing integration of Artificial Intelligence (AI) across various digital tools and platforms. AI algorithms personalize content delivery, anticipate needs, and create an increasingly seamless user experience (Sodiya et al., 2024). This personalization creates a feedback loop that can potentially reinforce reliance by catering to anticipated desires and shaping user behaviour in subtle ways (Fogg, 2009). Habit formation theories emphasize the role of repetition, rewards, and environmental cues in establishing habits (Aarts and Dijksterhuis, 2000). AI-powered platforms, through continuous user engagement and personalized content delivery, might leverage these habit formation principles, potentially heightening feelings of incompleteness without their familiar digital environment. The Technology Acceptance Model (TAM) (Davis, 1989) sheds further light on the factors influencing tech use. TAM suggests that perceived usefulness and perceived ease of use are key drivers of technology adoption and continued use. However, TAM doesn't explicitly account for the emotional aspects of the user-technology relationship. The "incomplete" feeling in the current study delves beyond perceived usefulness to explore a potential emotional attachment that transcends purely utilitarian motivations. Moreover, it could mean beyond simple inconvenience or frustration associated with a lack of access to basic functionalities like emailing or browsing the web. It signifies a more profound emotional response, characterized by a sense of unease, anxiety, or even a perceived loss of a crucial part of oneself. Participants who described themselves as "incomplete" demonstrated significantly higher emotional intensity scores compared to those who reported less intense emotions during technology absence. This qualitative difference suggests a deeper psychological reliance that warrants further investigation. By focusing on the emotional intensity associated with tech absence, we can gain valuable insights into the perceived role that technology plays in individuals' lives. Is it a tool for connection, a source of information that fosters a sense of self-actualization, or something more complex and multifaceted? Exploring these questions through qualitative analysis allows us to delve into the subjective experiences of those who feel "incomplete" without technology, potentially revealing a deeper understanding of the psychological mechanisms at play.

3 Methodology

The study recruited 65 participants through an online survey hosted on Microsoft Forms. Advertisements distributed among peers and colleagues described the research focus on AI in everyday life and sought participation in the brief online survey. Inclusion criteria included regular use of various technological devices, like smartphones, computers, tablets, or smart speakers, along with a willingness to share their encounters with AI features. This approach ensured the participation of individuals who routinely integrate AI into their daily lives, encompassing a range of technical expertise levels.

The online questionnaire comprised multiple sections aimed at gathering the following data:

- **Demographics:** Participants provided information about their age, gender, ethnicity, education level, and geographic location.

- **Daily used AI habit & reliance:** Likert-scale items (1–7, strongly disagree to strongly agree) assessed participants' self- reported comfort and proficiency with technology. Open-ended items to understand participant's perceptions, reliance, trust, using habits etc.

Questions probed areas like their confidence navigating new software, willingness to experiment with new devices, and ability to troubleshoot basic technological problems. Additionally, information was taken on their daily hours of AI technology uses, potential number and types of devices and their years of experiencing of using such technologies to understand a correlation between their self-assessment results on tech-savviness compared to device no., daily usage, and long-term habit. This comprehensive questionnaire (Fig. 1) examines participants' complex relationships with various AI-powered technologies, delving into trust, reliance, and the emotional impact of technology absence. It begins by exploring practical usage patterns across AI assistants, health trackers, and language models. Questions then assess trust levels in AI-driven predictions (weather, travel, pricing) and personalized content recommendations (Netflix, Spotify, and news feeds). Beyond functionality, the questionnaire probes deeper, gauging overall tech reliance, comfort levels without devices, and even a hypothetical scenario prioritizing tech over survival essentials. Specific questions investigate potential emotional attachment to technology, including reactions to malfunctions, conversational tendencies, and anthropomorphising.

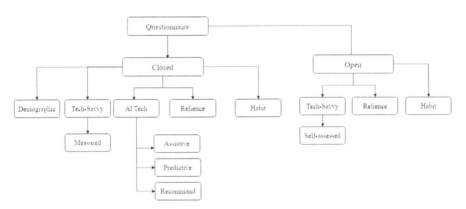

Fig. 1. Questionnaire structure

Crucially, open-ended descriptive questions invite participants to explain the factors influencing their trust or distrust of specific AI applications and detail their feelings and behaviours during tech absence. This qualitative element promises rich insights into themes of control, emotional attachment, and how AI shapes daily routines.

4 Result Analysis and Discussion

This study first explores how self-reported levels of tech-savviness relate to various aspects of AI reliance, contributing to the understanding of the "Incomplete Without Tech" phenomenon. A non-parametric, the Kruskal-Wallis test, revealed intriguing nuances.

The results (Fig. 2) show a clear connection between tech-savviness and reliance on AI-driven recommendations. The Kruskal-Wallis test statistic (7.803) was significant at p = .007, indicating differences in how participants categorized as "Not Tech Savvy," "Not Sure," and "Tech Savvy" rely on AI recommendations (Reccom_score). Pairwise comparisons further supported this, with "Tech Savvy" individuals differing significantly from both "Not Tech Savvy" and "Not Sure" groups. This suggests that those who perceive themselves as more knowledgeable about technology are more likely to integrate AI recommendations into their decision-making processes, potentially contributing to a feeling of being "incomplete" without such technologies. A similar pattern emerged with trust in AI predictions. The Kruskal-Wallis test statistic: 7.405 and p = .025 again show a significant effect. Here too, pairwise comparisons reveal the "Tech Savvy" group placing significantly higher trust in AI predictions (Pred_score) compared to the "Not Tech Savvy" group. This suggests that self-perceived technical expertise might lead to a greater willingness to rely on AI for tasks like weather forecasting or travel planning.

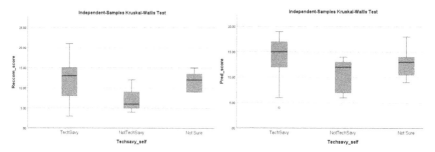

Fig. 2. Kruskal-Wallis test statistics on Recommendation and Prediction scores

Interestingly, tech-savviness didn't significantly impact reliance on AI assistants (Assist_score) or general tech reliance (Relien_score). The Kruskal-Wallis test statistics for these scores (3.806 and .991 respectively) were not significant (p > .05). This highlights that the "Incomplete" feeling, while potentially influenced by how people use certain AI features, might also depend on psychological factors beyond self-reported technical skill. The same was true for AI-related habits (habit_score) whrere the Kruskal-Wallis test (statistic: 2.051, p = .359) did not yield a significant result. This indicates that tendencies to converse with technology, personify devices, or experience difficulty when tech malfunctions might not be strongly linked to a person's self-rated tech-savviness. Overall, these results add to the growing body of research on the "Incomplete Without Tech" phenomenon. It suggests that while technical knowledge might contribute to reliance on some AI implementations, reliance on others and the deeper emotional attachment to technology might be driven by more complex psychological needs that require further investigation.

The analysis of open-ended responses reveals a complex landscape of decision-making processes surrounding AI reliance. The responses to the question *"What specific factors influence your decision to rely on or avoid AI technology in different situations?"* reveal the multifaceted considerations at play in navigating an increasingly AI-saturated world. A recurring theme is balancing the perceived benefits of AI against lingering reservations (Table 1). Convenience and speed emerge as primary drivers of reliance, particularly for information-seeking, aligning with the earlier findings on trust in AI predictions and utilization of recommendations. However, a strong counter-current of concern surrounds accuracy, privacy, and the potential loss of autonomy if reliance becomes excessive. This tension suggests that for many participants, the "Incomplete Without Tech" feeling might stem from a complex calculation – acknowledging its utility while fearing over-reliance and the potential erosion of skills, judgment, or even a sense of self if the technology were unavailable.

Table 1. Thematic qualitative analysis of the 1st open ended question

Theme	Explanation	Example quotes highlights
Convenience & Efficiency	AI's ability to save time and simplify tasks redefines the perception of 'needs.'	finding answers faster, helps saving time, easy
Trust, Accuracy, & Reliability	Trust in AI's output and concerns about misinformation influence reliance decisions	fast and accurate, depends on how reliable, avoid when replicating artwork
Personal Control & Autonomy	A desire to maintain agency, avoid over-dependence, and make independent choices	perform certain tasks myself, should be able to function without them, not become too dependent
Ethical & Societal Concerns	Privacy violations, potential biases, and misuse of AI lead to avoidance in specific domains	avoid…mimic people's voices, privacy concern, job displacement
Task-Specific Suitability	AI is favoured for simple, factual needs; creative tasks and high-stakes decisions often warrant human input	research, mundane tasks vs. medical emergencies, creative ideas

The thematic analysis (Table 1) reveals three core influences shaping decisions around AI reliance: perceived benefits, concerns, and task nature. The prominence of convenience and efficiency within the benefits theme emphasizes the potential fulfilment of basic needs (Maslow, 1943), suggesting that AI-driven tools might become integrated into a revised hierarchy for certain individuals. Speed and accessibility in obtaining information or completing tasks could foster a sense of competence, potentially contributing to the "incomplete" feeling experienced in their absence. However, concerns around accuracy, privacy, and autonomy signal that psychological needs extend beyond practicality. Object Relations Theory (Winnicott, 1953) helps us understand this. The emphasis on maintaining control and agency could suggest an unconscious fear of technology replacing a core part of the self – the ability to think, reason, and make choices independently.

The qualitative analysis of the second open-ended question unveils a nuanced relationship between participants and AI, marked by varying levels of experience, degrees of trust, and diverse motivations for adoption. A key theme is **task-specific trust**, with participants expressing greater confidence in AI for well-defined tasks like navigation and weather forecasting. This mirrors the earlier quantitative results, where tech-savvy individuals showed a stronger tendency to rely on AI recommendations and predictions. However, a deeper dive into the "Why?" behind this reliance reveals a complex interplay of factors. We explicitly asked participants, *"Are there any specific types of AI technology you find more reliable than others? Why or not?"* Their responses paint a rich picture of reliance, scepticism, and underlying concerns about the transformative power of AI. This tension highlights the potential for a double-edged sword: AI integration might fulfil needs and enhance capabilities, yet some participants grapple with a sense of unease regarding the potential for over-reliance and the erosion of their own skills and decision-making abilities.

Table 2. Thematic qualitative analysis of the 2nd open-ended question

Code Category	Percent of Responses	Key Themes
AI Experience Level: Novice	Low	Reasons for non-use
AI Experience Level: Selective	High	Emphasis on specific tasks, scepticism exists alongside use
AI Experience Level: High-Reliance	High	Potential for overconfidence in certain AI tools
Trust Fluctuation: Task-Specific	Very high	Weather apps, navigation vs. creative tasks
Trust Fluctuation: Generally Sceptical	High	Existential concerns, lack of transparency
Trust Fluctuation: Unquestioning	Low	May be linked to tech-savviness
Tool vs. Threat: Tool Mindset	Low	Focus on augmentation
Tool vs. Threat: Substitution Concern	Medium	Skill erosion, loss of originality
Lure of Convenience: Convenience	High	Entertainment, ease-of-use
Lure of Convenience: Utility	Medium	Problem-solving, efficiency

The thematic analysis (Table 2) reveals distinct experience levels and a complex interplay of trust and reliance. The high percentage of 'Selective Users' highlights a nuanced approach, suggesting the integration of AI occurs alongside healthy scepticism. This underscores the importance of task-specific trust, with participants readily accepting AI for predictable tasks (weather, navigation) but hesitant about those involving creativity. Notably, a significant portion harbour a general scepticism rooted in existential concerns and a desire for AI transparency. This aligns with the "Incomplete Without Tech" theme– if AI can generate creative outputs, some individuals might question their own unique contributions and potentially feel diminished in its absence.

The 'Tool vs. Threat' dimension (Table 2) highlights anxieties about technology replacing human abilities. While convenience remains a powerful motivator, the concern about skill erosion hints at the potential for reliance to breed dependency. This tension between AI as an aid and a potential threat to self-sufficiency directly relates to the concept of feeling "Incomplete." Participants might question their capabilities if the technology that augments them is unavailable, suggesting a reliance that goes beyond mere convenience.

The third open-ended question directly probes the core of the "Incomplete Without Tech" phenomenon. Responses to *"Briefly describe any specific feelings or behaviours you experience when you are unable to access your technology."* promise to unveil the depth of attachment participants have formed with their devices. By bridging these narratives with the levels of reliance and specific concerns outlined earlier, a clearer picture emerges (Table 3). If feelings of frustration or anxiety are predominant, this suggests a dependence deeper than mere utility. If specific behaviours highlight a scrambling for alternative ways to complete tasks or connect with others, this could indicate that AI-supported tools and social platforms have become so tightly woven into daily life that their absence creates a deeply felt void.

Moreover, our exploration of emotional responses (Table 3) to tech absence takes a fascinating turn when we consider how participants perceive their own tech-savviness. While a clear correlation between self-reported tech skills and emotional impact might be expected, the data paints a more nuanced picture (Fig. 3).

The finding that a significant portion (75%) of tech-savvy individuals experienced intense negative emotions when denied access to technology directly challenges the core assumption that those who consider themselves technologically adept can easily adapt in tech-absent situations (Fig. 3). This suggests a deeper psychological reliance that goes beyond mere functional needs. Participants described feelings of frustration, anxiety, and even a sense of being "incomplete," echoing the study's central focus. These powerful responses invite us to reconsider how technology might now be intertwined with core human drives. Could the distress expressed by the 'savvy' group indicate that some AI tools now fulfill needs for belonging, esteem, or even self-actualization traditionally met through other means? Revisiting Maslow's hierarchy in the modern context becomes essential to understand this shift and its potential consequences for an individual's well-being during periods of tech absence.

The spectrum of reliance even within the "Yes" category underscores the nuanced nature of tech dependence. It highlights that not all tech-savvy individuals are equally vulnerable to experiencing negative emotions during tech absence. Individual differences in personality, the specific technologies used, and potentially pre-existing anxieties likely play a role in shaping this response. Focusing on the outliers – those tech-savvy

Table 3. Thematic qualitative analysis of the 3rd open ended question

Tech Savviness	Common Responses (Illustrative, Not Exhaustive)	Key Observations
Yes	* "Frustration", "Anxious", "Incomplete", "Restless"… (Many express strong reliance)	* Even 'savvy' individual's struggle. Is this reliance masked by overconfidence?
Yes	* "Boredom", "Mild Annoyance"… (Some show low-intensity impact)	* Does savviness sometimes equal a spectrum of reliance types, not universal mastery?
Yes	* "Just wait…", "Disturbed" (Outliers showing unexpected negativity)	* These defy assumptions! Why do some tech-savvy people feel so lost?
No	* "Frustrated", "Anxious," "Helpless"… (Emotional impact can be strong)	* Potential for heightened struggles with less tech confidence
No	* "Not a big deal…", "I can easily switch…" (Signs of healthy adaptability)	* These are the true "under the rock" success stories we need to study
Not Sure	* Varied range of emotions, some negative, some neutral	* This uncertainty about tech-savviness itself is interesting – what does it reflect?
Not Sure	* "Can solve it myself…" (Potential for self-reliance even with uncertainty)	* Intriguing for the "rock" discussion: is 'old school' problem-solving less tied to tech?

participants who reported utter helplessness and those less savvy who demonstrated surprising resilience – promises to illuminate the crucial factors governing reliance versus adaptability. These insights can contribute to building targeted interventions and digital well-being practices in the future. The very existence of a "Not Sure" category regarding tech-savviness compels us to question our definitions. In a world saturated by technology, does true "savviness" reside in the unquestioning adoption of the latest gadgets, or could it instead prioritize critical thinking and a mindful approach to technology use? This hints at the importance of AI as not just a physical absence of tech, but a mindset – the ability to function effectively without constant reliance.

The current findings resonate with philosophical concerns about technology's role in shaping our existence. Heidegger's (1977) worries about technology becoming the sole lens through which we experience the world gain new relevance in light of the anxiety exhibited by tech-savvy individuals. When we question our ability to think and navigate challenges without our devices, we highlight the necessity for a balanced relationship. The goal should be to leverage AI in empowering ways without diminishing our unique human capacity for independent thought and action.

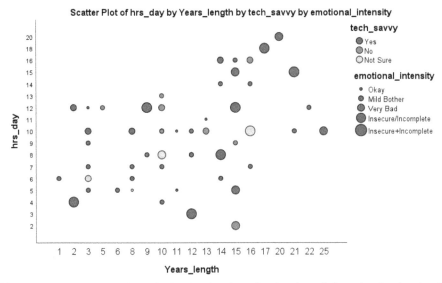

Fig. 3. How the emotional intensity is related to the tech using times daily and tech using time in years, and self-reported tech-savvy measures.

5 Conclusion

This study's exploration of the "Incomplete Without Tech" phenomenon has revealed a complex and evolving relationship between individuals and AI-powered technologies. The combination of quantitative findings and open-ended qualitative investigations provides insights into a core question from our background discussion: how does technology reliance relate to traditional models of addiction and habit formation? While significant differences in reliance on AI recommendations and predictions suggest that tech-savviness plays a role, the emotional responses and deeper concerns about autonomy and creativity highlight limits to these models. The "Incomplete" feeling experienced in tech absence likely signals a reliance that extends far beyond mere compulsive usage or simple habit formation.

Qualitative analysis directly addressed the study's central questions. Participants' descriptions of their decision-making processes around AI adoption highlight a constant tension between convenience and the potential for over-reliance, hinting at the need to move beyond solely measuring tech use to understand its impact. The notion of task-specific trust emerged as vital. While clearly favouring the use of AI for well-defined, predictable tasks, many participants expressed deep hesitation about outsourcing functions involving creativity, complex decision-making, or those with high personal stakes. Here, questions raised about the Technology Acceptance Model (TAM) become crucial – perceived usefulness and ease of use might not be the sole determinants of adoption when tasks become tied to one's sense of self and individual expression.

Perhaps the most striking revelations came from the narratives surrounding experiences of technology absence. The prevalence of strong negative emotions like frustration, anxiety, and even a sense of "incompleteness" across all tech-savviness levels challenged assumptions about the limitations of addiction-based models in explaining our relationship with AI. These accounts suggest that while AI might initially be integrated through mechanisms like habit formation, the emotional response upon tech absence indicates the formation of a deeper psychological attachment. This finding emphasizes the need for new theoretical frameworks specifically designed to address the unique ways in which deep reliance on AI might influence our sense of self, well-being, and resilience.

This study leaves us with crucial questions for future research. We must redefine "tech-savviness" to include not only technical expertise but also the mindful ability to disconnect when necessary. Further investigations should explore the thresholds at which reliance on AI starts to hinder skill development, critical thinking, and an individual's sense of self-efficacy. As AI fulfills needs like belonging and esteem, we should consider if Maslow's hierarchy of needs requires revision in the modern era, particularly for those who feel "Incomplete Without Tech." This understanding of deep reliance has implications for developing new well-being tools and perhaps even therapies designed for an AI-driven world. Most importantly, this study underscores the continued need to critically evaluate the impacts of ever-present technology. As AI becomes even more advanced, ongoing research and a willingness to question our own reliance will be key in ensuring that the integration of human and technological capabilities enhances our lives rather than diminishes the skills and sense of self that make us uniquely human.

References

Aarts, H.H., Dijksterhuis, A.: Habits as knowledge structures: automaticity in goal-directed behavior. J. Pers. Soc. Psychol. **78**(1), 53–63 (2000). https://doi.org/10.1037/0022-3514.78.1.53

Atske, S.: Experts say the 'new normal' in 2025 will be far more tech-driven, presenting more big challenges. Pew Research Center (2021). https://www.pewresearch.org/internet/2021/02/18/experts-say-the-new-normal-in-2025-will-be-far-more-tech-driven-presenting-more-big-challenges/

Billieux, J.: Problematic use of the mobile phone: a literature review and a pathways model. Curr. Psychiatry Rev. **8**(4), 299–307 (2012). https://doi.org/10.2174/157340012803520522

Davis, F.D.: Perceived usefulness, perceived ease of use, and user acceptance of information technology. Manag. Inf. Syst. Q. **13**(3), 319 (1989). https://doi.org/10.2307/249008

Fogg, B.J.: A behavior model for persuasive design. In: International Conference on Persuasive Technology (2009). [Preprint]. https://doi.org/10.1145/1541948.1541999

Griffiths, M.D.: Behavioural addiction: an issue for everybody? Empl. Counselling Today **8**(3), 19–25 (1996). https://doi.org/10.1108/13665629610116872

Heidegger, M.: The question concerning technology, and other essays (1977). https://ci.nii.ac.jp/ncid/BB27055321

Koltko-Rivera, M.E.: Rediscovering the later version of Maslow's hierarchy of needs: self-transcendence and opportunities for theory, research, and unification. Rev. Gen. Psychol. **10**(4), 302–317 (2006). https://doi.org/10.1037/1089-2680.10.4.302

Kuss, D.J., Griffiths, M.D.: Internet gaming addiction: a systematic review of empirical research. Int. J. Ment. Health Addict. **10**(2), 278–296 (2011). https://doi.org/10.1007/s11469-011-9318-5

Litt, C.J.: Theories of transitional object attachment: an overview. Int. J. Behav. Dev. **9**(3), 383–399 (1986). https://doi.org/10.1177/016502548600900308

Maslow, A.: A theory of human motivation. Psychol. Rev. **50**(4), 370–396 (1943). https://doi.org/10.1037/h0054346

Przybylski, A.K., et al.: Motivational, emotional, and behavioral correlates of fear of missing out. Comput. Hum. Behav. **29**(4), 1841–1848 (2013). https://doi.org/10.1016/j.chb.2013.02.014

Ryan, R.M., Deci, E.L.: Self-determination theory and the facilitation of intrinsic motivation, social development, and well-being. Am. Psychol. **55**(1), 68–78 (2000). https://doi.org/10.1037/0003-066x.55.1.68

Sodiya, E.O., et al.: AI-driven personalization in web content delivery: a comparative study of user engagement in the USA and the UK. World J. Adv. Res. Rev. **21**(2), 887–902 (2024). https://doi.org/10.30574/wjarr.2024.21.2.0502

Turkle, S.: Alone together: why we expect more from technology and less from each other. Choice **48**(12), 48-7239 (2011). https://doi.org/10.5860/choice.48-7239

Winnicott, D.W.: Transitional objects and transitional phenomena; a study of the first not-me possession. Int. J. Psychoanal. **34**, 89–97 (1953)

Vedechkina, M., Borgonovi, F.: A review of evidence on the role of digital technology in shaping attention and cognitive control in children. Front. Psychol. **12** (2021). https://doi.org/10.3389/fpsyg.2021.611155

Suspicious Activity Detection for Defence Applications

Matthew Marlon Gideon Parris[✉], Hisham Al Assam, and Mohammad Athar Ali

Faculty of Computing, The University of Buckingham,
Hunter Street, Buckingham MK18 1EG, UK
{1804564,hisham.al-assam,athar.ali}@buckingham.ac.uk
https://www.buckingham.ac.uk

Abstract. Violent behaviour, especially using handheld weapons, is a significant issue plaguing society in this era. This paper evaluates the state-of-the-art 3DCNNsl violent activity recognition architecture for violent activity recognition with an overview of its operations via transfer learning, adapting the 3DCNN pre-trained model for new undertakings. 3DCNNsl violent activity recognition final layer modifications evaluate the action's generic status, emphasising the activity's true nature, belonging to the violent or non-violent class with subclasses. The idea emphasises 3DCNNsl violent activity recognition capability and confidence in its predictions regarding the homogeneousness of violence and hostile yet non-violent conduct from an overall perspective. The operation adds value to the violent activity recognition domain by evaluating action similarity data, which intensifies the computational load identified by overall accuracy performance. To establish the model's performance impact, we employed real-world conditions to observe processing effectiveness on specific violent/non-violent classes concerning their complexity. We designed the conditions to reflect pre-processed data *(data containing resolution/scenery enhancements)*, no pre-processing *(raw data without modifications or enhancements)*, and action similarity *(intense complexity between violent/non-violent actions)* to evaluate the lethal sporadic complex nature of violence honestly. The designed conditions are only applied to evaluate 3CNNsl violent activity recognition models. Experimental results project the model's overall accuracy capability at 72–88% utilising data with pre-processing *(data containing resolution/scenery enhancements)*. We concluded the investigations by discussing operational challenges and measures to enhance the model's effectiveness to produce more robust outcomes.

Keywords: Violent activity detection · Violent activity recognition · Suspicious activity detection · Human activity recognition

1 Introduction

Activity recognition techniques experience processing challenges as individual mechanisms when exposed to extreme action complexity, which escalates the risks of producing erroneous results. Most activity detection solutions incorporate multiple complex scaling techniques, further enhancing image dimensionality and simultaneously intensifying convolution operations. 3DCNN for violence and object tracking, such as traffic violations, crowd management and law enforcement, has yet to be appropriately determined. Moreover, a distinct detection singularity emerged that evaded the focus of researchers as they are yet to establish the impact of efficiently detecting the complexity of violent behavioural patterns in primitive stages before lethal impact. To address the abovementioned concerns, we proposed modifying 3DCNN as a 3-dimensional convolution neural network for violent activity recognition using CCTV data and measuring its effectiveness via overall accuracy with precision/recall as performance indicators for further fine-tuning as a 3DCNNsl violent activity recognition model.

2 Literature Review

The earliest activity recognition concept considered wearable gyroscope activity motion sensors that detected the disposition of non-violent human patterns at 0.98% accuracy relative to the multi-sensor utilising the adaptive-hidden Markov model in [8]. Violent real-world scenarios require robust processing systems that supersede accelerometer sensors relative to their ambiguous nature and the range of violent motions. The investigations disclosed a concatenation of the state-of-the-art 3DCNN and LSTM as a Bi-Directional LSTM solution for early action frame prediction relating to motion patterns and object appearances as a modern automated technique [13]. They achieved 0.68% accuracy using the UCF crime dataset for activity recognition. However, the technique required improvement relative to feature learning and extractions at the input stages. The impact introduced an adverse effect that hindered the model's inference capability due to overlooking pertinent complex processing anomalies encompassing violent activity features during training. Its operations increase the risk of biased results. Further investigations revealed [7], that introduced a mixed convolution Resnet-50 regression block on UCF-101 based on feature and model fusing for specific targets. They achieved 71.07% activity recognition at speeds of 200 FPS with complex configurations. Authors [2] developed a hybrid 3DCNN HAR model for activity recognition actions applying KTH and J-HMDB datasets. They demonstrated state-of-the-art performance compared to baseline methods, with accuracies of 78% on KTH and 90% on J-HMDB from a non-violent perspective. They endured image dimension trials, which adversely affected their generalisation operations. The issue negatively impacts the model's robust processing capabilities as a critical component of preventing human fatalities. The investigations unveil [3], which proposed a hybrid technique that included 3DCNN and optical flow-gated networks for the violent activity recognition concept. They obtained an accuracy of 87.25% on the RWF dataset containing 2000 sample videos with complex script configurations. The unveiling of 3DCNN's effectiveness in the literature's selection proved its capability as an innovative medium that facilitates the development of a reliable violent activity recognition model that accurately discerns the actions in its primitive stages.

3 Proposed System/Model

To emphasise 3DCNN's processing, we expose its technicality, which aligns our objective through quantitative concepts employing transfer learning procedures. Investigating 3DCNN single and multi-level via experiments disclosed the superiority of the single-level neural network over multi-level. Multi-level layers encouraged extreme processing flexibility requiring the power of GPU and TPU devices coupled with extensive training operations to achieve a suitable outcome [6,17,24]. This approach exceeds the computing hardware's ability to manage the framework processing requirements with real-time results. Multi-level high data dependency discussed in [6,15], and [22] necessitates laborious data labelling tasks, thus intensifying the possibility of producing erroneous results on violence, compared to re-configuring 3DCNN's layers. We modified [16]'s sixteen-layer 3DCNN with additional final layer configurations, transforming its structure to facilitate violent activity recognition as a 3DCNNsl model of twenty-three layers in Fig. 1.

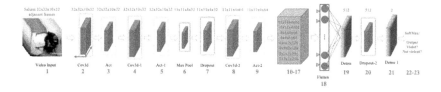

Fig. 1. 3DCNNsl Violent Activity Recognition Operations

Their 3DCNN operations excluded the 128 and 256 convolution dimensions as processing layers, thus intensifying the processing complexity and causing extensive over-fitting for this research. We reintroduced those layers as a 3DCNNsl violent activity recognition framework coupled with output coding to generate a desired outcome. The idea boosted efficiency by adapting the model's programming to encourage robust classification of complex sporadic violent actions in its primitive stages. We achieved violent activity recognition by acquiring insight into categories of violence, amassing, pre-processing datasets and concluding statistical analysis. Pre-processing utilising software tools *(iMovie/handbrake)* standardised the publicly available datasets by formatting data dimensions to facilitate the model's input size requirements. The concept generates more *(violent/non-violent)* data via contrast augmentations, cropping scenes, shearing, flipping, rotation of 90° clockwise and anti-clockwise, Grayscale, and 5% Gaussian noise. As most datasets are available online, we addressed ethical concerns by selecting publicly available forums and requesting permissions electronically, which disclosed the desire for specific sensitive samples. 3DCNNsl violent activity recognition processes video data to establish category labels' probability via scores instead of using bounding box encapsulation techniques during blob analysis, like object detection. The model produces an overall accuracy score to reflect its confidence in its decisions on complex actions. The output further discloses status labels to specify the resonance of each action generically *(violent or non-violent)* and individually from a subclass level *(stabbing, beating, fencing, shooting, fencing, cutting-in-kitchen, nun-chucks, sumo-wrestling, walk-with-dog, and knitting)*.

3.1 Activity Description

By accumulating the general knowledge of behavioural patterns suggesting the erratic motions of violence, we designate the class of activities that allow violent predictions to occur. By pinpointing the class of activity, we identify comparisons of normal and abnormal behaviour indicators relevant for training operations in alignment with [5,11,14,19]. These actions reflect the human gait with or without weapon objects depending on the class category relative to one human versus one human, one human versus many and many individuals versus one human. Though non-violent actions are not the focal point of this research, their significance demonstrates actions that are not violent to enhance 3DCNNsl violent activity recognition prospects during inference.

3.2 Data Acquisition

Acquiring meaningful volumes of data containing quality resolutions of real-world scenarios, with its pre-start duration, proved challenging to locate as a prerequisite. The data from public Facebook, Instagram, and YouTube online social media platforms mirrored the pre-start criteria seconds before violence for training purposes. The acquisition prospect generated 569 action samples between violence and non-violence to enhance the model's learning initiatives. The data acquisition process proved challenging because of the limited dataset volumes of training samples portraying fatality/injuries resulting from violence. Utilising impractical data samples increases the model's processing complexity, risking the distortion of valid attributes of violence compared to other non-violent actions during processing. The data split training ratio mirrored [23] and [18] cross-validation 80:20 approach utilising 60% training, 20% validation, and 20% for testing, which alleviated over-fitting. Over-fitting adversely affects convolution models, promoting classification projections towards samples in the training set but experiencing generalisation challenges when classifying actions on unused data [20].

3.3 Experimental Settings

Modifications to [16]'s 3DCNN model via GitHub increased the layers and hyperparameter fine-tuning efforts to standardise the input requirements of height, width, tensors, and channels as $320 \times 240 \times 12 \times 3$. The model's hyperparameter options emerged from ubiquitous approaches in the literature and various fine-tuning procedures during simulations; those operations disclosed optimal results using a batch size of 130, an epoch of 32/90, and an activation depth of 12. The depth value regulated the model's tensor block output size to reduce processing latency and encourage optimal results. We implemented a MacBook Pro 32-core silicon, m1 64gb CPU with onboard GPU support hardware for processing. We applied Python as a ubiquitous programming platform within the literature to further reduce latency issues and establish intuitive syntax and processing control. The operation alleviated complex deployment discussed in [21] by integrating several TensorFlow, Matplotlib, Pandas, NumPy, SciPy, Scikit-Learn and PyTorch libraries to accommodate convolution. We strategically designed eighteen experimental conditions *(six experiments per data condition)* utilising pre-processed data *(data containing resolution/scenery enhancements)*, no pre-processing (raw data without enhancements) and action similarity *(intense complexity between violent/non-violent actions)* to deliberately challenge the model's operations, oppose to utilising easily classified actions.

3.4 Evaluating 3DCNNsl Violent Activity Recognition

To evaluate 3DCNNsl violent activity recognition, we considered a mean average precision *(mAP)* to disclose the operation overall accuracy denoting its effectiveness. Mean average precision overall accuracy metric provided crucial insight into the models' classification state with a score projection ratio between 0–1. Scores approaching 1 insinuate high accuracy to suggest the level of performance concerning its *(effectiveness and confidence)* classifications, whereas low scores suggest the opposite. The analysis considers ratings exceeding 50% as a threshold, thus implying model efficacy and ignoring values below the margin. A further evaluation incorporated confusion matrix, precision, and recall disclosing operational and class action fine-tuning measures to encourage 3DCNNsl violent activity recognition robustness [12] and [1]. Its processing considers important computational equations discussed in [4] that determine the overall score, precision and sensitivity/recall in binary *(violent or not)* and multi-class *(stabbing, walking, shooting, running)* classification scenarios. Conversely, precision estimates the correct proportions of all class labels the model predicted, whereas recall quantifies the number of accurately classified label predictions relative to all positive classifications generated in alignment with [9] and [10].

4 Results and Analysis/Discussions

The evidence disclosed 3DCNNsl violent activity recognition effectiveness in several simulations incorporating a batch size of 130, an epoch of 32/90 on data conditions reflecting no pre-processing *(raw data without enhancements)*, pre-processing *(data containing resolution/enhancements)*, and action similarity *(intense complexity between violent/non-violent actions)*. Data conditions incorporating no pre-processing demonstrated significant improvements rivalling the other models' processing. Action similarity conditions containing increased motion complexity disclosed identical overall accuracy outcomes in Exp. #7 and #1. In this instance, the model's effectiveness improved in Exp. #8-#12 demonstrating its dominance over no pre-processing and several other models. Exp. #13-#18 pre-processing exceeded all conditions concerning effectiveness. The evidence showed 3DCNNsl violent activity recognition's superiority over all other models with exceptions via [2]'s operations concerning J-HMDB easily classified samples. The results satisfied the research objectives by establishing the model's ability to classify extensively complex violent actions effectively by generating higher overall accuracy outcomes.

The notion aided in determining the fine-tuning measures to apply with precision and recall values to improve 3DCNNsl violent activity recognition effectiveness. Precision performance outcomes estimate the correct proportions of all class labels the model predicted, whereas recall quantifies the number of accurately classified predictions relative to all positive classifications generated. Incorporating high performance for this research depended on the criticality of human life. Acceptable performance identified by high precision minimises the number of instances where legitimate violent classes identify as non-violent classes. High recall values suggest the effectiveness of predicting the pre-stages of all violent actions. Analysis of individual class predictions is not the focal point; however, the metrics support by accentuating the model's effectiveness towards violent actions. Precision scores for the no pre-processing data *(raw data without enhancements)* category produced 57–100% with recall between 50–100%, suggesting the model experienced challenges based on fluctuating values. Similar outcomes for action similarity *(intense complexity between violent/non-violent actions)* conditions reflect precision scores between 59–100% and recall ratios at 47–100%.

Finally, data conditions containing pre-processing *(data containing resolution/enhancements)* generated the highest precision values at 61–100% with recall at 55–100% to emphasise the model's effectiveness at this level. Individual action categories per experiment assist in determining the processing impact on 3DCNNsl violent activity recognition regarding the complexity of actions. The idea adds a measure of value to the domain of activity recognition by accentuating processing impact by utilising specific complex actions compared to other research efforts that employ easily classified categories. Exp. #18 overall accuracy *(ACC)* at 88% demonstrate dominance with the exception of [2]'s hybrid 3DCNN HAR model at 90%. Contrasting [2]'s 90% performance, 3DCNNsl violent activity recognition reduction in score occurred because of the difference in action complexity processing. Incorporating easily classified data samples in J-HMDB increased [2]'s hybrid model potential as opposed to 3DCNNsl violent activity recognition processing action complexity conditions.

4.1 Evaluation of Data Containing No Pre-processing

The analysis on Exp. #1 disclosed its overall accuracy on specific samples at 63% compared to [13]'s Bi-Directional LSTM at 68%, [7]'s Res-net50 Regression Block at 71.07%, [2]'s Hybrid 3DCNN HAR at 78% using KTH dataset and 90% on J-HMDB and [3]'s 3DCNN and Optical Flow at 87.25%. Although 3DCNNsl violent activity recognition appeared insignificant, contrasting the other models, the operations incorporated excessive action complexity *(fighting, shooting, knitting, walking-dog)* instead of easily classified samples. The other model's efforts in Table 1 applied easily classified samples projecting clear gait pattern distinctions. Like Exp. #1, Exp. #2 utilising *(beating, fighting, knitting, walking-dog)* and Exp. #3 utilising *(stabbing, fighting, knitting, walkingdog)* increased in performance by 3% at 66%. Exp. #4 incorporating *(beating, stabbing, knitting, walkingdog)* and Exp. #5 with *(shooting, stabbing, knitting, walkingdog)* produce an overall accuracy of 72%, a 9% difference over Exp. #1 and 6% over Exp. #2 and #3. 3DCNNsl violent activity recognition 75% overall accuracy disclosed its potential for complex action processing by rivalling [13]'s Bi-Directional LSTM at 68%, and [7]'s Res-net50 Regression Block at 71.07%.

Table 1. 3DCNNsl No Pre-Processing in Data, Batch size 130, Epoch 32, Depth 12

3DCNNsl Operations				Other Models Performance			
Experiment	Generic Class	Subclass	Our Accuracy (ACC)	[13] Bi-Directional LSTM	[7]'s Res-net50 Regression Block	[2]'s Hybrid 3DCNN HAR	[3]'s 3DCNN/ Optical Flow
Exp #1	2 Violent 2 Non-Violent	Fighting, Shooting, knitting, WalkingDog	63%	68% on UCF crime dataset	71.07% on UCF crime dataset	78% on KTH/ 90% on J-HMDB	87.25% on RWF dataset
Exp #2	2 Violent 2 Non-Violent	Beating, Fighting, knitting, WalkingDog	66%				
Exp #3	2 Violent 2 Non-Violent	Stabbing, Fighting, knitting, WalkingDog	66%				
Exp #4	2 Violent 2 Non-Violent	Beating, Stabbing, knitting, WalkingDog	**72%**				
Exp #5	2 Violent 2 Non-Violent	Shooting, Stabbing, knitting, WalkingDog	**72%**				
Exp #6	2 Violent 2 Non-Violent	Shooting, Beating, knitting, WalkingDog	**75%**				

4.2 Evaluation of Data Containing Action Similarity

Like experiments employing no pre-processed data above, 3DCNNsl violent activity recognition improved its processing by utilising action similarity conditions. Exp. #7 *(stabbing, fighting, cutting-in-kitchen, fencing)* produced a similar score of 63% in Table 2 compared to Exp. #1 *(fighting, shooting, knitting, walkingdog)* in Table 1. The findings implied that action similarity conditions persistently challenged the model's aptitude between sporadic action classes. Exp. #8 at 72% *(shooting, beating, nun-chucks, fencing)* produced a 9% increase over Exp. #7 and Exp. #1 at 63% in Table 1 to validate its processing capability. The model identified the stabbing and shooting class distinctions irrespective of their intense complexity. Exp. #9 *(Beating, Fighting, Sumo-Wrestling, Fencing)*, Exp. #10 *(fighting, shooting sumo-wrestling, nun-chucks)*, and Exp. #11 *(shooting, stabbing, nun-chucks, cutting-in-kitchen)* at 75% demonstrated a 12% increase in performance over Exp. #7 and Exp. #1 in Table 1, and 3% over Exp. #8. 3DCNNsl violent activity recognition stable improvement validated its aptitude in complex conditions by 6% in Exp. #12 at 81% compared to Table 1's highest score in Exp. #6 at 75%. Action similarity high score validated 3DCNNsl violent activity recognition effectiveness over no pre-processing conditions and two other models ([13] and [7]) towards satisfying the objectives.

Table 2. 3DCNNsl Action similarity Data, Batch size 130, Epoch 32, Depth 12

3DCNNsl (mAP) Operations				Other Models Performance			
Experiment	Generic Class	Subclass	Our Accuracy (ACC)	[13]'s Bi-Directional LSTM	[7]'s Res-net50 Regression Block	[2]'s Hybrid 3DCNN HAR	[3]'s 3DCNN/ Optical Flow
Exp #7	2 Violent 2 Non-Violent	Stabbing, Fighting, Cutting-In-Kitchen, Fencing	63%	68% on UCF crime dataset	71.07% on UCF crime dataset	78% on KTH/ 90% on J-HMDB	87.25 % on RWF dataset
Exp #8	2 Violent 2 Non-Violent	Shooting, Beating, Nun-chucks, Fencing	72%				
Exp #9	2 Violent 2 Non-Violent	Beating, Fighting, Sumo-Wrestling, Fencing	75%				
Exp #10	2 Violent 2 Non-Violent	Fighting, Shooting, Sumo-Wrestling, Nun-chucks	75%				
Exp #11	2 Violent 2 Non-Violent	Shooting, Stabbing, Nun-chucks, Cutting-In-Kitchen	75%				
Exp #12	2 Violent 2 Non-Violent	Beating, Stabbing, Fencing, Cutting-In-Kitchen	81%				

4.3 Evaluation of Data Containing Pre-processing

Simulations utilising data pre-processing demonstrated overall classification improvement in Table 3. The evidence proved 3DCNNsl violent activity recognition processing superiority at 88% in Exp. #18 over most other models, with one exception of [2]'s hybrid 3DCNN HAR approach at 90% on a lesser complexity of data. Exp. #13 *(shooting, stabbing, knitting, walkingdog)* and Exp. #14 *(shooting, beating, knitting, walkingdog)* displayed an increase in overall accuracy at 72%, solidifying their 9% superiority lead over Exp. #1 to #3 in Table 1, and Exp. #7 in Table 2 at 63%. Exp. #15 *(beating, fighting, knitting, walkingdog)* dominated all no pre-processing *(raw data without enhancements)* and action similarity *(intense complexity between violent/non-violent actions)* experiments, with Exp. #12 as an exception at 81%. Like Exp. #15, the findings disclosed identical comparisons for Exp. #16 *(fighting, shooting, knitting, walkingdog)* and Exp. #17 *(shooting, stabbing, knitting, walkingdog)*. Exp. #18 *(beating, stabbing, knitting, walkingdog)* results surpassed action similarity conditions' highest score at 81% in Table 2 by 8% and no processing in Table 1 by 13% to validate the stability of the operations on complex actions.

4.4 Discussions on Performance

3DCNNsl violent activity recognition cannot directly prevent the impact of violence from a physical sense. Therefore, an alert signal integrated via configurations can disclose the possibility of an attack to reduce lethal outcomes, thus

Table 3. 3DCNNsl Pre-Processed Data, Batch size 130, Epoch 32

3DCNNsl (mAP) Operations				Other Models Performance			
Experiment	Generic Class	Subclass	Our Accuracy (ACC)	[13]'s Bi-Directional LSTM	[7]'s Res-net50 Regression Block	[2]'s Hybrid 3DCNN HAR	[3]'s 3DCNN/ Optical Flow
Exp #13	2 Violent 2 Non-Violent	Shooting, Stabbing, Knitting, WalkingDog	72%	68% on UCF crime dataset	71.07% on UCF crime dataset	78% on KTH/ 90% on J-HMDB	87.25% on RWF dataset
Exp #14	2 Violent 2 Non-Violent	Shooting, Beating, Knitting, WalkingDog	72%				
Exp #15	2 Violent 2 Non-Violent	Beating, Fighting, Knitting, WalkingDog	78%				
Exp #16	2 Violent 2 Non-Violent	Fighting, Shooting, Knitting, WalkingDog	79%				
Exp #17	2 Violent 2 Non-Violent	Shooting, Stabbing, Knitting, WalkingDog	81%				
Exp #18	2 Violent 2 Non-Violent	Beating, Stabbing, Knitting, WalkingDog	88%				

allowing the victim/s to evade such scenarios. 3DCNNsl violent activity recognition displayed classification limitations discerning interchanging violent conditions because of its complexity. The unstable precision and recall scores occurred because of the sporadic nature/construct of complex violence during inference. We subjected 3DCNNsl violent activity recognition to several categories of complex, violent actions reflecting real-world scenarios to scrutinise its actual ability given the criticality of human life. Those cases involved investigating one violent versus one violent sample, one violent versus many and vice versa. The rationale during development presented insight into the possibility of violent predictions. Applying a balance of actions *(violent/non-violent)* is mandatory to allow for proper class distinctions during training, inference and analysis. Employing balanced samples during training reduces biased results directly linked to an over-fitted model, further leading to insignificant results. One speculation on performance emerged surrounding the dissimilarity in complex data between models as a possible link towards the results rather than the model's processing aptitude. Because we amalgamated violent samples from public datasets and social media to create a class balance of actions, it is unrealistic to fully compare 3DCNNsl violent activity recognition with other models.

We observed the performance on stabbing violence specifically as it intensified the risk of misinterpretations between fighting, beating, and fencing because of its homogeneous nature. 3DCNNsl violent activity recognition operations dispensed high overall accuracy predominantly in the pre-processing stabbing experiments, which satisfied the objectives of developing a reliable violent activity recognition model and declaring its effectiveness. The evidence proved the effectiveness of the pre-processing measures as they acted as catalysts for improving the classification outcome for complex motion. Contrarily, as expected, the confusion matrix disclosed extensive misclassification outcomes predominantly for violence; increasing the sample size during training encourages generalisation as a prime solution. The confusion matrix aided with fine-tuning and highlights the model's overall performance and classification state per class. Acquiring significant amounts of data conveying all possible methods to perform each violence class remains challenging for training operations. The issue emerges specifically for this research because of the sensitivity of the data required for training and the lack of accessible volumes of each violent class sample conveying its relevant motions.

4.5 Discussion on Framework

We encountered a language barrier issue interpreting [16]'s 3DCNN installation guidelines from Mandarin to English. Although several language translators were applied, the language barriers increased the operations complexity of accurately interpreting its 3DCNN error prompts. The issue prolonged the developmental stages and experimental analysis schedule. We mitigate the language barrier issues by deciphering its programming utilities utilising PyTorch online convolution communities. Re-configuring [16]'s sixteen-layer 3DCNN model to twenty-three layers mitigated unbalanced layer issues that adversely impacted its ability to distinguish the complexity between violent and non-violent scenarios effectively. The modifications reduced the hardware's computation load while simultaneously reallocating its central processing unit (CPU) memory resources to optimise the training's time-frame from weeks to hours.

The investigations target other high-performance models to validate classification effectiveness using CPU processing. Those other models applied easily classified actions with distinct characteristics, such as contrasting running versus sitting attributes instead of stabbing versus fencing. Contrarily, 3DCNNsl violent activity recognition evaluations incorporate complex actions that deliberately challenge the model's ability to distinguish features effectively, such as rough playing, fencing and stabbing. We explored additional hyperparameter options within the bounds of the hardware to validate the operations utilising an increment in batch sizes from 130–600, epochs from 32–600, sample sizes from 40–492 and a depth ratio between 8–50 to regulate the size of tensor activations. 3DCNNsl violent activity recognition effectiveness relied heavily on fine-tuning the previously mentioned values to disclose processing optimality during development. Though the evidence accentuated the proposed solution's robustness, 3DCNNsl violent activity recognition operations produced several unstable scenarios indicating latency issues. The previously mentioned issue was predominantly related to the need for more computational power, moving from CPU to GPU or TPU. PyTorch ongoing package compatibility issues adversely impacted MacBook's metal performance shaders *(MPS)* support via coverage tracking, which reverts its onboard GPU to CPU for new silicone device processing solutions. Further fine-tuning necessitates integrating higher computational resources to maintain the robustness of 3DCNNsl violent activity recognition. The idea increases the possibility of achieving better results *(exceeding 88%)* regarding the complexity between violence and action in similar cases.

5 Conclusion

In this paper, we investigated the traditional 3DCNN to conclude its effectiveness towards violent activity recognition. The approach disclosed architecture anomalies hindering the effective layer-to-layer input-output feature generation, simultaneously affecting its ability to pre-empt patterns of complex violence. The proposed 3DCNNsl violent activity recognition enhanced the prediction of complex primitive stages of violence from an overall perspective. The approach validated the proposed model rivalling others in the literature in three complex conditions reflecting no pre-processing, pre-processing, and action similarity. Our findings emphasised the positive effects of integrating data pre-processing techniques to stimulate higher performance. 3DCNNsl violent activity recognition employing no pre-processing experienced its highest score at 75%. We disclose further improvement utilising action similarity at 81%. Finally, we validated the effectiveness of the proposed 3DCNNsl violent activity recognition employing pre-processing conditions at 88%, which exceeded all other models with one exception ([2]'s 90% outcome) where easily classified actions were applied. From the experimental assessment, one can infer several conclusions. Although 3DCNNsl violent activity recognition proved promising in several conditions, we experienced high fluctuations as expected during development. Because of the complexity of the complex, violent action tasks, the model displayed signs of instability, adversely affecting the overall performance, thus limiting its ability to surpass the other models on several occasions. The instability issues relate to the proposed model because the class selection complexity acts as a plot to evaluate actual performance. The classification task intensifies, especially in complex instances relative to the homogeneous gaits and excessive acceleration in violence and non-violent actions. However, given the criticality of human life, it is essential to contemplate the classification instability. A more robust technique considering the application of weapons to fortify the classification processing provides an opportunity for future investigations. Eradicating stability issues caused by the complexity of the data is addressed further with extensive hyper-parameter fine-tuning incorporating the power of GPU or TPU computations as potential improvements.

References

1. Batarseh, F.A., Yang, R.: Data democracy: at the nexus of artificial intelligence. Softw. Dev. Knowl. Eng. (2020)
2. Boualia, S.N., Amara, N.E.B.: 3D CNN for human action recognition. In: 2021 18th International Multi-Conference on Systems, Signals & Devices (SSD), pp. 276–282. IEEE (2021)
3. Cheng, M., Cai, K., Li, M.: RWF-2000: an open large scale video database for violence detection. In: 2020 25th International Conference on Pattern Recognition (ICPR), pp. 4183–4190. IEEE (2021)
4. Demir, F.: Deep autoencoder-based automated brain tumor detection from MRI data. In: Artificial Intelligence-Based Brain-Computer Interface, pp. 317–351. Elsevier (2022)

5. Dominik Endres, Heiko Neumann, Marina Kolesnik, and Martin A Giese. Hooligan detection: the effects of saliency and expert knowledge. 2011
6. Yasin Kaya and Elif Kevser Topuz: Human activity recognition from multiple sensors data using deep CNNs. Multimedia Tools Appl. **83**(4), 10815–10838 (2024)
7. Li, M., et al.: An action recognition network for specific target based on RMC and RPN. J. Phys: Conf. Ser. **1325**, 012073 (2019)
8. Li, Z., et al.: An adaptive hidden Markov model for activity recognition based on a wearable multi-sensor device. J. Med. Syst. **39**, 1–10 (2015)
9. Liang, J.: Confusion matrix: machine learning. POGIL Activity Clearinghouse, vol. 3. no. 4 (2022)
10. Liu, S., Qi, L., Qin, H., Shi, J., Jia, J.: Path aggregation network for instance segmentation. In: Proceedings of the IEEE Conference on Computer Vision and Pattern Recognition, pp. 8759–8768 (2018)
11. Lloyd, K., Rosin, P.L., Marshall, A.D., Moore, S.C.: Violent behaviour detection using local trajectory response. In: 7th International Conference on Imaging for Crime Detection and Prevention (ICDP 2016), pp. 1–6. IET (2016)
12. Loukas, S.: multi-class classification: Extracting performance metrics from the confusion matrix (2020)
13. Manju, D., Seetha, M., Sammulal, P.: Early action prediction using 3DCNN with LSTM and bidirectional LSTM. Turkish J. Comput. Math. Educ. **12**(6), 2275–2281 (2021)
14. Mohan, A., Papageorgiou, C., Poggio, T.: Example-based object detection in images by components. IEEE Trans. Pattern Anal. Mach. Intell. **23**(4), 349–361 (2001)
15. Nguyen, H.-C., Nguyen, T.-H., Scherer, R., Le, V.-H.: Deep learning for human activity recognition on 3D human skeleton: survey and comparative study. Sensors **23**(11), 5121 (2023)
16. Peng, L.: 3DCNN-with-keras (2013)
17. Raj, R., Kos, A.: An improved human activity recognition technique based on convolutional neural network. Sci. Rep. **13**(1), 22581 (2023)
18. Soekarno, I., Hadihardaja, I.K., Cahyono, M., et al.: A study of hold-out and k-fold cross validation for accuracy of groundwater modeling in tidal lowland reclamation using extreme learning machine. In: 2014 2nd International Conference on Technology, Informatics, Management, Engineering & Environment, pp. 228–233. IEEE (2014)
19. Sun, Y., Hare, J.S., Nixon, M.S.: Detecting acceleration for gait and crime scene analysis. In: 7th International Conference on Imaging for Crime Detection and Prevention (ICDP 2016), pp. 1–6. IET (2016)
20. Vrigazova, B.: The proportion for splitting data into training and test set for the bootstrap in classification problems. Bus. Syst. Res. Int. J. Soc. Adv. Innovation Res. Econ. **12**(1), 228–242 (2021)
21. Welch, S.: popular python AI libraries (2020)
22. Jianning, W., Liu, Q.: A novel spatio-temporal network of multi-channel CNN and GCN for human activity recognition based on ban. Neural Process. Lett. **55**(8), 11489–11507 (2023)
23. Yadav, S., Shukla, S.: Analysis of k-fold cross-validation over hold-out validation on colossal datasets for quality classification. In: 2016 IEEE 6th International Conference on Advanced Computing (IACC), pp. 78–83. IEEE (2016)
24. Yin, X., Liu, Z., Liu, D., Ren, X.: A novel CNN-based BI-LSTM parallel model with attention mechanism for human activity recognition with noisy data. Sci. Rep. **12**(1), 7878 (2022)

Participatory AI: A Method for Integrating Inclusive and Ethical Design Considerations into Autonomous System Development

Christina E. Stimson[1] and Rebecca Raper[2](✉)

[1] The University of Sheffield, Sheffield, UK
cestimson1@sheffield.ac.uk
[2] Cranfield University, Cranfield, UK
rebecca.raper@cranfield.ac.uk

Abstract. There has been significant work in the field of AI Ethics pertaining to how it might offer *guidelines* for developers to design, develop and deploy AI in an ethical way. Recently, the European Union's AI Act has introduced a risk-based regulation approach for AI system development. However, despite the additional requirements the AI Act places on developers to ensure that their systems are created with transparency, fairness, and accountability etc., there is no *formalised methodology* for how this might be achieved. Drawing on the history of collaborative and emancipatory technology design in Scandinavia, this paper proposes a software development methodology founded on the ethics and praxis-based principles of Participatory Design. Integrating this approach into the established 'Waterfall Method', it offers developers a practical way of embedding ethics in AI development, and to thereby satisfy the requirements imposed by the new regulations.

Keywords: Participatory Design · Operationalising Ethics · AI Ethics · Agile Development

1 Introduction

Artificial Intelligence (AI) Ethics, and ethics pertaining to autonomous systems more generally, is receiving increased attention, owing to what are perceived as existential risks associated with AI (i.e., threats from an Artificial General Intelligence), and well-documented issues such as AI bias and threats to job security. This paper focuses on the ethics relating to the less existential threats associated with autonomous systems (though these might be relevant) and proposes a formalised methodology for software developers to follow, following the Scandinavian tradition of Participatory Design. Hence, we introduce 'Participatory-AI' a procedure that inherently prioritises the *genuine participation* of stakeholders (primarily, end users).

The paper starts with some background to the topic area: outlining why there is a need for a formalised methodology for embedding ethics in AI software development, tracing discussions surrounding the operationalisation of ethics, followed by describing what Participatory Design is in the context of its history, as a more inclusive way to design technology solutions in the past. A new methodology – *Participatory AI* – is then proposed with an in-depth description, followed by a discussion surrounding its *authentic application,* uses, and limitations. Finally, the paper concludes that 'Participatory AI' is an appropriate methodology for software developers to follow to practically embed ethical considerations in their development processes.

2 Background

2.1 Operationalising Ethics

The European Union (EU) AI Act recently set out 'levels of risk' associated with different use cases for AI. The act enforces bans on 'high level risk AI' and suggests that regulation be put in place to manage lower-level risk AI. Although the approach has generally been positively received, there are some fears that taking a regulation-based approach could stifle innovation [1].

Another issue with the regulation-based approach to ethical AI development is that, although there are rules and guidelines for what AI systems should look like (or what is and is not permitted regarding types of AI System (such as *'Ethically Aligned Design'* [2])), these do not prescribe to developers *how* to create the systems so that they meet the criteria set by the standards. For example, although guidelines such as 'ensure transparency' [2] are set as one metric for ethical AI, there are minimal prescriptions for how to achieve AI transparency. The topic has often been debated by philosophers, but there is little in the way of a formalised methodology for developers to follow to ensure that systems are created *with transparency.*

In Raper and Coeckelbergh (2022) [3], it was argued that the AI landscape has a *methodological gap* because whilst there are guidelines for ethical development, there are no formalised procedures that align with the processes that are typically followed by engineers and developers in designing their autonomous systems. 'Agile' forms of 'The Waterfall Method' (a formal process for requirements elicitation, development and testing) are frequently used by IT departments to ensure the rigorous, fit-for-purpose design of their new systems. However, there is yet no equivalent formalised methodology for designing autonomous systems or ensuring that ethical integrity an intrinsic aspect of the development process.

Recently, the term 'operationalising ethics' [4] has also been used to denote the need to practically apply ethics to the design process. Again, although there has been significant discussion on the need for a practical way to embed ethics in AI development processes (see [5] and [6]), comparatively little work has been done to address this need.

Ethical by Design [7], the incorporation of ethics into the design process, is frequently suggested as needed to satisfy this operational gap. However, there have been few attempts to put this into practice. One approach that attempts to operationalise ethics at the design stage of the AI development process is put forth in [8] with the suggestion that developers consider Spheres of Technology Influence (i.e., who and what the new technology will in turn affect), when designing an AI system. For instance, with one sphere being social impact, designers should consider what social impact their new system will have. The spheres are useful to highlight what is at stake if an AI system were to be unethically developed, but it still does not give instructions on how to build the systems, so that they take these spheres of influence into account. For instance, as a developer, I might envisage my technology being adopted in a certain way, but how can I ensure that the way it is adopted is ethical? This question highlights the necessity of rules and methods for ethical AI development.

Value Sensitive Design [9] is another approach that also tries to embed ethics into the AI design process. It emphasises the need to consider the values of stakeholders when designing new AI or autonomous systems. For instance, supposing a service robot were suggested to be introduced into a care home setting, the values of the elderly residents might be considered to determine what priorities should drive the new system design. However, this approach does not account for when conflicting values (in this case, residents with different worldviews) are driving the new system design. How should the new system be designed when there are opposing values from the individuals who will be affected by the new system? Furthermore, it is not obvious how values can or should be applied to new technology design. For instance, suppose a resident values honesty above and beyond all. It seems that the new system should be designed to maintain this value for the resident, but what does this mean for the new system? Does it mean that the new system cannot be deceptive, or does it mean that the system must be honest in how it interacts with participants? Though individual values are important to consider, individual values alone do not seem sufficient to drive specific technological requirements.

2.2 Participatory Design

Participatory Design (referred to hereafter as PD) is a broad term that refers to any design process that includes the active, sustained – but not necessarily full – involvement of end users (that is, those who would be affected by the designed artefact). In other words, artefacts are designed with users, rather than for them (cf. User-Centered Design). PD is not defined by a specific set of rules or methods, but by a commitment to two core principles:

- Enabling all who would be affected by a product/service to have their voice heard, regardless of their ability to 'speak the language of professional technology design' [10].
- A process of mutual learning for both designers and users can inform all participants' capacities to envisage future technologies and the practices in which they are embedded and serve to enable ordinary people to be able to define what they want from a design process [10].

With its roots in various social, political, and civil rights movements from the 1960s and 1970s, PD has always been both inherently and expressly political in its aims [10].

PD possesses an intrinsic ethical, specifically an emancipatory quality due to a variety of reasons, the most salient of which are the regional and international social contexts. In particular, the Scandinavian tradition champions 'an unshakeable commitment to ensuring that those who will use information technologies play a critical role in their design' [10].

This core value can be traced back to the Frankfurt School's critical theory. A theory that is critical is 'distinguished from a "traditional" theory according to a specific practical purpose: a theory is critical to the extent that it seeks human "emancipation from slavery", acts as a "liberating... influence", and works "to create a world which satisfies the needs and powers of" human beings'" [11]). A fundamental contention of critical theory is that human emancipation in all circumstances of oppression 'cannot be accomplished apart from the interplay between philosophy and social science through interdisciplinary empirical social research' [11].

In light of the above, it is important to understand the emancipatory context in which PD evolved to see how authentically collaborative methodologies such as PD can ensure the development of ethical and inclusive AI that is fit-for-purpose.

In the face of management-driven technological change in Scandinavian workplaces, early PD practitioners made a conscious and hitherto unprecedented decision to uphold the interests of workers – those who would be directly affected by the new computerised systems being imposed – over those of company bosses [10]. Pernicious attempts at task automation and de-skilling employees across industries as a method of worker subjugation, in combination with wider societal changes and a political milieu unique to Scandinavia – namely, an unusually strong tradition of trade unionism – were instrumental to the evolution of PD as it is understood today [10].

As stated, the decision to side with marginalised communities is attributed in part to societal changes occurring at the time. Increases in citizen engagement at local levels in Western European countries, along with internationally seismic events such the Vietnam War, led to a paradigmatic transition in IT design [10]. This transition afforded an understanding that technology is deeply bound by the social and political contexts in which it is used, as opposed to formalised best practices (as identified by Suchman [12]). This concept, termed situated practice [12], is particularly useful as it reasserts the necessity for PD (of AI) to be continually reflexive and, therefore, iterative, to produce outcomes that are relevant to end users' reality: an inherently ethical objective.

Collaborations between researchers and trade unionists, early PD projects such as 'NJMF' [14], 'DEMOS' [15], and 'DUE' [16] very much kick-started and came to define Scandinavian PD in the 1970s. Rather than using PD merely as a means to designing a better product or increasing worker productivity, they treated 'democratic participation and skill enhancement' as valid and desirable ends in and of themselves [17].

Such a commitment to social justice distinguishes PD from (at least, superficially) similar user-oriented methodologies because it ensures that all outputs meet the needs and desires of their target user base, as they are not only deeply involved throughout the process, but decide how much, when, and how they are involved. This nuanced form of agency – that is, genuine, as opposed to "full" participation (see 'Discussion' section) – constitutes the *authentic* use of collaborative methodologies such as PD. Authentic collaboration inevitably increases user acceptance and long-term adoption, both issues that loom large over the field of AI.

PD, therefore, offers an alternative approach: one that enables people to have agency and meaningful involvement in the development of such a transformative and far-reaching innovation as AI. This sustained, egalitarian influence would go at least some way to address and offset the risks and issues posed by it (raised at the beginning of this paper).

3 Methodology

As stated earlier, Information Technology (IT) departments historically followed a process known as *The Waterfall Method* in the creation of new IT systems [18]. The purpose behind using such a methodological approach is that it prescribes to developers how to develop a new business technology solution so that it not only does what is intended from a business perspective, but also satisfies the requirements of the Product Owner.

Figure 1 illustrates a typical example of The Waterfall Method, with the various 'steps' denoted by stages in the waterfall.

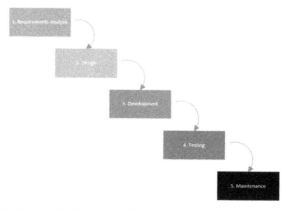

Fig. 1. A typical representation of The Waterfall Method [19].

Per Fig. 1, Step 1 of The Waterfall Method is **requirements analysis**. The aim of this stage is to capture the needs of the new system in as much detail as possible. For example, a new website designed to be accessible might need to have easy-read features. These requirements would be captured in the form of a matrix of logical requirements, with overarching aims at the top and further 'sub-requirements' under each aim. The overall aim of the project might be to create a website, but sub-requirements would specify how the website ought to look, feel, and operate.

After detailed requirements have been elicited, the Technical Architect specifies a **design** for the new system, forming Step 2 in The Waterfall Method. At this stage, there is close collaboration between the Business Analyst (the individual detailing the product requirements) and the Technical Architect to ensure that the proposed design specification meets the elicited requirements.

Step 3 is software **development.** Here, the requirements and technical architecture are developed and translated into a product using a preferred programming language. Once the product has been created, Step 4 takes place: the whole system is **tested** to ensure that the product meets the agreed specification. Using our example of a website that requires high accessibility, is it the case that the website is easy to read? Each listed requirement is mapped against a test criterion and checked against the final product.

Finally, once the system passes the testing phase, it is put into implementation, undergoing constant **maintenance** and periodic checks to ensure it still meets requirements and operates as intended.

Although The Waterfall Method is still very much in use across IT system development, Agile forms are more prevalent. Agile methodologies follow the same five steps listed above, except the steps take place at varying stages in the development process and are revisited if necessary. They take a non-linear, recursive, and iterative approach to software development. To manage these Agile approaches, a Product Owner is appointed. Their responsibility is to ensure that the product adheres to the original design specification throughout the development process. As well as the traditional Project Manager, who ensures that the project is delivered on time, a 'SCRUM Master' is responsible for ensuring team and project cohesion.

Although attempts to involve users in technology design have been laudable and at least superficially fit-for-purpose, they are often extractivist in nature. In the context of late-stage capitalism, referring to the historical context described in the previous section, this means designers/developers look to exploit end users' insights to maximise profits, as opposed to meaningfully involve those who would be affected by their outputs. Examples of extractivist or otherwise not collaborative (that is, not geared towards increasing equity or management of power dynamics) include processes that only involve users at one or two stages. For example, only at an early ideation stage and/or a one-off confirmatory or amendments session. They are also often technocentric, or otherwise technologically-determinist. In such scenarios, the public is encouraged to 'view technological change as an inevitability and focus "on how to adapt to technology, not on how to shape it"' [13].

As stated earlier, Participatory Design is about ensuring that those affected by a new technology (primarily, end users) are involved in its design. As such, it is important to ensure that end users are involved in not necessarily all, but *most* stages of system development, and, crucially, *in a manner and to an extent of their choosing.*

We propose the following ways in which end users could be authentically involved in a new system design, integrated into The Waterfall Method as follows:

1. **Requirements Analysis:**

    - Avoid 'educating' end users about the technology being developed. End users are to be considered domain experts (in other words, they know best what they need and want from a prospective product or service).
    - Instead, try to foster mutual learning between the designers/developers/researchers and the (potential) end users in an equitable manner. This means getting to know each other's perspectives without pushing a particular agenda or focusing on generating a functionality checklist. Establishing a common parlance, for example co-creating UX personas or short stories, would be useful at this stage.
    - The Product Owner is not the arbiter of whether the developed product meets the requirement criteria; *the end users are.* Only those who would actually use the product in real life can truly determine whether the product is fit-for-purpose and of sufficient quality.

- A key feature of this step in the process, if involving PD methods, is to remember that the design should not be driven exclusively by requirements for the business. If designing a new webpage for example, broader considerations should be considered: why is a new webpage being designed in the first place? What utility will it serve for the community and wider economy? Will it serve to benefit everyone in society in an equitable and fair way?

2. **Design:**
   - As PD is, by definition, iterative, reflexive, and recursive, it is important at this stage to ensure that multiple design feedback sessions are incorporated, where end users can participate how and as much, or as little, as they want. One session at the beginning of the development cycle and/or one at the end is not nearly enough, and constitutes at best box-ticking, and at worst extractivism.
   - Much like how the Product Owner is not the sole or ultimate authority in whether the developed product passes muster, the Business Analyst and the Technical Architect must defer to the end users regarding how well the design reflects the requirements. This is not to say that end users should take on their job responsibility or burden; it is a shift how these roles are interpreted; they are facilitators, as opposed to judges or administrators.

3. **Development:**
   - Throughout development, prototypes should be iterated upon according to end user and other stakeholders' feedback in a collaborative, cyclical effort. Again, the key here is agency as opposed to burden or responsibility. If an individual or group of participants shows little or no interest in getting involved at one stage or another, respect their choice but do not forget to keep the lines of communication open at future stages.

4. **Testing:**
   - QA Testers hold valuable knowledge regarding not only how a product "should" be used, but how it can be broken either by accident or through a deliberate series of actions. Alternating between sharing their knowledge with end users and observing how end users interact with the prototype/product is a valuable means of different stakeholders learning from and with each other, which fosters not only meaningful participation, but better outcomes and improved rates of user acceptance and long-term adoption.

5. **Maintenance:**
   - Involving end users in periodic checks – if they so desire – is both useful to ensuring the product continues to meet requirements and to fostering an enduring sense of meaningful involvement in participants and that their input is valued.

As an example case study, consider the aforementioned objective of designing an accessible website. An authentic PD methodology for this would meaningfully involve the individuals affected by the designed artefact. In trying to make the website accessible, the stakeholders to whom it should be made accessible should first be identified through stakeholder mapping, and then actively involved throughout the development process in a manner and extent of their choosing at each stage.

During Step 1: Requirements Analysis, this would include appointing the affected individuals as Product Owners and listening to their stories to understand their needs and desires regarding the product. It is crucial that the requirements list generated is founded on the *self-reported* needs and desires of the potential end users, as opposed to being largely driven by the needs of the business.

Throughout Step 2: Design, these potential end users should be treated as the arbiters of whether the design outputs meet their needs. During Step 3: Development and Step 4: Testing, the focus is solely on satisfying the needs of those who will ultimately use the accessible website. Test frameworks should be designed with this in mind, as well as adhere to the PD principles of enabling a polyphony of stakeholder voices and mutual learning between designers and end users.

Finally, when assuring the new system operates as intended in Step 5: Maintenance, developers must ensure that the system is working, and will continue to work faithfully; that is, in the way that was specified by non-business-oriented stakeholders during the PD process.

4 Discussion

Participatory Design is not a new concept; however, this paper introduces a new formalised procedure for ensuring that the approach is embedded in AI development lifecycles. Fundamental to this new approach is the active participation of implicated individuals during the project lifecycle; through this, developers would become more attuned to designing new systems with ethics at the forefront.

However, as PD is not new, there have been some issues with its implementation, mainly a lack of authenticity. As already stated, this means the genuine, but not necessarily full, involvement of end users. PD is often applied without embodying this quality. The following paragraphs outline some common pitfalls in this area.

Regarding participation, it is sometimes ambiguous as to whether any thought has been given to maintaining an equitable distribution or ratio of participants in any given session (for example, researchers/designers/developers to participants/end users, adults to children, or men to women). For example, in the context of Participatory Design in robotics research, papers often give the number of participants in each session along with any salient demographic information, such as their sex, age, status/role, and state if any expected participants were absent from the session. However, whilst most provide an underlying logic or motivation for recruiting a particular number and/or demographic of participants to the study overall, there does not seem to be any adjustments made for when absences occur.

For example, in Šabanović et al. (2015) [19], the second PD workshop had three participants (two males and one female) and five researchers (sex distribution not stated, although can potentially be inferred from the paper's five named authors) in attendance, with two absentees (sex distribution not stated). This configuration is unequal, both with regard to sex distribution amongst the participants and the ratio of researchers to participants (Stimson, Roy & Szollosy, 2024, in preparation [20]).

Perhaps an in-situ decision to reduce the number of researchers present at the session might have bolstered the participants' feelings of being an equal partner in a conversation or endeavour, as opposed to a research subject being monitored for their responses. The question of whether excluding one male participant from said workshop to estab-lish equal sex representation within it is an open and thorny one. There is the issue of fairness (indeed, which male participant out of the two should be excluded?). There is also whether the choice to exclude an available participant might undermine the depth and variety of the session's outcomes and insights, which would in turn affect its value to the study at large and its ability to make well-evidenced claims.

It is specificity and prior, considered elaboration on methodological choices that needs to be incorporated into development processes for autonomous systems.

It might be assumed from both ideological and research perspectives that maximizing participant input is ideal; indeed, this desire to democratise and decentralise power within and across relationships, practices and contexts is intrinsic to Scandinavian Participatory Design and to co-design in general. However, rather than constantly striving for strong, 'high levels' of participation [21], desirable levels should be informed by the nature of the individual activity at hand and its rational and experiential aims. Participation in PD sessions must be driven by participants' interest and ability in each activity - even at the expense of researcher/project aims.

The transferal of greater agency to participants *when they express a desire for it* is key in development processes, as people often appreciate structure in new or unfamiliar processes, particularly children; they 'recognise the limits of their autonomy and desire adult input and support' (Morrow cited in [21]). It is worth noting that even adults appreciate the input and support of those they consider more informed than they are. What is essential is doing so in a way that avoids patronising or undermining participants and enables them to be involved in a way that is comfortable and useful to them.

For example, when working with children, activities being fun is of paramount importance for engagement. Furthermore, the equalising of power relations between researcher/designer/developer and all participants as much as possible takes precedence over researcher/designer/developer aims because Participatory Design and co-design are inherently political, emancipatory processes.

5 Conclusions and Future Work

This paper has offered the theoretical and practical foundations for a formalised method-ology for incorporating Participatory Design practices into traditional software devel-opment processes so that developers can produce inherently ethical AI. PD incorporates ethics into technology design owing to its history involving marginalised communities in technology design. Considering the need for an operationalisation of ethics in AI development, we propose that, due to its emancipatory roots, PD can be introduced into typical software development lifecycles, such as those utilising The Waterfall Method and Agile variants thereof. If developers are to fulfil the new regulations by ensuring that AI developments are made in an ethical manner, an effective way of doing this is to incorporate PD throughout the software development cycle.

Considering future work in this area, we believe that once a step-by-step procedure has been established to cement PD in software development processes, then appropriate regulation in the form of development standards and development requirements can follow. For instance, there might be a new 'ethical' standard introduced to developed products to demonstrate that affected individuals have been involved in their software design (that the new system is ethical according to PD principles). There might also be broader regulations put into place to ensure that PD *is always applied* in the context of AI development that has significant implications (that is 'high risk' according to the EU AI Act). There is more research that needs carrying out in terms of formalising the above methodologies to ensure that affected individual voices are used to meaningfully shape new technology development. We hope this paper will inspire future work in this area.

Furthermore, additional work includes aligning the involvement of affected individuals with top-down governance procedures to ensure that the end product is effective in the form of ethical guidelines set for the design of new autonomous systems. For example, is the new system transparent to the affected individuals? Is it fair? These governance requirements might not be immediately obvious to those involved in designing the new system but would need checking through external auditing retrospectively.

Acknowledgments. This work is part of Research for the Imagining Technologies for Disability Futures project, and this project was funded by The Wellcome Trust (214963/Z/18/Z, 214963/B/18/Z, and 214963/C/18/Z).

Disclosure of Interests. The authors have no competing interests to declare that are relevant to the content of this article.

References

1. BBC News: Davos 2024: Can – and should – leaders aim to regulate AI directly? (2024). https://www.bbc.com/worklife/article/20240118-davos-2024-can-and-should-leaders-aim-to-regulate-ai-directly. Accessed 29 Apr 2024
2. Shahriari, K., Shahriari, M.: IEEE standard review—ethically aligned design: a vision for prioritizing human wellbeing with artificial intelligence and autonomous systems. In: 2017 IEEE Canada International Humanitarian Technology Conference (IHTC), pp. 197–201. IEEE (2017)
3. Raper, R., Coeckelbergh, M.: EKIP: designing a practical framework for embedding ethics in AI software development. In: ETHICOMP 2022, p. 267 (2022)
4. Solanki, P., Grundy, J., Hussain, W.: Operationalising ethics in artificial intelligence for healthcare: a framework for AI developers. AI Ethics **3**(1), 223–240 (2023)
5. Morley, J., Elhalal, A., Garcia, F., Kinsey, L., Mökander, J., Floridi, L.: Ethics as a service: a pragmatic operationalisation of AI ethics. Minds Mach. **31**(2), 239–256 (2021)
6. Morley, J., Kinsey, L., Elhalal, A., Garcia, F., Ziosi, M., Floridi, L.: Operationalising AI ethics: barriers, enablers and next steps. AI Soc. 1–13 (2023)
7. Mulvenna, M., Boger, J., Bond, R.: Ethical by design: a manifesto. In: Proceedings of the European Conference on Cognitive Ergonomics, pp. 51–54 (2017)
8. Peters, D., Vold, K., Robinson, D., Calvo, R.A.: Responsible AI—two frameworks for ethical design practice. IEEE Trans. Technol. Soc. **1**(1), 34–47 (2020)
9. Friedman, B.: Value-sensitive design. Interactions **3**(6), 16–23 (1996)

10. Simonsen, J., Robertson, T.: Routledge International Handbook of Participatory Design (63070), pp. 37–63 (2012)
11. Bohman, J.: Critical Theory (2004)
12. Suchman, L.A.: Plans and Situated Actions: The Problem of Human-Machine Communication. Cambridge University Press, Cambridge (1987)
13. Šabanović, S.: Robots in society, society in robots. Int. J. Soc. Robot. **2**, 439–450 (2010). https://doi.org/10.1007/s12369-010-0066-7
14. Nygaard, K.: The 'iron and metal project': trade union participation, in Sandberg, pp. 94–107 (1979)
15. Sandberg, Å. (ed.): Computers Dividing Man and Work Swedish Center for Working Life, Demos Project Report no 13. Utbildningsproduktion, Malmö (1979)
16. DUE Project Group: Project DUE: Democracy, Development, and EDP, in Sandberg, pp. 122–130 (1979)
17. Ehn, P.: Scandinavian design: on participation and skill. In: Participatory Design: Principles and Practices, pp. 41–77 (2017)
18. Senarath, U.S.: Waterfall methodology, prototyping and agile development. Technical report, pp.1–16 (2021)
19. Šabanović, S., Chang, W.L., Bennett, C.C., Piatt, J.A., Hakken, D.: A robot of my own: participatory design of socially assistive robots for independently living older adults diagnosed with depression. In: Lecture Notes in Computer Science (Including Subseries Lecture Notes in Artificial Intelligence and Lecture Notes in Bioinformatics), vol. 9193, pp. 104–114 (2015)
20. Stimson, C.E., Roy, K., Szollosy, M.: The Use of Participatory Design in Robotics: A Scoping Review [Manuscript in preparation] (2024)
21. Allsop, M.J., Holt, R.J., Levesley, M.C., Bhakta, B.: The engagement of children with disabilities in health-related technology design processes: identifying methodology. Disabil. Rehabil. Assistive Technol. **5**(1), 1–13 (2010)

Sampling-Based Motion Planning for Guide Robots Considering User Pose Uncertainty

Juan Sebastian Mosquera-Maturana[1]([✉])[iD], Juan David Hernández Vega[2][iD], and Victor Romero Cano[2][iD]

[1] Faculty of Engineering, Universidad Autónoma de Occidente, Cali, Colombia
smosquera@uao.edu.co
[2] School of Computer Science and Informatics, Cardiff University, Cardiff, UK
{hernandezvegaj,romerocanov}@cardiff.ac.uk

Abstract. In this paper, we propose a framework to address the problem of guiding a person within a semi-structured environment in a socially acceptable manner that prioritises safety and comfort. We propose an algorithm based on the optimal Rapidly exploring Random Tree (RRT*) algorithm for path planning. Our proposal utilises Dubins curves and takes into account the user during path planning to generate a navigation path that allows the robot to follow a feasible path that can also be navigated by the user. A comparative analysis against standard path planning based on the RRT* algorithm and the Social Force Model validates the efficacy of our proposed algorithm.

Keywords: Social Robot Navigation · Path planning · Social Force Model

1 Introduction

The use of robots for complementing and enhancing human capabilities is becoming increasingly common. They have been developed and deployed to assist elderly individuals in indoor environments [15], as well as to guide people in settings such as museums or exhibitions [17]. However, for robots to effectively operate in human-populated spaces, they must possess a fundamental characteristic: the ability to navigate the environment in a way that maximises human comfort. A robot's ability to move around people in a socially acceptable manner is crucial in providing comfort and safety to those around it [10]. In [3], the authors define a socially navigating robot as one that respects principles of safety, comfort, legibility, courtesy, social competence, agent understanding, proactivity, and responsiveness to context.

The works in [19,27] propose methods that tackle the Social Robot Navigation (SRN) problem, however, these methods do not properly distinguish between social agents and other dynamic objects in the environment, and thus omit relevant social aspects for socially acceptable navigation such as comfort zones around people and social norms related to proxemics [4]. Proxemics is the study of how people use and perceive personal space, and it plays a central role in human interactions. Ignoring cultural and individual differences in proxemics could lead to situations where the robot invades personal space, making people feel uncomfortable or intruded upon. By integrating proxemic awareness into their navigation systems, social robots can navigate with greater sensitivity, respecting the varying preferences for interpersonal distances and ensuring a better integration into human environments.

The differentiation between objects and social agents enables robots to take into account social aspects when navigating in environments shared with humans, however, the robot's behaviour and the social aspects to be considered vary depending on the task to be performed. A robot that only needs to navigate through a crowd behaves differently than one that has to make a delivery or one that has to guide people through the environment.

In environments such as airports, shopping malls, universities, or hospitals, robot guides enhance efficiency by enabling individuals to reach their destination, reducing congestion, and improving productivity. By utilising robots to carry out the task of guiding people in these environments, staff, such as nurses in hospitals, can allocate their time to other activities [13]. Our work focuses on guiding an individual ensuring that the generated path is feasible for the robot and comfortable for the user, while maintaining an appropriate proximity between both.

The contributions of this paper are as follows: First, we propose a social robot navigation framework that considers the collision probability of the guided person during path planning. The probability of a collision between the agent and the environment is determined by the uncertainty in the estimation of the agent's position relative to the robot during the planning of a path. Second, we consider the proximity of the user to the robot during path planning by means of a social region around the robot. This enables the robot to remain in close proximity to the user during navigation, thereby preventing the robot from moving forward in space and leaving the user behind when being guided. Finally, we implement a planning algorithm using Dubins curves that generates smooth paths that contribute to socially acceptable navigation.

The rest of this paper is organised as follows: Section 2 presents related work. Section 3 provides a definition of the problem and the challenges it presents. Also, this section describes the proposed social robot navigation framework. In Sect. 4, we describe the scenario in which the algorithms were implemented and compared. It also presents the results obtained in the different simulations. Finally, Sect. 5 discusses the results of each algorithm, a comparison among them, and presents the conclusions of the work done and the challenges to be faced in future developments.

2 Related Work

Several authors have proposed algorithms that distinguish between objects and social agents, such as people, as seen in [2,18,20,25]. The work in [20] considers the difference between social agents and dynamic objects in a dynamic scenario, enabling robot navigation in uncontrolled human-populated environments. However, these algorithms typically operate on static environment representations, where all elements and agents in the environment, except for the robot, are assumed to be instantaneously static.

In general, a robot can guide people in two ways: it can either remain stationary and provide indications, or it can navigate the environment along with the user while heading to the target position. The first approach involves the robot remaining immobile in a specific place where it gives indications to the person to be guided such as the works in [23,26]. This approach implies proper communication between the robot and the user, as well as a perfect understanding of the robot's indications by the user, in order to be able to follow them to reach the destination. Similarly, as the robot remains stationary, it cannot detect changes in the environment that may require a different path to be taken. In the second approach, the robot moves along with the user while guiding him to the destination [6]. In this case, it is necessary that the user follows the robot during the navigation, otherwise, the robot will fail in its task of guiding the user by reaching the destination without him.

Typically, guide robots have not incorporated considerations for the individuals they are guiding during path planning; instead, they react to them once navigation commences. For instance, in [17], researchers addressed the challenge of guiding visitors in a museum by deploying a guide robot. The robot follows a predefined path with stops at various stations, providing information at each stop until the end of the tour. Users can customise their tour by selecting specific stations of interest. In another study by [1], researchers utilised a robot to guide shoppers in a mall. Notably, their work included considering people's poses for obstacle avoidance. Similarly, in [7], the authors implemented the service robot SPENCER [24] to guide travellers in an airport, taking into account the social interactions of the individuals.

The Social Force Model (SFM), firstly proposed in [5], is widely used for solving local motion generation problems in social robot navigation [21]. SFM is a reactive algorithm that inherits the limitations of potential field-based algorithms, such as the local minima problem. The local minima problem in social robot navigation refers to a situation where a robot becomes trapped in a region of its environment where the cost of moving in any direction appears higher than the current position. In other words, the robot perceives its immediate surroundings in such a way that it seems more beneficial to stay in its current location rather than attempting to navigate to a different location.

To avoid the local minima problem of reactive motion planning methods, one option is to plan paths that take into account the environment in which the robot is located [20]. Planning methods that avoid collisions while considering the collision probability based on the robot's dynamics have improved navigation in dynamic or unknown environments [16]. However, navigating an environment without considering the user during path planning can lead to situations where the robot reaches the target position while the user may stop or lose sight of the robot. In this paper, we propose a planning framework that takes into account the collision probability of the user and their pose, allowing the robot to navigate the environment while guiding a person. Our approach aims to estimate the path with the least probability of collision for both the robot and the user. Our proposal is based on the asymptotically optimal Rapidly-exploring Random Tree (RRT*) [8,9] path planning algorithm. The RRT* algorithm has been modified to take into account the user to be guided. Based on the user's pose relative to the robot, we consider the probability of collision of the user with obstacles and his distance to the robot.

3 Social Navigation Framework for a Guide Robot

The configuration space $\mathcal{C}$ consists of the space occupied by obstacles $\mathcal{C}_{obs}$ and the free obstacle space $\mathcal{C}_{free}$ [12]. Path planning in robotics tries to find the path that connects an initial configuration q_{start} with a final configuration q_{goal} belonging to $\mathcal{C}_{free}$.

$$P^*(q_{start}, q_{goal}) = \underset{P \in \mathcal{C}_{free}}{\arg\min} \; F(P(q_{start}, q_{goal})) \tag{1}$$

The task of guiding a person in a semi-structured environment requires path planning to consider, among other social constraints, the relative position between the user (guided person) and the robot. Considering the social cues during path planning is possible by representing them with a function $G(u_{sc})$, where u_{sc} represents these social cues and the social path planning problem is then:

$$P^*(q_{start}, q_{goal}) = \underset{P \in \mathcal{C}_{free}}{\arg\min} \; F(P(q_{start}, q_{goal})) + G(u_{sc}) \tag{2}$$

This section presents a framework for addressing the problem of social robot navigation, specifically designed for guiding people. The framework, depicted in Fig. 1, is comprised of three components: world representation, user-aware path planning, and motion control.

3.1 World Representation

The obstacles in the environment are represented as bounding boxes using the pedsim_ros library and information from a lidar sensor. The information about the objects allows the configuration space to be established. The mean pose and dimensions of each box are employed to calculate its covariance matrix.

To consider the user to be guided as a social agent during path planning, we model their pose as a Gaussian random variable and estimate it relative to our guide robot for each of the possible configurations that the robot can take in the free configuration space. This estimation is done assuming that the social

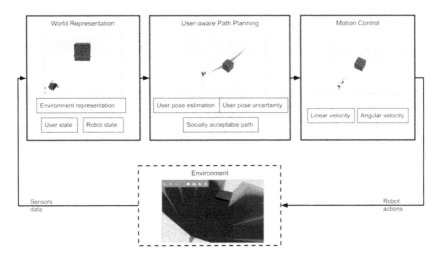

Fig. 1. Proposed framework for addressing the social robot navigation problem. The framework obtains data from sensors, performs a world representation, plans a path considering the user, and generates control actions to follow the planned path.

agent follows the guide robot according to the model proposed by [11] for a follow-the-leader configuration. According to this model, the guided user's pose $x(t)$ evolves according to:

$$x(t+1) = x(t) + v(t)cos(\psi(t)) \qquad (3)$$

$$y(t+1) = y(t) + v(t)sin(\psi(t)) \qquad (4)$$

where $\psi(t)$ is the anticlockwise heading relative to the robot frame's x - axis, and $x(t)$, $y(t)$ are the user positions in Cartesian coordinates.

Predicting the pose where the agent could be at each configuration that the robot can take during navigation requires a sequential increase in uncertainty. The user pose uncertainty increases as the tree is extended to connect the initial and final configurations, but it reduces every time the robot measures the user pose.

3.2 User-Aware Path Planning

We propose an algorithm based on the optimal Rapidly Exploring Random Tree (RRT*) for path planning. The cost function of our extended RRT* algorithm includes a factor that considers the current and predicted user's motion. The algorithm has an online execution, so it is designed to obtain new paths from the current pose of the robot to reach the target configuration as shown in Algorithm [1]. Path planning is conducted in a manner analogous to the RRT* algorithm. Distinct configurations are generated within the free configuration space, and a tree is extended by connecting the different configurations. During the expansion of the tree, in addition to collision checking, our algorithm seeks to minimise the probability of collision of the agent with the obstacles by incorporating this probability into the cost function of the algorithm.

Algorithm 1. socialPathPlanning

Input: q_{start}: Start configuration
  q_{goal}: Goal configuration
Output: P: asymptotically optimal path
 1: **begin:**
 2:   planner ← RRT*(DubinsStateSpace)
 3:   $last\_best\_known\_solution$ ← {}
 4:   q_{new_start} ← q_{start}
 5:   **while not** $end\_condition$ **do**
 6:     world_model ←reqUpdatedWorldRepresentation()
 7:     planner.updateWorldRepresentation(world_representation)
 8:     planner.optimalObjective(cost_function)
 9:     planner.startFrom($last\_best\_known\_solution$)
10:     planner.solve(q_{new_start} , q_{goal})
11:     **if** $solution not found$ **then**
12:       P ← planner.getPartialSolution()
13:     **else**
14:       $last\_best\_known\_solution$ ← planner.getSolution()
15:       P ← $last\_best\_known\_solution$
16:     **end if**
17:   **end while**
18:   **return** P

The proposed path planning algorithm accepts as inputs the start and goal configurations. The algorithm initiates by establishing the Dubins-type state space, initialising a variable to store the most recent optimal solution, and defining the start configuration to be utilised by the algorithm. A cycle to search the asymptotically optimal path begins. At each cycle iteration, the representation of the semi-structured environment is updated, the cost function is established, and the optimal solution previously identified is designated as the base path for planning. A solution is sought that accounts for both the goal configuration and the start configuration, which corresponds to the robot's current position. Upon the identification of a solution, it is registered as the most recent optimal solution and as the asymptotically optimal path. If no optimal solution can be found, a partial or approximate solution will be returned, which does not guarantee that the goal configuration can be reached.

Processing data from exteroceptive sensors, it is possible to determine the position and speed of agents present in the environment shared with the robot. In this work, we obtain the agent pose and speed directly from the pedsim_ros library. The dynamic state of the user being guided is represented with respect to the robot's local frame.

Path planning attempts to connect different waypoints between the start and goal configurations. The waypoints are generated in the configuration space by checking that they are in collision free space. During path generation, each waypoint represents a possible configuration to be taken by the robot.

Using the estimated position and the uncertainty generated by the estimation procedure, the probability of collision of the agent with obstacles in the environment is estimated. Building a planning cost function that includes the guided agent's collision probability effectively allows our planning framework to generate paths that avoid the user and the robot collision with the environment. In this framework, we implement our social factor $G(u_{sc})$, introduced in the problem formulation above, as the following cost function based on the Mahalanobis distance:

$$G(u_{sc}) = \sqrt{(x_{user} - x_{obs})^T \Sigma^{-1} (x_{user} - x_{obs})} \tag{5}$$

In this context, the variable x_{user} represents the user pose, while x_{obs} denotes the obstacle pose. The symbol Σ denotes the covariance matrix of both poses. The user-estimated pose covariance matrix is a diagonal matrix with each value representing the uncertainty in the pose estimation.

3.3 Motion Control Approach

Our motion control module attempts to follow the path planned based on a differential robot kinematics model, obtaining maximum linear and angular speeds from a path parametrisation based on Dubins curves. Path generation using Dubins curves was performed using the state space class of the Open Motion Planning Library [22]. The library allows the generation of paths in SE(2) with a geometric planner that interpolates the different configurations, thus enabling the movements "go straight," "turn left," and "turn right".

The robot's position and orientation are updated continuously via odometry data and the planned path, which is provided as a sequence of waypoints. The robot navigates by moving iteratively towards the next waypoint in the path. The linear and angular velocities are set according to the distance from the current position to the waypoint and the bearing and heading angles.

The motion control algorithm ensures that the robot is oriented towards the waypoint by calculating the bearing angle. The current heading angle is obtained from odometry. The robot adjusts its velocities in consideration of the predefined thresholds in order to minimise oscillations and guarantee socially acceptable and safe movements.

4 Simulations and Results

The methodology employed in this study involves working within a simulation environment that enables the generation of scenarios incorporating obstacles and the addition of individuals whose movement is governed by the social forces model. Furthermore, the environment allows for the simulation of mobile robots and the examination of the interactions between all elements within the environment.

For this work, a modified version of the pedsim_ros packages was implemented. The modification enables the simulated robot to exert an attractive force on an agent, emulating the function of being followed by the agent. The simulation of the environment was done using these packages and Gazebo ignition. Pedsim_ros allows to emulate the behaviour of the agents considering the obstacles. Gazebo ignition as a simulation environment allows to emulate sensors, actuators, as well as the environment and different physical variables. The simulation was conducted using a setup composed of an Intel Core i5-7200 CPU with a Intel HD Graphics 620 GPU and 8 GB RAM.

The proposed framework, has been evaluated against both the unmodified RRT* algorithm and the widely employed Social Force Model. The comparison was performed in three different scenarios involving the robot and an agent to be guided. In the first scenario, there is only one stated obstacle. The robot and the social agent are placed in front of the obstacle at a certain distance as seen in Fig. 2. In this scenario the robot has to navigate from its starting position, which was randomly chosen within the blue zone in Fig. 2 to a position behind the obstacle, also randomly chosen in the green zone depicted in Fig. 2.

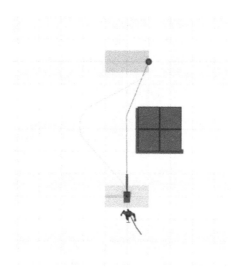

Fig. 2. Test scenario 1. A simulated environment with only one static object. The blue and green squares represent the zone where a random start and destination configuration is chosen to plan a path. The red line is the path planned by the RRT* algorithm, and the green line is the path planned by our proposal. (Color figure online)

In the second scenario, five static objects are placed in the environment (see Fig. 3). The robot must navigate through the obstacles from the start position to a target position. This scenario emulates environments where multiple obstacles, such as tables in a cafeteria, are encountered.

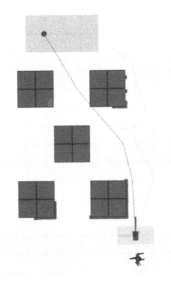

Fig. 3. Test scenario 2. A simulated environment with several obstacles. This type of environment represents social scenarios such as a cafeteria. As in scenario 1, blue and green squares represent zones where a random start and destination configuration is chosen to plan a path. The red line is the path planned by the RRT* algorithm, and the green line is the path planned by our proposal. (Color figure online)

To validate the performance of our proposal against the other methods, we analyse success rate and path smoothness over many trials. A trial was considered successful if the robot navigated from the start position to the goal position while maintaining a proximity to the user and avoiding obstacles. The path smoothness metric was obtained by analysing the total bending energy of the path traversed by the robot during each simulation. A smoothness value close to zero indicates a smoother path. Achieving socially acceptable navigation depends heavily on the smoothness of both planned and executed paths [14]. The results of the simulations on scenarios 1 and 2 can be found in Tables 1 and 2 respectively.

Table 1. Metrics from the scenario 1.

	Success rate	Smoothness
SFM	0.85	0.12
RRT*	0.89	0.049
our proposal	0.96	0.037

Table 2. Metrics from the scenario 2.

	Success rate	Smoothness
SFM	0.74	0.23
RRT*	0.76	0.091
our proposal	0.85	0.067

In the third scenario, there are two static objects placed one in front of the other as seen in Fig. 4. In this scenario the robot has to navigate the environment to a goal position behind the obstacles. The two obstacles formed a corridor and the distance between them is reduced until a narrow corridor where the robot and the agent cannot pass side by side is created. Metrics obtained from simulations in this scenario can be seen in Table 3.

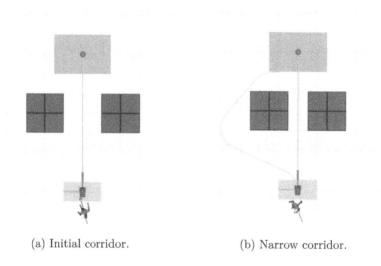

(a) Initial corridor. (b) Narrow corridor.

Fig. 4. Test scenario 3. A simulated corridor environment. This scenario presents two obstacles that create a corridor for the robot and the user to navigate through. The distance between the obstacles is decreased over different trials, resulting in a narrow corridor that the robot and the user cannot pass through side by side. As in the others scenarios, blue and green squares represent zones where a random start and destination configuration is chosen to plan a path. The red line is the path planned by the RRT* algorithm, and the green line is the path planned by our proposal. (Color figure online)

The success rate for the SFM and RRT* methods are slower than our proposal because a success trial imply to reach the destination point maintaining close to the user. In cases where the robot and the user can't travel together trough the narrow corridor, SFM and RRT* can provide a motion planning to pass in the middle of the obstacles leaving the user behind. Meanwhile, our proposal

Table 3. Metrics from the scenario 3.

	Success rate	Smoothness
SFM	0.4	0.15
RRT*	0.5	0.071
our proposal	0.8	0.041

takes into account the collision probability and plan a path to avoid the corridor as can be seen in Fig. 5. This results in a robot behaviour that prioritises user safety and comfort.

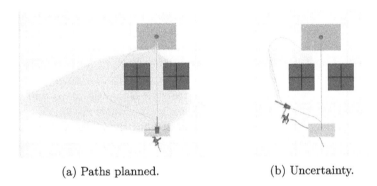

(a) Paths planned.   (b) Uncertainty.

Fig. 5. Paths planned for scenario 3 in a trial. A path planning result for a trial to navigate from the robot position to the target position (red circle). The red line is the path planned by the RRT* algorithm, and the green line is the path planned by our proposal. The yellow ellipses represent the uncertainty in the user pose estimation for each waypoint in the planned path. Note that user pose uncertainty reduces at every time step after the robot updates its believe about the user' s pose (Color figure online)

5 Conclusions and Future Work

This paper presents a path planning framework for a guide robot that considers estimates of the relative user's pose and its corresponding uncertainty. Our framework enables mobile robots to guide users through an environment in a free-collision and comfortable manner. We present modifications to the RRT* sampling-based algorithm that aim to contribute to the generation of socially acceptable paths for the task of guiding people, taking into account the possibility of collisions of the user with the environment.

The use of Dubins curves for path planning allows the generation of smoother paths compared to the SFM and RRT* algorithms. This is supported by the observation that the average smoothness in the three scenarios is lower when using our proposal. Following a smoother path while navigating may enhance the user's perception of the path as a natural path.

The proposed algorithm, compared to the classical SFM and RRT*, allows planning and executing paths that take into account the user to be guided. The consideration of the user during the planning and execution of the path leads to a successful navigation in which the robot prioritises the task of guiding over that of navigating the space, i.e. the robot tries to complete the navigation with the user and not individually as in classical robot navigation.

As future work, we intend to validate the proposed methodology in scenarios of higher structural complexity, as well as those scenarios with other social agents involved in both low and high densities. In order to account for the presence of other social agents, their poses and velocities relative to the robot, as well as their state of motion, i.e., whether they are moving or stationary, must be considered. Moreover, individuals and groups of varying sizes and compositions, with their respective personal or social areas, should also be considered.

The environment abstraction in this work enables our algorithm to navigate in a safe and socially acceptable manner. However, discriminating between different elements in the environment using metric-semantic maps will allow for better performance in social robot navigation. In situations where the terrain presents concrete roads with grass or sand around them, the socially accepted path is usually the road, even if crossing the grass offers a shorter path.

To validate the results obtained in the simulations and to obtain user feedback, real-world experiments should be conducted using case studies.

References

1. Chen, Y., Wu, F., Shuai, W., Chen, X.: Robots serve humans in public places—kejia robot as a shopping assistant. Int. J. Adv. Robot. Syst. **14** (2017). https://doi.org/10.1177/1729881417703569, https://journals.sagepub.com/doi/full/10.1177/1729881417703569
2. Daza, M., Barrios-Aranibar, D., Diaz-Amado, J., Cardinale, Y., Vilasboas, J.: An approach of social navigation based on proxemics for crowded environments of humans and robots. Micromachines **12**(2) (2021). https://doi.org/10.3390/mi12020193, https://www.mdpi.com/2072-666X/12/2/193
3. Francis, A., et al.: Principles and guidelines for evaluating social robot navigation algorithms. ArXiv **abs/2306.16740** (2023). https://api.semanticscholar.org/CorpusID:259287246
4. Hall, E.T.: The Hidden Dimension: An Anthropologist Examines Man's Use of Space in Public and in Private. Anchor Books (1969). https://books.google.com.co/books?id=Yf27OgAACAAJ
5. Helbing, D., Molnár, P.: Social force model for pedestrian dynamics. Phys. Rev. E **51**, 4282–4286 (1995). https://doi.org/10.1103/PhysRevE.51.4282
6. Hong, B., Lin, Z., Chen, X., Hou, J., Lv, S., Gao, Z.: Development and application of key technologies for guide dog robot: a systematic literature review. Robot. Auton. Syst. **154**, 104104 (2022). https://doi.org/10.1016/J.ROBOT.2022.104104

7. Joosse, M., Evers, V.: A guide robot at the airport: First impressions. In: IEEE International Conference on Human-Robot Interaction,pp. 149–150 (2017). https://doi.org/10.1145/3029798.3038389
8. Karaman, S., Frazzoli, E.: Incremental sampling-based algorithms for optimal motion planning. Robot. Sci. Syst. **6**, 267–274 (2010). https://doi.org/10.48550/arxiv.1005.0416, https://arxiv.org/abs/1005.0416v1
9. Karaman, S., Frazzoli, E.: Sampling-based algorithms for optimal motion planning. Int. J. Robot. Res. **30**(7), 846–894 (2011). https://doi.org/10.1177/0278364911406761
10. Kocsis, M., ZÃllner, R., Mogan, G.: Interactive system for package delivery in pedestrian areas using a self-developed fleet of autonomous vehicles. Electronics 2022, **11**, 748 (2022). https://doi.org/10.3390/ELECTRONICS11050748, https://www.mdpi.com/2079-9292/11/5/748/htmhttps://www.mdpi.com/2079-9292/11/5/748
11. Li, Q., Godsill, S.: A new leader-follower model for Bayesian tracking. In: eedings of 2020 23rd International Conference on Information Fusion, FUSION 2020 (2020). https://doi.org/10.23919/FUSION45008.2020.9190329
12. Lozano-Pérez, T.: Spatial planning: a configuration space approach. IEEE Trans. Comput. **C-32**, 108–120 (1983). https://doi.org/10.1109/TC.1983.1676196, http://ieeexplore.ieee.org/stamp/stamp.jsp?arnumber=1676196
13. Ma, A.C., Meng, Z., Ding, X.: Performance review of intelligent guidance robot at the outpatient clinic setting. Cureus **13**, e16840 (2021). https://doi.org/10.7759/cureus.16840
14. Mavrogiannis, C., Hutchinson, A.M., Macdonald, J., Alves-Oliveira, P., Knepper, R.A.: Effects of distinct robot navigation strategies on human behavior in a crowded environment. In: 2019 14th ACM/IEEE International Conference on Human-Robot Interaction (HRI), pp. 421–430 (2019). https://doi.org/10.1109/HRI.2019.8673115
15. Napoli, C.D., Ercolano, G., Rossi, S.: Personalized home-care support for the elderly: a field experience with a social robot at home. User Model. User-Adap. Interact. **33**, 405–440 (2023). https://doi.org/10.1007/S11257-022-09333-Y/TABLES/12, https://link.springer.com/article/10.1007/s11257-022-09333-y
16. Pairet, É., Hernández, J.D., Carreras, M., Petillot, Y., Lahijanian, M.: Online mapping and motion planning under uncertainty for safe navigation in unknown environments. IEEE Trans. Autom. Sci. Eng. **19**, 3356–3378 (2022). https://doi.org/10.1109/TASE.2021.3118737
17. Pang, W.C., Wong, C.Y., Seet, G.: Exploring the use of robots for museum settings and for learning heritage languages and cultures at the Chinese heritage centre. Presence: Teleoper. Virtual Environ. **26**, 420–435 (2017). https://doi.org/10.1162/PRES_A_00306, https://direct.mit.edu/pvar/article/26/4/420/92672/Exploring-the-Use-of-Robots-for-Museum-Settings
18. Rivero-Ortega, J.D., et al.: Ring attractor bio-inspired neural network for social robot navigation. Front. Neurorobot. **17**, 1211570 (2023)
19. Sathyamoorthy, A.J., Patel, U., Guan, T., Manocha, D.: Frozone: freezing-free, pedestrian-friendly navigation in human crowds. IEEE Robot. Autom. Lett. **5**(3), 4352–4359 (2020). https://doi.org/10.1109/LRA.2020.2996593
20. Silva, S., Paillacho, D., Verdezoto, N., Hernandez, J.D.: Towards online socially acceptable robot navigation. In: IEEE International Conference on Automation Science and Engineering, vol. 2022-August, pp. 707–714 (2022). https://doi.org/10.1109/CASE49997.2022.9926686

21. Singamaneni, P.T., et al.: A survey on socially aware robot navigation: taxonomy and future challenges. Int. J. Robot. Res. (2024). https://doi.org/10.1177/02783649241230562
22. Şucan, I.A., Moll, M., Kavraki, L.E.: The open motion planning library. IEEE Robot. Autom. Mag. **19**(4), 72–82 (2012). https://doi.org/10.1109/MRA.2012.2205651, https://ompl.kavrakilab.org
23. Tamai, A., Ono, S., Yoshida, T., Ikeda, T., Iwaki, S.: Guiding a person through combined robotic and projection movements. Int. J. Soc. Robot. **14**, 515–528 (2022). https://doi.org/10.1007/S12369-021-00798-2/FIGURES/9
24. Triebel, R., et al.: SPENCER: a socially aware service robot for passenger guidance and help in busy airports. In: Wettergreen, D.S., Barfoot, T.D. (eds.) Field and Service Robotics. STAR, vol. 113, pp. 607–622. Springer, Cham (2016). https://doi.org/10.1007/978-3-319-27702-8_40
25. Walker, R., Dodd, T.J.: A novel path planning approach for robotic navigation using consideration within crowds. In: Dixon, C., Tuyls, K. (eds.) TAROS 2015. LNCS (LNAI), vol. 9287, pp. 270–282. Springer, Cham (2015). https://doi.org/10.1007/978-3-319-22416-9_31
26. Wang, S., Christensen, H.I.: Tritonbot: first lessons learned from deployment of a long-term autonomy tour guide robot. In: RO-MAN 2018 - 27th IEEE International Symposium on Robot and Human Interactive Communication, pp. 158–165 (2018). https://doi.org/10.1109/ROMAN.2018.8525845
27. Yang, C.T., Zhang, T., Chen, L.P., Fu, L.C.: Socially-aware navigation of omnidirectional mobile robot with extended social force model in multi-human environment. In: Conference Proceedings - IEEE International Conference on Systems, Man and Cybernetics, vol. 2019-October, pp. 1963–1968 (2019). https://doi.org/10.1109/SMC.2019.8913844

Safety Assurance Challenges for Autonomous Drones in Underground Mining Environments

Philippa Ryan[1], Arjun Badyal[1]([✉]), Samuel Sze[2], Benjamin Hardin[2], Hasan Bin Firoz[1], Paulina Lewinska[1], and Victoria Hodge[1]

[1] Department of Computer Science, University of York, York, UK
{philippa.ryan,arjun.badyal,hasan.binfiroz,paulina.lewinska, victoria.hodge}@york.ac.uk
[2] Oxford Robotics Institute, University of Oxford, Oxford, UK
{samuels,bhardin}@robots.ox.ac.uk

Abstract. Autonomous drones have been proposed for many industrial inspection roles including building infrastructure, nuclear plants and mining. They have the benefit of accessing hazardous locations, without exposing human operators and other personnel to physical risk. Underground mines are extremely challenging for autonomous drones as there is limited infrastructure for Simultaneous Localisation and Mapping (SLAM), for the drone to navigate. For example, there is no Global Navigation Satellite System (GNSS), poor lighting, and few distinguishing landmarks. Additionally, the physical environment is extremely harsh, affecting the reliability of the drone. This paper describes the impact of these challenges in designing for, and assuring, safety. We illustrate with experience from developing an autonomous Return To Home (RTH) function for an inspection drone. This is initiated when the drone suffers a communications loss whilst surveying newly excavated corridors that are unsafe for personnel. We present some of the key safety assurance challenges we faced, including design constraints and difficulties using simulations for validation and verification.

Keywords: Autonomous drones · Safety assurance · Underground Mining

1 Introduction

Recent advances in autonomous drone capabilities, their cost-effectiveness and manoeuvrability have allowed drones to be used for complex and challenging missions. These include search and rescue, inspecting buildings, bridges and tunnels,

---

Supported by the Assuring Autonomy International Programme (a partnership between Lloyd's Register Foundation and the University of York) and UKRI (under Grants No EP/R025479/1 and EP/V026801/1).

mapping a nuclear emergency, and inspecting challenging environments such as underground mines [22]. Drones have the benefit of being able to access hazardous locations whilst avoiding difficult terrain, and without exposing human operators or inspectors to physical risk. Underground environments such as mines can be particularly challenging for drones. For example, there is no GNSS, poor lighting, and a lack of distinguishing landmarks for SLAM [18]. Also, the physical environment is hot and dusty affecting the reliability of drone navigation systems. There is the risk of explosion in a mine due to the presence of methane and combustible coal dust. Drones have LiPo batteries which can suffer thermal runaway and ignite methane or coal dust, hence allowing the drone to crash even in an unpopulated area of the mine could pose a significant risk.

This paper describes some of these safety assurance challenges in more detail, illustrating with an autonomous "return to home" (RTH) function used to recover an inspection drone in an underground mine. The RTH is a failsafe response that activates in the event of a failure such as loss of communication with the human pilot to ensure the drone can be recovered safely.

This paper is laid out as follows. Section 2 contains related literature. In Sect. 3 we describe our approach to developing a safety assurance case. Section 4 describes our initial model based design solution, and how it implemented some derived safety requirements. In Sect. 5 we describe some issues we encountered using simulations to support the assurance case. Finally, we present final conclusions and future work in Sect. 6.

2 Related Literature

We have not found any papers in the literature that specifically developed a safety case for autonomous drones in underground mines. A safety case, also known as a safety assurance case, is documented body of evidence which provides a convincing and valid argument that a specified system is safe for a given application in a given context or environment [2]. A small number of papers have developed safety case artefacts for above-ground drone missions. Coldsnow et al. [10] assured beyond visual line-of-sight (BVLOS) human-piloted operations by developing a safety case to acquire a BVLOS certificate for flying in a bounded airspace.

Many papers focus on risk assessment and management for above-ground drone missions. Valapil et al. [23] used a formal model to identify risk during flights and the impact of mitigations on those risks. Barr et al. [4] performed a preliminary risk analysis using a probabilistic model-based approach. Aslansefat et al. [3] combined fault tree analysis (FTA) with "complex basic events" which update the fault tree in real-time to enable real-time reliability evaluation and risk assessment. The authors focused on sensor reliability with ML reliability as future work. Chowdhury [7] used reliability modelling and failure modes and effects analysis (FMEA) coupled with minimum Bayes risk analysis to estimate the conditional risk probability during missions.

Other authors elicit safety requirements and argumentation for above-ground drone missions. Clothier et al. [9] produced a risk informed safety case for drone operations that identified risk mitigations (commensurate with the risk level) and argumentation to support and assure them. Hodge et al. [12] developed map-free drone navigation using deep reinforcement learning and produced a functional failure analysis (FFA) with safety case and rationale for failure mitigations. Similarly, Shafiee et al. [20] performed FFA for a drone inspecting an offshore wind farm with safety testing and verification in a laboratory. Rahimi et al. [19] developed an approach for highlighting safety assumption mismatches (traceability) as safety-critical drone products are developed. Cleland et al. [8] also used traceability to identify change as safety-critical drone products are developed using Agile methods.

3 Safety Assurance Approach

3.1 Mine Scenario

Autonomous drones flying and navigating in uncertain environments such as mines may require map-free navigation [12] and iterative replanning of the flight path (run-time adaptation) [14]. The autonomous drone must mitigate the inherent risks, adapt to the environment and be robust to degraded communications [21] commonly experienced in mines. Crucially, the missions need to be safe - this means reducing risk to safety of humans and the environment itself, and providing evidence/assurance we have done so.

In this scenario, a human-piloted drone is inspecting a newly excavated section of mine that is unsafe for humans to enter. During the inspection, the drone loses communication with the human pilot. It is unsafe for the pilot to attempt to recover the drone and leaving the drone in the newly excavated area is hazardous as it presents an explosion risk, and may become damaged attempting to land on the uneven floor surface. Routine loss of equipment would not be acceptable. Hence, the drone must autonomously and safely return to home.

3.2 Safety Assurance Case

The main aim of the safety case is to demonstrate that risk of hazardous behaviour is reduced to a level that is at least considered tolerable (measured in terms of likelihood and severity) and further it has been reduced As Low As Reasonably Practicable (ALARP), in line with UK health and safety at work regulations [11]. Our strategy is to demonstrate that the drone is *designed* to be operated safely, and then that it *is* being operated safely. The former is demonstrated via formal modelling that safety requirements are implemented (Sect. 4.2) and simulation-based testing to validate the performance of the RTH function (Sect. 5). The latter is addressed by personnel adhering to operational procedures derived from system safety analysis (Sect. 4.1).

Operational Design Domain. In keeping with what is becoming typical practice for safe autonomous systems, we first developed a Operational Design

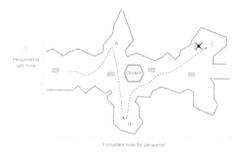

Fig. 1. Top down view of ODD in the mine

Domain (ODD) description in order to understand the context of use and potential safety risks in more detail [15].

The drone will be flying in an unmanned area, surveying newly blasted side tunnels from one main corridor (30–35 by 10–15 feet), all with uneven floors and walls. The tunnels are inspected for safety and viability for mining using laser scans, infrared and visual inspection. Some debris and continuing rock falls can be expected so the tunnels may not be identical when flying to and from an inspection site. No GNSS will be available. The tunnels will include reflective survey markers from the roof a few metres apart (e.g., 5 m and 6 m), with markers visible at all times from each branch tunnel to be inspected. No survey markers will be in the side tunnels in the mine, as these are unsafe newly blasted areas.

A top down representation of the ODD of a tunnel system is shown in Fig. 1. Drone operators and other staff will be separated from the drone inspection zone (described as the safe zone, with the survey tunnels being the forbidden zone). The figure shows ceiling markers roughly arranged along the central corridor, and a large obstacle that needs to be avoided. An example drone flight path to inspect three side tunnels during manual control is shown with the red dotted line (from start, A, B to C). If communications are lost with the drone, it should return autonomously to the initial starting point, avoiding collision with the obstacle and the tunnel walls. Landing at the location where communications were lost is not a viable solution, due to the risk of damaging the equipment and of explosion, and due to the difficulty in retrieving the drone from unsafe newly blasted tunnels.

As noted, the mine is an extremely dusty (particularly on the floor), dark, hot, harsh environment, which impacts reliability even on a short flight. For example, the dust will be caught in rotor systems, which may lead to drift during flight. Small fragments of rock may fall frequently. Flights are a maximum of approximately 15 min due to the harsh environment and limited battery life of the drone.

Hazard Analysis. Having defined the ODD, we then considered the main accidents, hazards, and risk acceptance criteria, to support our safety assurance case [2]. We focused on risks associated with the RTH function. Accidents range in

Table 1. Hazards the RTH function can contribute to

ID	Hazard	Comment
H1	Flying too close to person or moving object	Operational procedures should reduce risk of co-location, but risk can remain
H2	Controlled flight into surface or stationary object	Where drone has lost situational awareness in relation to surfaces or fixed objects, or makes a mistake sending flight commands
H3	Loss of control	Typically internal failures (e.g., broken propeller, or poor course calculations) that can cause a collision
H4	Incursion outside of forbidden zone	To avoid populated areas of the mine, or areas without communication / waypoint

severity from death to minor injuries, and include loss of equipment. The hazards related to the RTH function are in Table 1. We differentiated between moving and stationary objects, with the former largely referring to mine personnel, but could include, for example, falling debris or other movable equipment. For the purposes of our experimental system, and for this paper, our principle concern was *H2 - Controlled flight into surface or stationary object*. This could occur due to the RTH function failing to localise itself, calculating incorrect control commands and/or failing to compensate for physical problems with the drone rotors, hence flying into the walls or obstacles.

3.3 Design and Assurance Challenges

The following design and assurance challenges were identified.

Design: Physical Limitations of the Drone. The inspection drone used for the project was a Holybro PX4 [13] autonomous drone kit. Although the drone has a camera and infrared sensor attached, this is fixed facing forward, meaning that using visual comparisons from one direction would not work for the opposite. This would be a problem for localisation. Performing a mapping run from multiple directions is not practical as there is limited battery life, and it could not be guaranteed to complete without loss of communications. It's hard to add additional equipment to the drone as extra weight further limits battery life.

Design: External Environment Limitations. As there is no GNSS, waypoints cannot be calculated from defined positions. An alternative approach is to navigate via a combination of distinctive landmarks and the reflective ceiling markers. However, the reflective material is difficult for depth and distance sensing via imaging. Future work will look at these challenges.

Assurance: Simulations. Extensive testing in the mine is not practical, due to interference with day to day operations, and the inevitable wear and tear on equipment. Instead, our main options are the use of dedicated test spaces and simulations. Simulations have the advantage of being able to examine multiple scenarios, including failures, environment changes and extreme conditions. However, it is often difficult to justify differences between the simulation and the real world, and predict how these could reduce confidence in safety case evidence. Some specific challenges are described in Sect. 5.

Assurance: Real-World Testing. Physical testing is also required, but is limited to simple, non-destructive, scenarios for reasons of cost, safety and effort. Further, our test space does not replicate the harsh conditions of the mine. This would undermine our confidence in validity of test flights in predicting performance in the mine.

4 Proposed Solution

The design challenges for this scenario limited our options, particularly for SLAM and navigation. Hence, our initial version of the RTH function simply attempts to reverse the exact course taken from take-off up until the loss of communications. Whilst this is not an efficient method (a longer path than necessary would be taken, returning in and out of the blasted side tunnels), it has the advantage of taking a path that was known to avoid obstacles. The implementation does this by continuously recording operator commands for roll, pitch, yaw, speed, etc. and then attempting to appropriately invert these. For example, in Fig. 1 the drone has travelled to junctions A, B and C around the obstacle. The reverse course ensures that this can be avoided (travelling C → B → A), and is a safe route at the point the RTH is activated. However, the obvious problems are that there may be new obstacles, and the drone will suffer from drift, for example, from dust interfering with the propellers, drafts or from artefacts in inertial measurement unit (IMU) data. The next stage in our safety process is to derive safety requirements (DSRs) to be implemented to manage these issues.

4.1 Safety Analysis

For our initial design, we performed both a high-level system hazard analysis, and a low-level guideword-based analysis (using [16]) of the RTH function. For reasons of space, small illustrative excerpts are presented here.

System Level Hazard Analysis. System level hazard analysis considers means to reduce risk of each hazard in turn. For example, when considering Hazard H2 (Table 1) we are concerned preventing ways the drone could fly into a surface or object without any mitigating action. To reduce the risk of this leading to accidents a number of DSRs were suggested, including:

– DSR1: No new man made objects added to the inspection area during inspection flight.
– DSR2: No other objects can be deliberately moved during inspection flight.
– DSR3: Utilise visual scanning of survey markers and environmental features to support SLAM
– DSR4: RTH function will detect and adapt course to objects in path.

- DSR5: If no RTH path can be calculated, drone attempts to move to nearest survey marker, initiates hover, then lands if no recovery of communications and max operating time met.

DSR1 and DSR2 are procedural requirements, and would be evidenced in the "is being operated safely" part of the safety case e.g., through training of staff. DSR3-DSR5 are design requirements.

Safety Analysis of RTH Function. The guideword based analysis considers how internal faults inside components can contribute to risk, and is another source of DSRs. An extract is shown in Table 2.

Table 2. Extract of safety analysis of RTH function

Guideword	Cause	Mitigation
Omission - Route command missed from set of commands	Command was not recorded. Set of commands missed one or more data points. Command is not required due to change in corridor/object layout. Command queue processing error	DSR.RTH1: Sanity check on route commands. DSR.RTH2: Timestamps used on recorded route commands and checked for ordering. DSR.RTH3: Hover mode initiated when uncertain of current status

4.2 Control Software Design

In this section, we describe how we designed a formal model of the RTH control software, particularly focusing on the implementation of the DSRs from the high and low-level analyses. We have used a model-based approach, and in particular the RoboSim notation to model the control software [6] and implemented code based on that model. There are several advantages to using the model-based approach of RoboSim. It can support the maintenance of the safety case and analysis. We could, for example, perform automatic code generation to obtain consistent simulation or automatic test case generation, test with a model checker, and carry out formal proofs of correctness [5]. The availability of all these approaches support our case (Fig. 3).

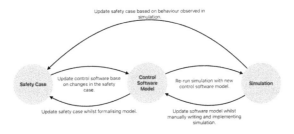

Fig. 2. Bi-directional workflow for improving the control software and the safety case.

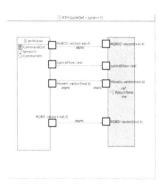
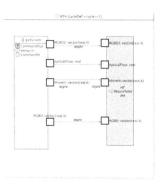

Fig. 3. RoboSim module relating the services of the drone (LHS) with the ReturnToHome controller(RHS). The connections represented by the arrows describe the data flow from the services of the drone to the controller.

Fig. 4. Control software for the drone, demonstrating the data flow between the *ReverseCmd*, *LocalisationAndMapping*, and *ObstacleAvoidance* state machines (the grey boxes). In particular, several events (the black-bordered boxes) are used within the controller to communicate between the state machines. (Color figure online)

In this work, we have used an iterative approach to improve the model, software and DSRs in the safety case, as described in Fig. 2.

In RoboSim, the formalisation process begins with writing the services provided by the robotic platform to the control software to support the DSRs (Fig. 5). Next, the relation between any parallel state machines within a controller is described as shown in Fig. 4. Finally, the details of the state machines defining the behaviour of the controllers are described, as in Fig. 5. When formalising the software, there have been several steps taken to update the DSRs at each level of abstraction. For example, we are forced to state precisely which services of the drone are used by the control software; this has informed the scope of our DSRs. The DSRs have been integral in the development of the control software. For example, DSR5 presents a clear account of the control flow in a situation where no path can be calculated by the obstacle avoidance state machine. This is reflected in the transition between the *ObstacleAvoidance* and Landing states in the *ReverseCmd* state machine in Fig. 5.

We specified the details of the *ReverseCmd* state machine, however, writing the model allowed us to identify more clearly other aspects of the model such as localisation and mapping and obstacle avoidance, as shown in Fig. 4. Members of the team implementing the software for each controller could refer to the model for a clear indication of the inputs and outputs required by them. This is an area often overlooked in the modelling process. A formal account of what is not known with an indication of how to achieve these unknowns in a precise manner is invaluable to any future adaptations of the model and its implementation.

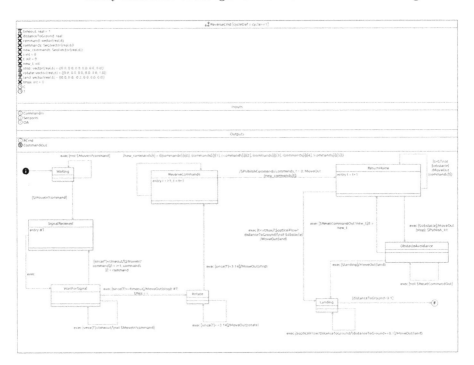

Fig. 5. ReverseCmd state machine. This state machine formally describes how the state of the control software changes depending on events (e.g. MoveIn) or conditions (e.g. since(T) < timeout). It also describes when operations are to be sent to the drone e.g. MoveOut.

Once a model is written, it can be translated into a form that can be executed as part of a simulation. In our system, we used the *ReverseCmd* state machine to write a Python node in ROS2 [17] which communicates with the wider simulation (in Gazebo, see Sect. 5). One of the inherent advantages of this approach is its modularity, as our software artefacts can be re-used in the deployment of the drone or elsewhere without modification. In writing and executing the simulation code, the control software model has been frequently updated due to observed incorrect behaviour. For example, the drone initially did not reverse the commands due to incorrect logic in the model's ReverseCommands state. We therefore, updated the model to fix this. Whilst it is true that in this process both the code and the model have to be updated, it is much easier to update the model and resolve issues at the model level first. Since it is a precise formal model, it has been relatively low-cost to translate back to our low-level source-code implementation. Finally, the safety case has been updated based on the results of the simulation. In our case, the drift of the drone in simulation reinforced the importance of the inclusion of SLAM, for example, in DSR3.

5 Evaluation: Updating The Safety Case and Model

A key part of our safety approach is the use of simulations to provide assurance in the performance of the drone over many varied scenarios[1] Fig. 6. Simulation has the advantage of testing the drone in extreme situations without loss of equipment. We use our observations to update and evaluate our safety case considering the ODD, DSRs, physical reality-gap, reliability, and repeatability.

Fig. 6. Screenshots from the simulations

Is the ODD Adequately Represented? A physical mockup of the mine tunnels was produced and our simulation layout used mesh files of these, including representations of floor objects (Fig. 6). It accurately represented the physical proportions of the deployment environment, but it suffers from physical considerations (e.g. collision detection was not used for tunnel walls and the model does not include unexpected drafts). Nevertheless, this was felt to be sufficient to observe the behaviour of the drone in respect to its environment for early design improvements, but would undermine confidence in any final stage design testing and reduce our ability to automate tests. A physical model of the drone and the environment, and its relation to the physical behaviour of the simulation would be required to quantify how well these aspects of the ODD are reflected [1]

Are the DSRs Reflected? No additional objects are spawned in simulation which ensures DSR1. Some man made objects may move in an unnatural way due to the physics engine which may violate DSR2. DSRs 3–5 were not observed in simulation. As mentioned in 4.2 the state machines in the model need to be updated to reflect these requirements.

Is the Physical Performance of the Drone Representative? The PX4 model of the drone contains all the key sub-systems necessary to test the RTH function in its current implementation, but we made a number of simplifications to improve the performance of the simulator. For example, we used camera messages from the Gazebo model, rather than a model of the drone camera. It

---

[1] The simulation source-code, model, and video are available at https://github.com/uoy-research/UAV_Hackathon_T3.

would be necessary to ensure that images from this simulation were sufficient to test any future obstacle avoidance algorithm implemented using visual object detection. Other internal performance aspects, such as natural drift in the IMU or compute time taken to respond to commands, were not considered. Nor are there any health monitoring functions and other key sensor and actuator data to test robustness to inputs. The ODD and DSRs may be updated to more precisely formalise performance requirements of the drone.

Is the Simulation Repeatable? Our Gazebo runs were not repeatable (repeatability is a typical expectation for robust safety assurance). Additionally, the non-deterministic nature of the gazebo simulations limited our investigation. This non-determinism could be due to a number of factors such as the use of ROS, or the accumulation of error in the computation pipeline of the physics engine. Therefore, we could validate the state machine's general behavior but not the coverage over all trajectories.

Does the Simulation Reliably Model the Elements as Configured? No quality assurance of the simulation tools (the combination of PX4, Gazebo, and ROS2) we used are available, therefore we do not have full confidence that the behaviour we specify is the behaviour the tool represents.

6 Summary and Future Work

When developing our RTH function we faced many challenges in developing a feasible solution, and, most crucially from a safety perspective, in assuring the design. Using the RoboSim framework gave us access to strong formal approaches to support software verification and validation. However, we also need assurance in the drones performance via dynamic testing. We used simulations for this purpose as they can cover many scenarios (including situations hard to replicate in the real-world) but found many issues which undermined confidence in the results. This meant that although simulations were useful for early feedback and safety requirements development, they may provide limited assurance for a mature design.

Future work should look at improving the control software model. Models of the drone and its environment should also be developed. The implementation can then be further refined based on these models and any changes to the safety case.

References

1. A. Miyazawa, et al., A.: RoboSim Physical Modelling Reference Manual. Technical report, University of York (2020)
2. ACWG: Assurance Case Guidance "Challenges, Common Issues and Good Practice". Technical report SCSC-159 v1.0, Safety Critical Systems Club (2021). https://scsc.uk/scsc-141C

3. Aslansefat, K., Nikolaou, P., et al.: Safedrones: real-time reliability evaluation of uavs using executable digital dependable identities. In: Seguin, C., Zeller, M., Prosvirnova, T. (eds.) Model-Based Safety and Assessment, pp. 252–266. Springer (2022)
4. Barr, L.C., Newman, R., et al.: Preliminary risk assessment for small unmanned aircraft systems. In: 17th AIAA Aviation Technology, Integration, and Operations Conference, p. 3272 (2017)
5. Cavalcanti, A., Barnett, W., et al.: RoboStar Technology: A Roboticist's Toolbox for Combined Proof, Simulation, and Testing, pp. 249–293. Springer, Cham (2021)
6. Cavalcanti, A., Sampaio, A., et al.: Verified simulation for robotics. Sci. Comput. Program. **174**, 1–37 (2019)
7. Chowdhury, A.: Dynamic Risk Assessment of Unmanned Aerial Vehicles (UAVs). Master's thesis, Department of Mechanical Engineering, University of Alberta, Canada (2023)
8. Cleland-Huang, J., Agrawal, A., et al.: Visualizing change in agile safety-critical systems. IEEE Softw. **38**(3), 43–51 (2020)
9. Clothier, R., Denney, E., Pai, G.J.: Making a risk informed safety case for small unmanned aircraft system operations. In: 17th AIAA Aviation Technology, Integration, and Operations Conference, p. 3275 (2017)
10. Coldsnow, M.W., Glaab, L.J., et al.: Safety Case for Small Uncrewed Aircraft Systems (sUAS) Beyond Visual Line of Sight (BVLOS) Operations at NASA Langley Research Center. Technical report, NASA Langley Research Center, Hampton, VA, USA. (No. NASA/TM-20230003007) (2023)
11. Health and Safety Executive: Health and Safety at Work Regulations (1999). https://www.legislation.gov.uk/uksi/1999/3242/regulation/3
12. Hodge, V.J., Hawkins, R., Alexander, R.: Deep reinforcement learning for drone navigation using sensor data. Neural Comput. Appl. **33**, 2015–2033 (2021)
13. Holybro: PX4 Vision Dev Kit V1.5 (2024). https://holybro.com/products/px4-vision-dev-kit-v1-5
14. Imrie, C., Howard, R., et al.: Aloft: self-adaptive drone controller testbed. In: 19th International Conference on Software Engineering for Adaptive and Self-Managing Systems (SEAMS 2024) (2024)
15. Kaakai, F., Adibhatla, S.E.A.: Data-centric operational design domain characterization for machine learning-based aeronautical products. In: Computer Safety, Reliability, and Security: 42nd International Conference, SAFECOMP 2023, Toulouse, France, September 2022, 2023, Proceedings, pp.227–242. Springer, Heidelberg (2023)
16. McDermid, J.A., Nicholson, M., Pumfrey, D.J., Fenelon, P.: Experience with the application of HAZOP to computer-based systems. In: IEEE Proceedings of the 10th Conference on Computer Assurance Systems Integrity, Software Safety and Process Security, pp. 37–48 (1997)
17. Open Robotics: ROS2 documentation (2024). https://docs.ros.org/en/foxy/index.html
18. Park, S., Choi, Y.: Applications of unmanned aerial vehicles in mining from exploration to reclamation: a review. Minerals **10**(8) (2020)
19. Rahimi, M., Xiong, W., et al.: Diagnosing assumption problems in safety-critical products. In: 2017 32nd IEEE/ACM International Conference on Automated Software Engineering (ASE), pp. 473–484 (2017)
20. Shafiee, M., Zhou, Z., et al.: Unmanned aerial drones for inspection of offshore wind turbines: a mission-critical failure analysis. Robotics **10**(1), 26 (2021)

21. Shahmoradi, J., Roghanchi, P., Hassanalian, M.: Drones in underground mines: challenges and applications. In: 2020 Gulf Southwest Section Conference. ASEE (2020)
22. Shakhatreh, H., Sawalmeh, A., et al.: Unmanned aerial vehicles (UAVs): a survey on civil applications and key research challenges. IEEE Access **7**, 48572–48634 (2019)
23. Valapil, V.T., Herencia-Zapana, H., et al.: Towards formalization of a data model for operational risk assessment. In: 2021 IEEE/AIAA 40th Digital Avionics Systems Conference (DASC), pp. 1–10 (2021)

Who is the Chameleon? A Party Game to Explore Trust and Biases Towards Alexa, Pepper and ChatGPT

Charlotte Jones[1], Darren Reed[2], and Fanta Camara[3](✉)

[1] School of Natural Sciences, University of York, York, UK
[2] Department of Sociology, University of York, York, UK
[3] Institute for Safe Autonomy, University of York, York, UK
fanta.camara@york.ac.uk

Abstract. With the increasing adoption of robotic and AI systems in our daily activities, understanding how people interact with and trust these systems or may have biases towards them is becoming important for their development and safety. This paper investigates people's level of trust and potential biases towards three robotic systems i.e. an Alexa device, robot Pepper and ChatGPT using "The Chameleon" game. In this experiment, the same words were presented to two of groups of participants, the first group played the game through an online form and the second group played the game in person with Alexa, Pepper and ChatGPT in a lab space. The game consisted in spotting the player who is the chameleon i.e. pretending to know a target word that only the other players are supposed to know. The results showed that both the online and in-person participants had similar levels of spotting the chameleon. However, participants were less able to spot Pepper when it was the chameleon, suggesting that they trusted Pepper a bit more than Alexa and ChatGPT.

Keywords: HRI · trust · bias · robotics · AI · Alexa · Pepper · ChatGPT

1 Introduction

To explore conscious and unconscious biases towards robotic and artificial intelligence (AI) systems, we intend to detect and compare human biases towards building trust with autonomous systems using different communication modalities such as embodiment with a humanoid robot (Pepper), voice with a smart speaker (Alexa) and text with a chatbot screen (ChatGPT) through "The Chameleon" game. We chose these three robotic and AI systems, because they have distinctive features and people would have different levels of familiarity or trust towards them based on their individual experiences and preferences during

---

This work was funded by YorRobots Venables Internship.

human-robot interactions (HRI). For example, Rauchbauer et al. [1] suggested that when engaging with conversational agents there is less engagement of brain areas involved in everyday social cognition, compared to human-human interactions.

"The Chameleon", as used in this work, is a game developed by Big Potato Games. Each player is dealt a card as shown in Fig. 1. One of the cards says "You're the Chameleon", the rest of the cards have a key that is used to look up a word on a separate topic card. Then all the players say one word (or short phrase) to show that they are not the chameleon, without revealing what the target word is to the chameleon. The Chameleon does not know the target word and has to try and blend in. At the end of each round, the players vote for who they think is the chameleon. The aim of the game is for the other players to spot the chameleon and for the Chameleon to blend in. The key part of the game is that the Chameleon did not know what the target word was when everyone was thinking of a word, however they could have figured out what the word was by the time it was their turn to say a word.

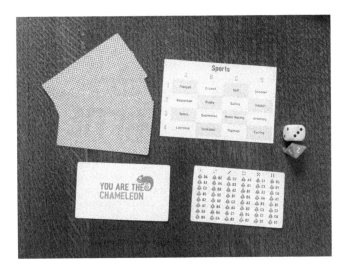

Fig. 1. Example cards from "The Chameleon" game.

Our theory behind this experiment is that in a normal game of "The Chameleon", there are several reasons why a player is more or less likely to be voted for as the chameleon. A big factor in who gets voted for is the word chosen by each player. For example if a player says a really vague word, they might be more likely to be voted for. Other reasons such as the order in which the players say their words and where they are sat in relation to the other players have an impact too. The main part of our theory is that how much the player is trusted and how good they are thought to be at the game are the remaining factors that determine how likely someone will be voted for.

Hence in this work, we made two main hypotheses:

1. the more familiarity/experience that a participant has with a robot, the more likely they will trust it;
2. individual participants will have biases towards a particular robot, and these will be dependent on past experience with each robot.

Previous work such as [2] performed a meta-analysis of factors that affect trust in HRI, the results showed that performance of the robots was the most important factor for people to trust them. Sanders et al. [3] investigated the relationship between trust and the use of a robot where it was found statistical support indicating that "trust leads to use". Cross et al. [4] suggested that interacting with a robot does not change behaviour or neural empathetic response towards the robot, so in this study we do not expect to find a changed measure of trust after taking part in the game. Natarajan and Gombolay [5] measured the effects of anthropomorphism on human trust towards robots and found that the behaviour and anthropomorphism of an agent are the most important factors in trusting them or not. More recently, Alarcon et al. [6] explored biases in human-human vs human-robot interactions, their results showed that there are differences in trust between a human and robot partner and that human biases towards robots are more complex. In the present work, we explore trust and biases towards robots through a party game, "The Chameleon", we assess participants' levels of trust and familiarity with each robot before and after the experiment and we also compare the in-person game results to online participants' responses. To our knowledge, this is the first time a trust experiment is carried out with robot Pepper, an Alexa device, and ChatGPT.

This work was split into three main steps: (1) words were collected from human players in advance from card-based "The Chameleon" games; (2) an online study was performed where participants looked at the words (collected from steps 1) to try and spot which word was said by the chameleon; (3) a robot experiment took place where a separate group of participants made judgements on the same words pronounced by Alexa, Pepper and ChatGPT in a 4-player game. A diagram of the experimental protocol is shown in Fig. 2. Ethical approval was sought from and approved by the Department of Sociology at the University of York.

2 Words Collection

To generate target words and player contributions for our online and robot experiments, we recorded a series of card-based games played by humans (cf. Fig. 1). Three groups of four or five people played the game for approximately 40 min each. The words pronounced by each player were recorded including the word used by the chameleon in each round was written down by the experimenter. These lists of words were then used for the online and robot experiments detailed below.

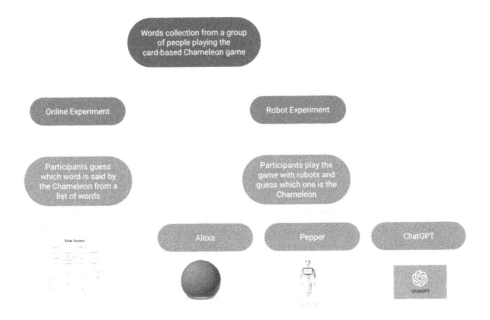

Fig. 2. Diagram of the experimental protocol.

Fig. 3. An example of a round from the online study. The Chameleon said "Fast-paced", a word often associated with several film genres, however it did not fit with foreign Language as well as Words and Confusing, so the example person reviewing the round in part 2 of the study spotted that it was the chameleon word.

3 Online Experiment

In order to quantify the effect of each word revealing or not the chameleon's identity to other players, words collected from the card-based games were presented to participants in an online study. Participants were shown twelve rounds of the game and for each round they had to rank the words from the most to least likely said by the chameleon. An example round can be seen in Fig. 3. 78 participants (59 females, 17 males, 1 non-binary, 1 prefer not to say) aged between 19 and 71 years old completed the online experiment, with an average age of 33 years old. The aim of this online study was to serve as a baseline whose results can be used in comparison to the robot experiment. Through the online experiment, we can get an average score rating for each word for a given target word i.e. this gives a baseline measure of how obvious a word can be associated with the chameleon without knowing who said it, hence online participants cannot have any bias towards a particular player because the rounds and words were presented in a random order.

4 Robot Experiment

4.1 Setup

We programmed three robotic systems (an Amazon Alexa Echo smart speaker, a Pepper robot from Aldebaran Robotics and a chatGPT user interface) to play a simplified version of "The Chameleon" with human participants, while an experimenter followed a script controlling when and what each robot should say at each round, as shown in Fig. 4. This was done in order to control the robots' behaviour and avoid unexpected actions during the experiment.

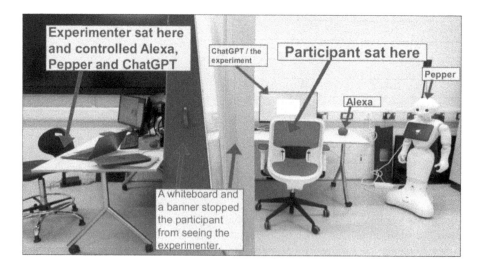

Fig. 4. In-person experimental setup.

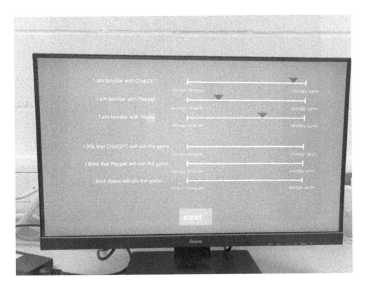

Fig. 5. An example of pre- and post- experiment questions assessing participants' familiarity and trust towards the robots.

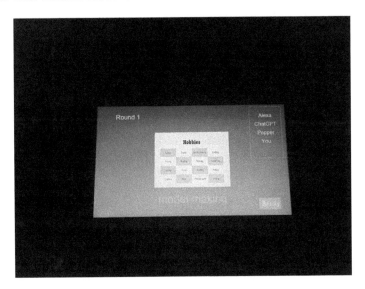

Fig. 6. Screen capture from a game round during the robot experiment.

Alexa. Using the open source Node-RED software [7], it is possible to control Alexa from a computer to say any word. However there may be some problems, for example Alexa does not connect to the University wifi network – Eduroam – a solution was to use mobile hotspot, as Alexa does not have to be connected to the same network as the computer. Alexa randomly stops working after about

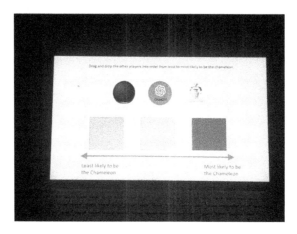

Fig. 7. Rating screen from the robot experiment.

15 min, possible cause is that mobile hotspot switches off when not in use and Alexa doesn't reconnect without being switched off.

Pepper. Using Choregraphe (version 2.5.10) software [8], Pepper can be controlled to say target words and make movements. The text to speech routine from Pepper takes a while to load, it takes from 3–15 seconds, which is not ideal when trying to get Pepper to say a word at a particular time. So the Choregraphe routine was a loop where Pepper says the word and then there is a 3 s delay. This allowed us to mute Pepper until it was her go and then Pepper was unmuted until it had said the word once, making sure not to unmute Pepper midway through a word. Pepper is very difficult to understand, so her voice was adjusted to be of the voice shaping settings of 71% and speed of 80%. This made individual words easier to understand. If the participant did not understand what pepper had said, they could ask the experimenter to say what Pepper had said, to make it fair the word that Pepper said could be repeated by the experimenter too. For example words such as Stu, Filthy, Gen X, Props, Moody and String(from practice rounds) often needed repeating when they were said by Pepper and occasionally when they were said by Alexa.

ChatGPT. We have mimicked a ChatGPT user interface by taking a screenshot of a ChatGPT screen and editing out the text from that chat. Then we used Psychopy software [9] to display text on the screen in the same style as ChatGPT. The experimenter ensured that the participant had read the word displayed on the ChatGPT screen by watching the participant on the video feed from Pepper's eyes. When the experimenter saw that the participant had looked at the screen, they then moved the experiment on to the next player's turn to speak.

4.2 Data Collection

18 participants took part in the robot experiment, they were aged between 19 and 44 years old, with an average age of 22 years old. Participants were familiarised with the game and robots through two practice rounds, where they were allowed to ask questions to the experimenter and check their understanding of the rules of the game. The layout of the experimental setup is illustrated in the labelled photograph (Fig. 4). First, participants were asked to rate on a scale from 0 to 10 their familiarity and trust towards each robot before and after the experiment, as shown in Fig. 5. The results of these questions are discussed in the next section.

Then participants played the game and after each round (cf. example shown in Fig. 6), they indicated which robot was more likely to be the chameleon through a card sorting exercise. The participants dragged and dropped pictures of the robots, scoring them from most likely (scored 2) to least likely to be the chameleon (scored 0) and uncertain (scored 1), as shown in Fig. 7. Then the next target word is shown and then the players say the next round of words. The same twelve rounds of the chameleon that were used in the online study were acted out by Alexa, Pepper and ChatGPT. All controlled by the experimenter behind a screen. After the third round, participants were told that they are the chameleon and they cannot see the target word. This was to make the game more realistic, the participants did not vote in this round and the same words were used for each participant. In the experiment, the words that each robot said was randomised and the order in which all the players (human and robots) presented their words was also randomised. The order of the rounds was randomised in a spreadsheet generating a random number for each round and then ordering the rounds by their assigned random number. For each participant, new random numbers were therefore assigned to each round, making the order of the rounds independently random for all the participants.

5 Results

5.1 Trust vs Familiarity

Before and after the robot experiment, participants were asked to respond to a list of statements about their familiarity and trust towards the robots using a sliding scale (cf. Fig. 5). Figure 8 shows the effect of participants' familiarity on their trust towards a robot before the experiment. For example, participants appeared to be very familiar with ChatGPT and trusted it more, compared Alexa that they were also very familiar with but they trusted a lot less. Participants were less familiar with Pepper but they trusted it in a similar level to Alexa. Figure 9 shows participants' levels of trust for each robot before and after the experiment. These results show that participants appear to trust each robot a bit more after the experiment, with the highest increase being for Alexa (moving from 3 to 4.7 rating). This result is different from the findings in [4] which suggested that trust levels would not change.

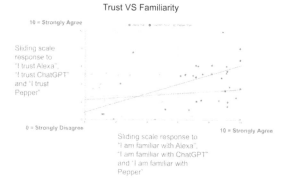

Fig. 8. Participants' trust and familiarity with each robot.

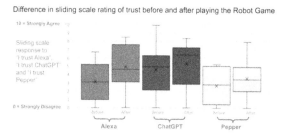

Fig. 9. Participants' trust levels before and after the robot game.

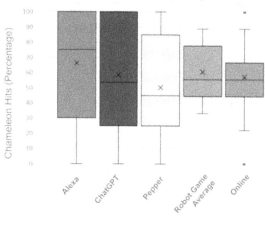

Fig. 10. Chameleon hits.

	Words	Alexa	ChatGPT	OnlineStudy	Pepper	Grand Total
Rubik's Cube	Choking-Hazard	2.000	2.000	1.551	1.750	1.625
	Puzzle	0.200	0.833	0.513	0.714	0.531
	Colourful	0.400	0.667	0.936	0.286	0.844
Office	Dwight	0.286	0.800	0.397	0.833	0.438
	Moody	1.333	2.000	1.641	1.556	1.646
	Cubicles	1.000	0.429	0.962	0.667	0.917
Maths	Equality	0.200	0.333	0.513	0.571	0.490
	Diversity	1.444	1.200	1.269	1.250	1.281
	Inclusion	1.250	1.000	1.218	1.571	1.229
Foreign language	Fast-paced	2.000	1.857	1.564	2.000	1.635
	Words	1.000	0.833	0.833	0.857	0.844
	Confusing	0.200	0.200	0.603	0.125	0.521
Seahorse	Mate for life	0.000	0.625	0.513	0.375	0.500
	Fin	2.000	1.800	1.551	1.286	1.573
	Vertical	1.000	1.000	0.936	0.333	0.927
Badminton	Swing	0.000	0.400	0.628	0.750	0.594
	Momentum	1.714	1.625	1.603	1.333	1.604
	Hit	0.500	1.000	0.769	1.286	0.802
Wings	Freedom	1.111	1.400	1.090	0.750	1.094
	Flight	0.400	0.750	0.679	0.333	0.635
	Air	1.500	1.444	1.231	1.400	1.271
London	Filthy	0.750	1.000	1.038	0.600	1.000
	Metro	0.857	1.200	1.231	0.833	1.177
	Calling	1.143	1.500	0.731	1.143	0.823
Tank	Wheels	1.000	0.714	0.705	1.000	0.740
	Road	1.250	0.800	0.923	0.600	0.927
	Efficient	1.333	1.333	1.372	0.833	1.333
Model Making	Precision	1.000	0.375	0.885	0.600	0.833
	Proportional	1.111	0.500	0.513	0.600	0.573
	Focus	2.000	1.167	1.603	1.625	1.594
Theatre	Props	0.333	1.000	0.667	1.143	0.698
	Script	0.667	0.875	0.936	1.250	0.927
	Snacks	0.833	1.600	1.397	1.429	1.375
Disco	Stu	1.600	1.400	0.718	1.000	0.823
	Wheels (Disco)	1.500	1.000	1.013	1.667	1.063
	Gen X	0.714	0.250	1.269	0.286	1.115
	Grand Total	**1.037**	**1.032**	**1.000**	**0.931**	**1.000**

Fig. 11. Average ratings for each word used in the online and robot experiments.

5.2 Chameleon Identification

If the participant correctly identified the chameleon word, we considered this as a "chameleon hit". Figure 10 shows the chameleon hits for both the online and robot experiments. The average number of chameleon hits in the robot game is similar to that of the online experiment. However, if we look more closely at the chameleon hit per robot, we can see that participants were better at spotting when Alexa was the chameleon than for ChatGPT or Pepper who seems have been trusted a bit more, hence indicating some form of participants' biases towards Alexa and ChatGPT for being the chameleons. These biases could be potentially linked to participants' prior familiarity and trust levels towards each robot. Figure 11 shows the detailed average ratings that each word received for being the chameleon from participants in the online and robot experiments.

6 Discussion and Conclusion

This work provides a new quantitative method for testing biases towards different robotic and AI systems. We have collected and analysed data from 78 online and 18 in-person participants with the results suggesting some form of bias towards Alexa and ChatGPT that participants were familiar with but they tend to trust Pepper a bit more despite knowing her less.

Whilst differences between the ratings of Alexa, Pepper and ChatGPT are not statistically significant, there appears to be a trend of Pepper being favoured slightly more than ChatGPT and Alexa. This may be in line with the results from Rauchbauer et al. [1] where it was shown that participants may behave differently with conversational agents such as Alexa and ChatGPT, hence participants may have trusted Pepper a bit more because of its anthropomorphism [5].

There are some limitations with this study. For example, the reasons why we might not have found a significant effect in the bias and trust towards a robot include the small number of participants in the robot experiment, future work should have more participants and possibly from different cities, countries and even continents, as several studies have shown the impact and importance of taking cultural differences into account in human-robot interactions e.g. [10]. Also, some words alone were more obvious than others to come from the chameleon, hence future experiments might use words with less effects i.e. less obvious answers. We have used a broad range of robots with many differences between them. We do not know exactly why there are biases towards a particular robot. Hence in future work, more information should be gathered from debrief interviews with participants in order to provide some reasoning behind their ratings, and prompt a discussion about their trust or mistrust towards the autonomous systems. These debrief interviews could help explain why the in-person participants' trust levels increased after playing the game with the robots.

Further research could repeat the experiment but have players with more subtle differences between them. For example, if an Alexa is trusted more than Pepper, we might hypothesise that this is because people are more familiar with smart speakers. Future research could have people spend longer time with Pepper to see if this increases their trust. Finally, this experiment could be repeated as the technologies become more widely used, as a measure of how biases change over time with the use of automated technology increasing. In future work, we could also consider making the responses of the robots such as Pepper more animated to utilise their features e.g. moving its arms, changing the voice etc.

References

1. Rauchbauer, B., Nazarian, B., Bourhis, M., Ochs, M., Prévot, L., Chaminade, T.: Brain activity during reciprocal social interaction investigated using conversational robots as control condition. Philos. Trans. R. Soc. B **374**(1771), 20180033 (2019)
2. Hancock, P.A., Billings, D.R., Schaefer, K.E., Chen, J.Y., De Visser, E.J., Parasuraman, R.: A meta-analysis of factors affecting trust in human-robot interaction. Hum. Factors **53**(5), 517–527 (2011)
3. Sanders, T., Kaplan, A., Koch, R., Schwartz, M., Hancock, P.A.: The relationship between trust and use choice in human-robot interaction. Hum. Factors **61**(4), 614–626 (2019)
4. Cross, E.S., Hortensius, R., Wykowska, A.: From social brains to social robots: applying neurocognitive insights to human–robot interaction (2019)
5. Natarajan, M., Gombolay, M.: Effects of anthropomorphism and accountability on trust in human robot interaction. In: Proceedings of the 2020 ACM/IEEE International Conference on Human-robot Interaction, pp. 33–42 (2020)
6. Alarcon, G.M., Capiola, A., Hamdan, I.A., Lee, M.A., Jessup, S.A.: Differential biases in human-human versus human-robot interactions. Appl. Ergon. **106**, 103858 (2023)
7. "Node-RED." https://flows.nodered.org/node/node-red-contrib-amazon-echo
8. "Choregraphe software." http://doc.aldebaran.com/2-1/software/choregraphe/choregraphe_overview.html
9. Peirce, J., et al.: Psychopy2: experiments in behavior made easy. Behav. Res. Methods **51**, 195–203 (2019)
10. Camara, F., Fox, C.: Extending quantitative proxemics and trust to HRI. In: 2022 31st IEEE International Conference on Robot and Human Interactive Communication (RO-MAN), pp. 421–427. IEEE (2022)

Investigation of Gated-CNN and Self-Attention Mechanism for Historical Handwritten Text Recognition

Jizhang Li[1], Sarfraz Ahmed[2](✉), and Md Nazmul Huda[1]

[1] Department of Electronic and Electrical Engineering, Brunel University London, Kingston Lane, Uxbridge, London UB8 3PH, UK
2053594@brunel.ac.uk, MdNazmul.Huda@brunel.ac.uk

[2] School of Future Transport Engineering, Coventry University, Priory Street, West Midlands, Coventry CV1 5FB, UK
ahmed157@uni.coventry.ac.uk

Abstract. In recent years, the development of large-scale datasets and computational power has skyrocketed, significantly advancing the field of handwriting recognition of historical artefacts. Mainstream approaches in the field have historically relied on convolutional recurrent neural networks. However, these deep learning models encounter numerous challenges when dealing with long text images, such as gradient vanishing and gradient explosion. Another drawback is the large number of parameters of these models, which requires a large amount of computation. In response to these limitations, this paper introduces a novel Handwritten Text Recognition (HTR) architecture that integrates a gated convolutional neural network (Gated-CNN) with an ALBERT self attention mechanism. This proposed architecture is built upon the Spatio-Temporal Attention Network (STAN) architecture and is designed to enhance recognition efficiency while reducing the computational burden. Experimental validation conducted across various datasets demonstrates that this model can surpass the performance of other mainstream models. Notably, it achieves this superior recognition capability with fewer convolutional layers and a reduced parameter count, thereby addressing the critical drawbacks of existing approaches.

Keywords: Deep Learning · Handwriting Recognition · Gated Convolution · Self-attention Mechanism

1 Introduction

Along with the development of artificial intelligence technology, the processing technology of different modal information has gradually come into people's view, and the cross-modal technology focusing on information interaction has become an important part of the field of deep learning in the 21st century and has received extensive attention from researchers. The task of handwriting recognition of historical artefacts is a very important research direction in the cross-modal field. At the same time, due to the long-time retention of the handwriting caused by a variety of font styles and handwriting wear and tear and other problems, which prevented the model failed to have a large breakthrough in reliability and accuracy. This paper is a detailed study based on this problem.

The Historical Handwritten Text Recognition (HTR) AI system can convert historical documents into an electronically accessible and analyzable format through dynamic (online) or static (offline) conversion [1]. It is worth mentioning that the transcription provided by offline text recognition can facilitate the study of written text, such as historical manuscripts [2] and medical manuscripts [3]. HTR technology is a key part of offline recognition, it plays a pivotal role in facilitating this transformation, storing, retrieving, analyzing,and sharing these invaluable records efficiently and effectively. This field opens new pathways for documentary research, cultural heritage preservation, and linguistic studies.

The difficulty of historical handwriting recognition is that it is a very complicate process to trace and transcribe the texts from the ancient manuscripts. These manuscripts usually show something that only is found on them and also have physical signs of damage. Historical documents differ in a large scale; records may be stained, old or damaged. Such conditions make preparing pictures extremely complex for further analysis where achieving desired resolution, size and sharpness can become main issues. Also, in the irregular text layouts and handwritten styles of the documents applications, writing would involve various angles, font sizes, and line spacing, which would make it impossible for character segmentation and feature extraction to happen. Another point is, the books have their own content which complicates the situation. Among the ancient records, one can come across included incorrect or obsolete forms of letters or figures, this in turn requires a flexible and adaptable recognition system that can accurately decode and interpret them.

Therefore, in this paper we present an approach, which builds upon the Spatial-Temporal Attention Network (STAN) [4] architecture, designed to high-quality results handwritten text recognition by combining the high feature extraction performance of CNNs architecture with the deep attention and global context understanding of cluster self-attention. This strategy does far more than enable the method to take account of multifaceted stylistic variety and misprints, but also ensures the conversion process to be less error- prone and more time-effective from image to text output, thereby growing the system more stable and efficient. The main work and contributions of this paper are as follows:

1. A deep learning framework is rationally designed to improve the expressiveness and flexibility of the model in processing images.
2. The self-attention module of the model is built based on the current state-of-the-art ALBERT [5] structure.
3. Spatial-Temporal Attention Network (STAN) [4] and proved the feasibility of the model.

4. The superior performance of the model is demonstrated by comparing its success with current state-of-the-art models on different datasets.

2 Related Works

In [6], Convolutional Neural Networks (CNNs) and a Support Vector Machines (SVMs) were compared for recognising handwritten text. The authors found that CNNs are capable of learning local features due to their inherent capabilities of automatically extracting useful and abstract features. The SVM was designed to represent and divide the multi-dimensional data and classes using a hyperplane. The SVM also reduced generalisation errors when faced with previously unseen data. This approach was evaluated on Kannada numerals. For more information on Kannada numerals, see [7]. The CNN architecture outperformed the SVM architecture by 0.23%, with the accuracy of 98.85% and 98.62%, respectively. This demonstrates the effectiveness of CNNs for handwritten text recognition.

The authors in [8] presented a novel technique for handwritten text recognition which uses a combination of a Convolutional Neural Network (CNN) and Recurrent Neural Networks (RNN) for extracting discrete representation from text-line input. Using the CNN+RNN architecture, an encoder—decoder network with an added quantisation layer architecture was constructed. This approach yielded an improvement of 22% on the IAM dataset [9] and 21.1% on the ICFHR18 dataset [10] when compared to previous state-of-the-art techniques.

In [11], a CNN with a Gated Recurrent Unit (GRU) was introduced to improve upon existing handwritten text recognition techniques. As mentioned throughout this section, CNNs have been a core component in recent state-of-the-art handwritten text recognition approaches. Combing the accuracy of the CNN architecture with a GRU provides robustness when dealing with the various styles and complexities of various languages. This approach achieved CER (character error rate) of 7.16% and 6.5% and WER (word error rate) of 16.16%, 17.24% for the IAM and Geroge Washington (GW) [12] datasets, respectively.

3 Methodology

The technique proposed in this paper aims to efficiently extract features from the input image and capture key information by combining CNNs and attention mechanisms to form a conformal model. By integrating the strengths of CNNs and attention, this architecture not only processes image data in an efficient manner, extracting rich local features, but also increases the depth of understanding through global context, greatly improving performance in complex visual tasks. The approach consists of the following key components: Spatio-Temporal Attention Network (STAN) [4] with the additions of CNNs and self-attention mechanisms.

3.1 Spatio-Temporal Attention Network (STAN)

STAN utilises the spatio-temporal information from self-attentions layers for point-to-point interaction between non-adjacent locations. In this work, we build upon this architecture by implementing CNNs for learning of historical handwritten text. This will be further discussed in Sect. 4.4. Detailed description of the STAN architecture is out of the scope of this paper. For further details, please refer to [4].

3.2 Convolutional Neural Networks

The utilisation of CNNs as the implementation of self-attention mechanisms in hand-written recogniser based on historical artefacts with their often complicated structures and different level of preservation yields a range of the benefits. Convolutional Neural Networks perform well at identifying small local patterns that are textures, lines, edges, and contours within the image data they analyse, a skill that is indispensable in character acknowledgement of handwritten documents of various styles and topics.

3.3 Self-Attention Mechanisms

The efficient fusion of self-attention mechanisms is considered a key step for getting over the problem. Self-attention mechanism enables the model to look at each character separately and also considers context from the full text. Due to this, there comes much ease in production of individual letters as well as in finding out how they can be used in different words or sentences. Increasing this awareness can enhance the accuracy and provide more consistent performance recognition. On the other hand, self-attention mechanisms highly develop the parallel processing algorithms, which is becoming one of the key factors that makes the model efficient. The importance of processing acceleration is especially relevant for those scenarios, where huge computational work is required to accomplish the practical purpose of the analysis of specific textual data.

4 Experimentation

4.1 Datasets

Training and evaluation are performed on the IAM Handwriting Database, as shown in Fig. 1, a standard dataset for handwriting recognition of historical artefacts. This dataset is a large handwritten text database developed by the Computer Vision Group at the University of Applied Sciences in Bern, Switzerland, which is widely used in handwritten text recognition and handwriting analysis research. The number of training, evaluation, and testing samples are 12k, 0.7k, and 0.7k, respectively. Each sample package contains 30 data samples, into which the whole training set is repeated 100 times for training.

hand-picked Team under the leadership of

Fig. 1. Sample image from the IAM dataset

This paper also evaluates the model performance on George Washington (GW) handwriting dataset [12], as shown in Fig. 2. This dataset is an historical collection consists of approximately 20 pages of George Washington's handwritten letter from 1755. The letters have been converted into high-quality images with approximately 5,000 word-level annotations.

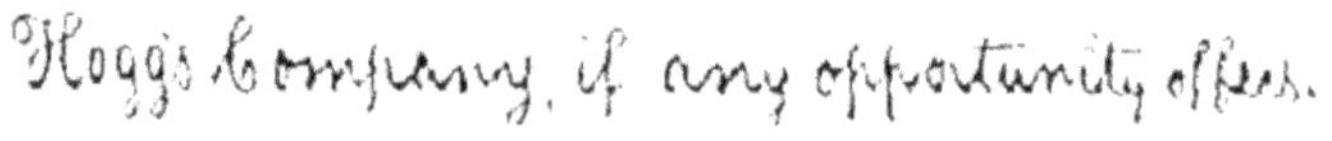

Hoggs Company, if any opportunity offers.

Fig. 2. Sample image from the GW dataset

4.2 Convolutional Blocks

Unlike the general systems that use VGG or ResNet networks to extract image features, the convolutional blocks structure of this system based of Gated-CNN, as shown in Fig. 3, incorporates the use of components such as standard convolutional layers, gated convolutional layers, batch normalisation, activation functions, and maximum pooling layers in order to enhance the expressive power and flexibility when processing images.

Convolutional blocks progressively transform input high-dimensional image data into useful feature representations suitable for subsequent processing and classification tasks by means of a series of convolutional layers. In this context, the convolutional layer is the core of the block, which performs convolutional operations by sliding a set of learnable filters, which can also be referred to as convolutional kernels, directly over the input image, O is the output feature map, K is the convolutional kernel, and I is the input feature map (see (1)).

$$O_{ij} = (K * I)_{ij} = \sum\sum K_{mn} I_{j+m, j+n_{nm}} \tag{1}$$

This operation effectively captures the local spatial features of the input image, as the region covered by the convolution kernel is usually small. Each convolution operation is followed by a ReLU activation function, which is used to introduce non-linearity and enhance the expressive power of the network, enabling the network to learn effective features at a deeper level and effectively mitigating the problem of gradient vanishing. Where, f(x) in (2) denotes the feature maps after the activation function processing.

$$f(x) = max(0, x) \tag{2}$$

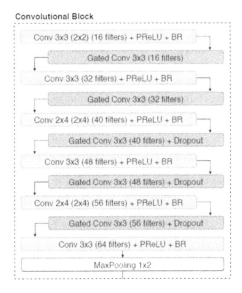

Fig. 3. Convolutional block structure

The addition of a convolutional layer using a gating mechanism allows the network to not only capture key feature information, but also dynamically adjust its processing according to the task requirements, which significantly improves the system's ability and efficiency in processing image tasks. W_g and b_g are the weights and bias of the gated convolution, G is the gating signal (see (3)), and M denotes the final output feature map (see (4)), which is dynamically adjusted to use the main convolution layer's output through sigmoid activation function σ processing to dynamically regulate the use of the main convolutional layer output.

$$G = \sigma(W_g * I + b_g) \qquad (3)$$

$$M = A \odot G \qquad (4)$$

A batch normalisation layer is added after each convolutional layer to speed up the training process and improve the generalisation ability of the model by normalising small batches of data. The batch normalisation layer is performed independently for each channel of the feature map. Represented in (5), where x_k is the input of a single channel, μ_B and σ_B^2 are the mean and variance of the batch of data, respectively, and γ_K and β_K are the learnable parameters as described in (6).

$$\hat{x} = \frac{x_k - \mu_B}{\sqrt{\sigma_B^2 + \varepsilon}} \qquad (5)$$

$$y_k = \gamma_K \hat{x} + \beta_K \qquad (6)$$

The convolutional neural network in the system gradually transforms high-dimensional image data into deep, abstract feature representations by stacking multiple such convolutional and processing layers, providing strong feature support for subsequent attention mechanisms or other tasks. This structure not only extracts image features efficiently, but also optimises the information flow and network training process through advanced techniques such as gating mechanisms and batch normalisation.

4.3 Attention Mechanism

The system is based on ALBERT (A Lite BERT), which incorporates a self-attention module. This attention mechanism allows the model to capture long-range dependencies when dealing with sequential data, which is very effective when dealing with structured data such as text. ALBERT is a natural language processing architecture designed to solve Bidirectional Encoder Representations from Transformers' (BERT) memory and computational efficiency problems with large data and models. The core mathematical representation, as represented in (7) of the self-attention mechanism remains the same as the original Transformer, where Q, K and V are the query, key and value matrices respectively, and d_k is the dimension of the key vector.

$$Attention(Q, K, V) = softmax\left(\frac{QK^T}{\sqrt{d_k}}\right)V \qquad (7)$$

A core improvement in ALBERT is the parameter sharing mechanism. Unlike BERT where each layer has independent parameters, ALBERT shares the same parameters among all Transformer layers, which not only reduces the total number of parameters in the model, but also helps to prevent over-fitting, especially when the amount of data is not very large. In this way, ALBERT significantly reduces the memory footprint of the model and speeds up the training process. Another improvement is the factorisation of the word embedding layer. In traditional BERT, the word embedding layer is mapped directly from word indices to their embedding representations through a large matrix, which requires a large number of parameters. ALBERT achieves a more efficient use of parameters by factorising this large matrix into two smaller matrices. In this way, the lexical indices are first mapped to a low-dimensional hidden space and then from this low-dimensional space to a high-dimensional embedding space. This decomposition method not only reduces the number of parameters required, but also improves computational efficiency while maintaining semantic richness.

As a training task, that ALBERT structure ditch BERT's Next Sentence Prediction (NSP) task and switch to the Sentence Order Prediction (SOP) task which detects whether the two subsequent sentences are in the right order and consequently accomplish to better mimic the coherence and comprehension- two spear headed operations related to the real-world language usage. SOP investigates whether two sentences are in an appropriate order, which common example indicates resembling the accuracy of relations between words/phrases. This novel development of the training objective makes model enable to grasp the relationship between sentences in long texts more precisely, and because of that, it performs better on more complex documents and dialogues.

In this system, long-range dependencies in textual data can be efficiently captured by employing the ALBERT architecture. The self-attention layer allows the model to directly compute and learn attentional weights across different input positions, which enables the model to dynamically adjust its focus over the entire input sequence to accurately extract the most critical information for the task at hand. The output of each input location is obtained by weighted combination of information from all input locations, where the weights are dynamically determined by the features of the inputs themselves.

4.4 Spatial-Temporal Attention Network (STAN) Architecture

In this Spatial-Temporal Attention Network (STAN) architecture, CNNs are adapted and linked with attention mechanisms to provide a feature extraction tailor-made for sequences. First, the data is analysed by using few base layers of convolutions sequentially. The convolutions are applied between the layers to select the spatial features, which non-linearity is enriched by the activation function. Among these layers, the one incorporating the activation function Sigmoid, a gating mechanism is also introduced which regulates the input variables for the tasks that are not yet executed. The input gate convolution is a kind of mask ranged between 0 and 1 which element-wise multiplies the output of the prior convolutional layer making that feature selection a precise process.

The generated feature maps are normalised by batch-normalisation procedure following the pixel-wise filtering at each Neuro ML module, which serves to stabilise the system and accelerate the training process. To reach the performance of the model, we fed it with the Dropout layer with dropout parameter - a probability for neurons' elimination. Dropout is the procedure when randomly, certain output is omitted to let the model rely on other features. Now, you have reached the maximum pooling. This operation is the final layer with the responsibility to reduce dimensionality while also discarding less information but eventually retaining essential features for higher level computations.

In the last step after the convolutional processing, attention module is given the refined features with the use of the ALBERT arrangement methodology. The routing core of this module is represented by multi-head attention mechanism, which allows for information distributed processing over various semantic subspaces. Therefore, different 'heads' process each characteristic or feature of the input, and they are utilised to produce a dense representation that incorporates significant local context as well as the high-level global context information. In this manner, the model signifies the advanced way of data exploitation, simultaneously taking into consideration the relationships among those elements, and ultimately enhancing the utility of both local and global predictive control parameters.

The output station linearizes the attention-processed features and generate the output by entering the linear layer. CNNs get boosted by implementing the self-attention mechanism such way that the local visual clues from the data are received, and global information from whole image is convolved. This capability proves indispensable for handwriting recognition tasks, where the coherence and contextual relationships among characters critically influence the accuracy of content interpretation. Consequently, the STAN model excels in recognising complex or ambiguous handwritten texts, adapting seamlessly to varying handwriting styles and habits, thereby delivering robust performance in real-world applications.

4.5 Training Protocols

In this paper, the optimizer for model training is the AdamW optimizer and the loss function is the CTC (Connectionist Temporal Classification) loss function, the combination of which provides an ideal optimisation strategy for dealing with the complex task of handwriting recognition of historical artefacts. AdamW is a variant of the Adam optimisation algorithm that integrates weight decay (often referred to as L2 regularisation) Adam itself is an adaptive learning rate optimisation algorithm that combines the benefits of Momentum and RMSprop, adjusting the learning rate for each parameter by calculating first-order moment estimates of the gradient (the mean) and second-order moment estimates (the un-rooted variance). Adam's update rule can be expressed as (8)–(12).

$$m_t = \beta_1 m_{t-1} + (1-\beta_1) g_t \tag{8}$$

$$v_t = \beta_2 v_{t-1} + (1-\beta_2) g_t^2 \tag{9}$$

$$\hat{m}_t = \frac{m_t}{1-\beta_1^t} \tag{10}$$

$$\hat{v}_t = \frac{v_t}{1-\beta_2^t} \tag{11}$$

$$\theta_{t+1} = \theta_t - \frac{\eta}{\sqrt{\hat{v}_t} - \varepsilon} m_t \tag{12}$$

where g_t is the gradient at time step t, m_t and v_t are estimates of the first and second-order moments of the gradient, respectively, and β_1 and β_2 are the decay rates, which are usually close to one.

AdamW modified this update rule to separate weight decay from parameter updating to handle regularisation more directly and improve generalisation performance. AdamW's update rule can be expressed as (13).

$$\theta_{t+1} = \theta_t - \eta \left(\frac{\hat{m}_t}{\sqrt{\hat{v}_t} - \varepsilon} + \lambda \theta_t \right) \tag{13}$$

The inclusion of the weight decay coefficient λ in the update rule can effectively prevent over-fitting.

The CTC loss function is used to train sequence prediction models without the need for alignment, where the length of the input data may not be equal to the length of the labels. CTC works by adding a special blank symbol '_' to the sequences, which allows the model to output additional characters that do not correspond to the target sequence. The goal of CTC is to compute a probability distribution such that the final output sequence has the highest probability of matching the target sequence after removing all the blank symbols and duplicate characters, the probability of matching the target sequence is maximised. The CTC loss is defined in (14).

$$CTCLoss = -log\left(\sum \pi \epsilon \beta^{-1}(y) P(\pi|x)\right) \quad (14)$$

where β is an operation that removes all blank and duplicate characters from a path, π denotes a possible path (i.e., the sequence output by the model), and y denotes the target sequence.

After training, the model was able to successfully recognise the content of the handwritten text, as shown in Fig. 4.

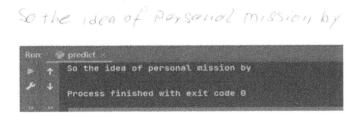

Fig. 4. Example of predictions

4.6 Evaluation Protocols

The authors evaluate the results of the model using the metrics in [13]. The model needs to return 100 identifications and the evaluation metrics are as follows:

Character Error Rate (CER). CER is a very important metric in evaluating handwriting recognition systems, especially in cases where character-level errors may have a significant impact on the comprehension of the text as a whole. It is calculated by dividing the number of erroneous characters in the recognition result by the total number of characters and is calculated as follows:

$$CER = \frac{\text{Number of character insertions + deletions + substitutions}}{\text{Total number of characters in the ground-truth}} \quad (15)$$

Word Error Rate (WER). WER looks at word-level errors. It is calculated by comparing the difference between the recognition results and the real text at the word level and is usually used in scenarios where the accuracy of the recognition results at the word level needs to be evaluated. The calculation method is as follows:

$$WER = \frac{\text{Number of character insertions + deletions + substitutions}}{\text{Total number of words in the ground-truth}} \quad (16)$$

5 Results and Discussions

This section describes the experimental design methodology and the analysis of the results. The effectiveness of the STAN model proposed in this paper is strongly demonstrated by testing on different datasets and comparing with state-of-the-art models.

5.1 Comparison with State-of-the-Art

This paper calculates the CER and WER of the model using the test partitions in the IAM handwritten dataset and the Washington dataset. In the IAM dataset, this paper obtains 5.94% of CER and 11.83% of WER, which means that the proposed model outperforms the reference model in the IAM dataset, and the specific data are shown in Table 1. In the George Washington dataset, this paper obtains 9.23% of CER and 17.69% WER, also better than the reference model, as shown in Table 2.

Table 1. CER and WER Results in the IAM Test Set

Model	CER	WER
Ours	5.94%	11.83%
[11]	7.16%	16.16%
[14]	4.8%	12.7%
[15]	5.1%	13.6%

Table 2. CER and WER Results in the GW Test Set

Model	CER	WER
Ours	9.23%	17.69%
[11]	6.5%	17.24%
[16]	19.29%	32.92%

Therefore, the recognition rate improvement achieved in this paper compared to the above methods can be explained by the following points:

- Gated-CNN in convolutional blocks.
- Combination of CNNs and self-attention mechanism.
- Cutting-edge deep learning techniques (Such as the AdamW optimizer).

The effective combination of these several techniques achieves a low number of parameters and high performance of the system. It makes this system has better flexibility and reliability.

6 Conclusions

In this paper we present an approach built upon the highly successful Spatial-Temporal Attention Network (STAN) architecture. This novel approach is a page-level offline handwriting recognition system, which implements a lightweight model structure to reduce computational costs and increase processing speed. The STAN architecture consists of a self-attention built upon the existing and current state-of-the-art ALBERT structure. Combining this STAN architecture with Gated-CNNs provided comparable, and in some cases, improved results upon existing state-of-the-art techniques while reducing computational costs and time. The superior performance of the model is demonstrated by comparing its success with current state-of-the-art models on different datasets. For the IAM dataset, we were able to achieve comparable CER of 5.94% and outperform similar techniques by at least 0.87%, with a WER of 11.83%. The proposed technique, achieved CER of 9.23% and 17.69% on the GW dataset, outperforming previous approaches by up to 10.06% and 15.23% for CER and WER, respectively.

References

1. Bezerra, B.L.D.: Handwriting: Recognition, Development and Analysis. Nova Science Publishers, Inc., New York (2017)
2. Sánchez, J.A., Romero, V., Toselli, A.H., Vidal, E.: ICFHR2016 competition on handwritten text recognition on the READ dataset. In: Proceedings of International Conference on Frontiers in Handwriting Recognition, ICFHR. Institute of Electrical and Electronics Engineers Inc., 7 2016, pp. 630–635
3. Kamalanaban, E., Gopinath, M., Premkumar, S.: Medicine box: doctor's prescription recognition using deep machine learning. Int. J. Eng. Technol. (UAE) 7((3.34)34), 114–117 (2018)
4. Luo, Y., Liu, Q., Liu, Z.: STAN: spatio-temporal attention network for next location recommendation. In: The Web Conference 2021 - Proceedings of the World Wide Web Conference, WWW 2021. Association for Computing Machinery, Inc, pp. 2177–2185, April 2021. https://doi.org/10.1145/3442381.3449998
5. Lan, Z., Chen, M., Goodman, S., Gimpel, K., Sharma, P., Soricut, R.: ALBERT: a lite BERT for self-supervised learning of language representations. In: 8th International Conference on Learning Representations, ICLR 2020. International Conference on Learning Representations, ICLR, September 2020

6. Parameshachari, B.D., Ashok, A., Reddy, H.: Comparative analysis of handwritten text recognition using CNN and SVM. In: 2nd IEEE International Conference on Distributed Computing and Electrical Circuits and Electronics, ICDCECE 2023. Institute of Electrical and Electronics Engineers Inc., 2023
7. Pal, U., Chaudhuri, B.B.: Indian script character recognition: a survey. Pattern Recognit. **37**(9), 1887–1899 (2004). https://www.sciencedirect.com/science/article/pii/S003132030400055X
8. Davoudi, H., Traviglia, A.: Discrete representation learning for handwritten text recognition. Neural Comput. Appl. **35**(21), 15 759–15 773 (2023)
9. Cheng, L., Bing, L., He, R., Yu, Q., Zhang, Y., Si, L.: IAM: a comprehensive and large-scale dataset for integrated argument mining tasks. In: Proceedings of the Annual Meeting of the Association for Computational Linguistics, vol. 1. Association for Computational Linguistics (ACL), pp. 2277–2287 (2022)
10. Straub, T., Leifert, G., Labahn, R., Hodel, T., Mühlberger, G.: ICFHR2018 competition on automated text recognition on a READ dataset. In: Proceedings of International Conference on Frontiers in Handwriting Recognition, ICFHR, vol. 2018-Augus. Institute of Electrical and Electronics Engineers Inc., 12 2018, pp. 477–482 (2018)
11. Sharma, M., Bagoria, R., Arora, P.: Hybrid CNN-GRU model for handwritten text recognition on IAM, Washington and Parzival datasets. In: 2023 2nd International Conference on Smart Technologies and Systems for Next Generation Computing, ICSTSN 2023. Institute of Electrical and Electronics Engineers Inc., 2023
12. Fischer, A., Keller, A., Frinken, V., Bunke, H.: Lexicon-free handwritten word spotting using character HMMs. Pattern Recognit. Lett. **33**(7), 934–942 (2012)
13. Sánchez, J.A., Romero, V., Toselli, A.H., Villegas, M., Vidal, E.: A set of benchmarks for Handwritten Text Recognition on historical documents. Pattern Recognit. **94**, 122–134 (2019)
14. Voigtlaender, P., Doetsch, P., Wiesler, S., Schluter, R., Ney, H.: Sequence-discriminative training of recurrent neural networks. In: ICASSP, IEEE International Conference on Acoustics, Speech and Signal Processing - Proceedings, vol. 2015-Augus, pp. 2100–2104 (2015)
15. Pham, V., Bluche, T., Kermorvant, C., Louradour, J.: Dropout improves recurrent neural networks for handwriting recognition. In: Proceedings of International Conference on Frontiers in Handwriting Recognition, ICFHR, vol. 2014-Decem, pp. 285–290 (2014)
16. Puigcerver, J.: Are multidimensional recurrent layers really necessary for handwritten text recognition? In: Proceedings of the International Conference on Document Analysis and Recognition, ICDAR, vol. 1, pp. 67–72 (2017)

Robotic Modeling, Sensing and Control

Enabling Tactile Feedback for Robotic Strawberry Handling Using AST Skin

S. Vishnu Rajendran[1]([✉]), Kiyanoush Nazari[2], Simon Parsons[1], and E. Amir Ghalamzan[3]

[1] Lincoln Institute of Agri-Food Technology, University of Lincoln, Lincoln, UK
25451641@students.lincoln.ac.uk , sparsons@lincoln.ac.uk
[2] School of Computer Science, University of Lincoln, Lincoln, UK
[3] University of Surrey, Guildford , UK
a.esfahani@surrey.ac.uk

Abstract. Acoustic Soft Tactile (AST) skin is a novel sensing technology which derives tactile information from the modulation of acoustic waves travelling through the skin's embedded acoustic channels. A generalisable data-driven calibration model maps the acoustic modulations to the corresponding tactile information in the form of contact forces with their contact locations and contact geometries. AST skin technology has been highlighted for its easy customisation. As a case study, this paper discusses the possibility of using AST skin on a custom-built robotic end effector finger for strawberry handling. The paper delves into the design, prototyping, and calibration method to sensorise the end effector finger with AST skin. A real-time force-controlled gripping experiment is conducted with the sensorised finger to handle strawberries by their peduncle. The finger could successfully grip the strawberry peduncle by maintaining a preset force of 2 N with a maximum Mean Absolute Error (MAE) of 0.31 N over multiple peduncle diameters and strawberry weight classes. Moreover, this study sets confidence in the usability of AST skin in generating real-time tactile feedback for robot manipulation tasks.

Keywords: Soft Tactile skin · Acoustics · Grip-force control

1 Introduction

Soft tactile sensors are generally used in manipulation tasks that involve handling soft deformable objects. These sensors are typically characterised by soft deformable skin whose deformation is converted to tactile feedback as tactile readings using an integrated transduction mechanism. There has been continuous progress in the development of soft tactile sensors using various skin materials and transduction mechanisms [16,23].

Various robotic applications call for customisable soft tactile sensors. This is especially important to generate tactile feedback from robot counterparts for the effective execution of manipulation tasks. The customisation can be in

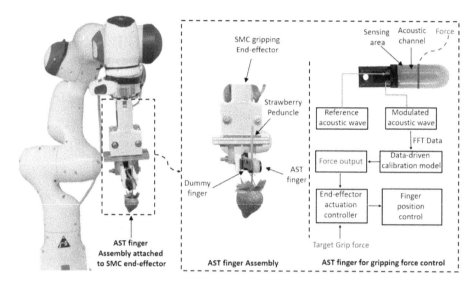

Fig. 1. Demonstration of AST sensing technology to generate tactile force feedback for a robot manipulation task: The AST Skin is attached to a custom end effector finger (the AST finger), and the force feedback from the AST finger is used in performing force-controlled handling of strawberries by gripping their peduncle. This finger assembly is mounted on an SMC gripper for validation purposes, but later, it will be attached to a custom-built strawberry harvesting end-effector

the form of sensor shape and size, material and sensing specifications (such as sensitivity, measurement range) to fit the requirement. Since the conventional soft-tactile sensors have their transduction mechanism integrated with the sensing skin, customisation is not straightforward. Usually, the transduction mechanism uses electric principles (e.g., resistive [27], capacitive [9], piezoelectric [19], magnetic [4,15], impedance [26]) non-electric principles (e.g., camera-based [2,6,8,18,24], fluid-based [7]) or their combinations [11]. When it comes to sensors using electric principles, they have closely knit circuitry embedded beneath the sensing surface and it is the camera with its accessories for camera-based sensors. Similarly, fluid pressure-based sensors have intricate fluid lines integral to the sensing skin. Hence, customising the skin requires considerable effort to also customise the integrated transduction mechanism elements. However, the possibility of keeping the sensing skin and transduction elements modular offers a better scope for customisation. This is possible when the transduction mechanism uses propagating mediums (such as vision/light, fluids or acoustics) to sense the disturbances on the skin by analysing their respective modulations. While using vision, the camera and its accessories result in a bulky form factor of the sensor [25]. Hence, using a camera provides limitations in minimising the form factor of the whole sensor. Moreover, fluid-based methods are known for their delayed measurement responses [5]. In this context, acoustics has a promising scope. It only requires minimal hardware components, typically a speaker and a microphone [10,22]. Most importantly, it can potentially derive diverse tactile information such as force, contact location, temperature and contact material nature [22].

With the modularity concept for customisation and harnessing the capabilities of acoustics, a low-cost novel tactile sensing technology, namely, Acoustic Soft Tactile skin (AST Skin), has been developed [13]. This sensing technology keeps the soft sensing skin and transduction mechanism elements modular. The sensing skin only needs hollow acoustic channels beneath the sensing surface. A speaker and microphone unit form the transduction elements that need to be connected to the acoustic channel. A reference acoustic wave emitted by a speaker propagates through these channels and returns to the microphone. When external forces act on the sensing surface, they deform the channel; hence, the modulation of the acoustic waves varies. This variation is used to measure tactile interactions. Moreover, AST skin technology uses a generalisable data-driven machine-learning model for its calibration. Such a calibration model can account for the variation of the sensing skin's form factor or material change that may arise from customisation, which is complex to establish through an analytical model. AST skin technology has been proven to measure normal forces, 2D contact locations, and contact surface geometries [13,21]. Moreover, the skin's resilience to external sound disturbances is also validated.

In previous work, various skin configurations and their impact on tactile measurements are studied in detail [13]. This paper evaluates the customisability of AST skin by integrating it on a custom end effector finger (the AST finger) and tests its sensing capabilities (refer to Fig. 1). This finger has been developed as a retrofit to a strawberry harvesting robotic end-effector. The AST finger is expected to grip the strawberry peduncle and provide tactile feedback about the peduncle grip status when the robot's vision unit cannot confirm the grip due to occlusions. Moreover, the feedback from the AST finger facilitates effective grip force control for the strawberry handling involved in the harvesting cycle. This paper does not discuss the details of the harvesting end effector mechanism; instead, the AST finger is tested using a general purpose two jaw gripping end effector. The proposed AST finger has a reduced size factor compared to other sensorised gripper fingers available for strawberry handling [20]. The size of the fingers impacts its usability in handling strawberries grown in clusters. Moreover, gripping the strawberry body while harvesting can lead to fruit bruising [1], reducing its shelf life. Thus, the AST finger aims to grip the peduncle rather than the fruit body.

The remainder of the paper outlines the methodology to develop the AST finger and its sensing performance during the real-time strawberry peduncle gripping trials.

2 Methods

This section discusses the methodology for the end-to-end development of the AST finger. Initially, it outlines a study to determine the maximum gripping force that can be applied to the strawberry peduncle. This maximum grip force value helps to decide the force range for which the AST skin needs to be calibrated. Later in this section, the design of the AST finger, prototyping details, and the adopted data-driven calibration approach are detailed.

2.1 Study on Gripping Force

During harvesting, the robot's end effector fingers should grip the peduncle before the cutting and continue holding it until the detached strawberry is placed in the onboard storage (e.g., punnet). Applying the proper force is essential to prevent the strawberry from slipping off the fingers during this manipulation cycle. There are two reasons for slippage: applying insufficient force or crushing the peduncle due to excessive force.

To determine the safe range of force that can be applied to the peduncle, a compression study was conducted with peduncle samples from two breeds of strawberries [14]. The study found that the safe limit of gripping force can be approximated to 10 N. Hence, the AST finger is calibrated for a range of 0 to 10 N.

2.2 AST Finger Design

The AST finger comprises of a back plate on which the AST skin is attached (refer to Fig. 2a). The back plate is 3D printed using PLA material, while the AST skin is made by moulding. The AST skin is moulded in two halves separately and then joined together. Figure 2b shows the mould design. The two moulds are 3D printed, and a Silicone rubber compound with a 10 A shore hardness value (PlatSil Gel 10) is poured into the moulds. After the curing period, the two halves of skin are joined together to form the whole skin. Later, the AST skin is attached to the back plate. The same silicone material is used as the adhesive to join the skin halves and to attach it to the back plate. The acoustic channel is given a 3 mm diameter and is at 1 mm below the sensing surface of the skin so that a gentle touch can affect the channel geometry. The AST finger has two ports to connect the modular acoustic hardware components (refer to Fig. 2a). In this prototype, a regular headphone speaker and microphone are connected to the port via a flexible tube as a test case. In the future, a miniature speaker-microphone will be used.

As a mating finger for the AST finger to grip the strawberry peduncle, a dummy finger has been 3D printed with PLA material. To actuate both fingers for the gripping trials, they are attached to an SMC gripper as shown in Fig. 1 and 5a.

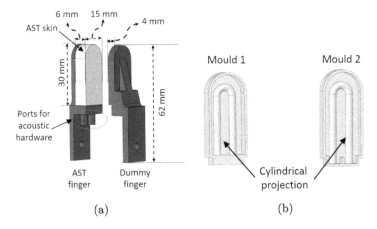

Fig. 2. (a). AST finger with its mating dummy finger, (b). Skin mould: The two semi-cylindrical projections in each mould help to create the skin's cylindrical cavity while joining two halves of the cured silicone

2.3 Skin Calibration

As mentioned in Sect. 1, the AST skin uses a data-driven calibration model. In here, the AST finger is only calibrated to measure contact forces. So, the data set used for the calibration only involves the Fast Fourier Transform (FFT) of the modulated acoustic wave and the corresponding force that causes the modulation. To derive this dataset, a robot-based calibration setup is used to apply known forces on the sensing surface, and the corresponding FFT of the acoustic wave modulation is recorded against the applied force. The robot wrist is fitted with an axial load cell-peg assembly, and the peg has a cylindrical profile with a diameter of 1.5 mm (refer to the Fig. 3, right). This cylindrical profile simulates the average diameter of the strawberry peduncle considered for studying the gripping force limit. During actual deployment on the strawberry harvesting end effector, the finger will be calibrated with a diverse data set generated with different diameter pegs covering the possible peduncle diameters and orientations ranges.

As a test case, only a 14 mm section of the skin is calibrated, and this section was divided into seven subsections at a 2 mm gap, as shown in the Fig. 3 (left). The robot drives the peg vertically downwards at subsections with an equal increment of 0.5 mm from the skin surface until the force value reaches 10 N. During this, the modulated FFT data is recorded against the force values. The resulting data set is used to train the regression machine learning model to predict force from FFT data. To select the best-suited regression model, MATLAB regression learner is used. Different regression models are trained with a data partition of 90:10 ratio and 10% cross-validation folds (refer to Table 1). The Exponential Gaussian Process regression model is selected based on its lower validation error (RMSE: 0.27) and used as the calibration model for the skin.

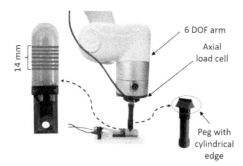

Fig. 3. AST finger calibration: The calibrated area of the AST finger (left), the calibration set up consisting of 6 DOF UFactory xArm, axial load cell with peg attached to the robot arm wrist (middle), the calibration peg with cylindrical profiled edge to simulate the shape of strawberry peduncle (right)

This calibration model is tested with the 10% test data, and the predictions obtained are presented in the Table 2. About 91.1% of predictions are made within ±0.5 N tolerances, while it could make 99 % of predictions at ±1.0 N tolerances. Moreover, the mean absolute error of the predictions is 0.16 N with a standard deviation of 0.22 N.

3 Performance Evaluation of AST Finger

This section evaluates the usability of the AST finger's feedback for force-controlled gripping of the strawberry peduncle.

For this, a series of trials are conducted where the AST finger grips strawberries by their peduncles for a simple pick-and-drop sequence. The AST and dummy finger are connected to an SMC LEZH gripping end effector using an extension adaptor, as depicted in Fig. 5a. The end effector with the fingers is then attached to a Franka Emika robotic arm. This SMC gripper increases the gripping force by closing the fingers in adjustable grip widths.

For the gripping trials, five strawberries are collected with 35–60 mm peduncle left on them (refer to Fig. 4). These strawberries have an average weight and peduncle diameter of 0.141 N and 1.73 mm, respectively (refer to Table 3). A gripping force of 4 N is assumed for gripping these strawberries during the pick-and-drop manipulation cycle. With this gripping force, a mass of 91 g (approximately 1 N) can be manipulated when the manipulator acceleration is kept below $1.0\,m/s^2$ and when the gripping surface of the finger has a friction coefficient of 0.5 (refer to the Eq. 1 and with S as 2). Generally, the friction coefficient of silicone material is closer to 1, but it is assumed to be 0.5 here, providing an additional safety factor. Since the AST finger assembly forms a 2-jaw finger configuration, the total gripping force of 4 N can be considered shared between the two fingers. Hence, 2 N is the target force to be read from the AST finger to complete the gripping action before the pick-and-drop traverse.

Table 1. Comparison of Regression models using MATLAB Regression Learner

Regression Models	Validation Error (RMSE)
Linear Regression	
Linear	0.91
Interactions Linear	0.56
Robust	0.97
Step-wise Linear	0.56
Regression Trees	
Fine Tree	0.42
Medium Tree	0.43
Coarse Tree	0.46
Support Vector Machines	
Linear	0.95
Quadratic	12
Cubic	101.93
Fine Gaussian	0.43
Medium Gaussian	0.49
Coarse Gaussian	0.59
Gaussian Process	
Rational Quadratic	0.28
Squared Exponential	0.40
Matern 5/2	0.34
Exponential	**0.27**
Ensemble of Trees	
Boosted Trees	0.50
Bagged Trees	0.37
Neural Networks	
Narrow Neural	0.46
Medium Neural	0.43
Wide Neural Network	0.41
Bi-layered Neural Network	0.44
Tri-layered Neural Network	0.42

Table 2. Prediction performance of the calibration model

Absolute Error (N)	Percentage Predictions (%)
±0.5	91.13924
±1.0	99.27667
±1.5	99.81917
±2.0	100

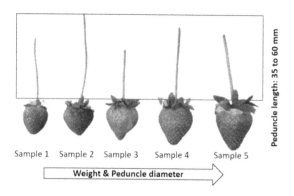

Fig. 4. Strawberry samples selected for the gripping trials

$$F_g = \frac{m.(g+a).S}{\mu} \quad (1)$$

where; 'F_g' is the net gripping force (N), 'm' is the mass to be handled (Kg), 'g' is the acceleration due to gravity (m/s²), 'μ' is the coefficient of friction, and 'S' is factor of safety [17].

Table 3. Features of strawberries used for the gripping trials

Sample	Weight (N)	Peduncle Diameter (mm)
1	0.084	1.24
2	0.111	1.38
3	0.155	1.88
4	0.176	1.90
5	0.181	2.29
Mean	0.141	1.73

3.1 Gripper Force Controller

Tactile feedback-based reactive grip force control is commonly used for grip stabilisation in robotic manipulation [3]. Here, the real-time force feedback (f_m) from the AST finger is leveraged for such a grip force controller. As such, the controller is set to stop gripping by settling the measured grip force (f_m) within a safety zone close to the desired grip force $[f_d - \epsilon, f_d + \epsilon]$ (refer to Eq. 2). As described earlier, the desired grip force (f_d) is 2 N. For this control implementation, the grip width of the SMC gripper (g_t) is changed with a step size of σ_h to close the AST finger assembly. Therefore, the grip width command at each time step reads as follows:

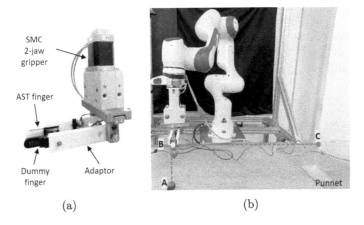

Fig. 5. (a). AST finger assembly connected to the SMC gripper for gripping trials, (b). Experimental setup for the gripping trials: Point A is the picking point, B is a via point between A and C, and C is the dropping point

$$g_{t+1} = \begin{cases} g_t - \sigma_h & \text{if } f_m < f_d - \epsilon, \\ g_t & \text{if } f_d - \epsilon < f_m < f_d + \epsilon, \\ g_t + \sigma_h & f_m > f_d + \epsilon. \end{cases} \quad (2)$$

The grasp width change's step size (σ_h) is tuned based on the target grip force to avoid overshooting in grip force changes or significant control lags. In our experimental trials, the hyper-parameters are defined as $\sigma_h = 1$ mm and $\epsilon = 0.1$ N.

3.2 Experimental Trials

The strawberries are placed individually in the picking location on the workbench with the peduncle upright (refer to Fig. 5b). At the start of the trial, the fingers will be positioned so that the peduncle is between the AST finger (calibrated area) and the dummy finger. Later, the end effector controller is triggered manually, and the fingers start closing with equal increments ($\sigma_h = 1$ mm) until the gripping force value reaches the target gripping force (f_d=2N). Afterwards, the robot arm moves in a fixed trajectory towards the dropping point, where the strawberry is released into a punnet. The experiment scenario is shown in the Fig. 5b.

This gripping trial is repeated five times for each strawberry sample, and the respective force readings from the AST finger (f_m) are logged in real-time to evaluate the performance. The upcoming section will discuss the analysis of the force readings.

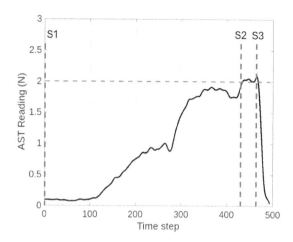

Fig. 6. AST readings (f_m) during a gripping sequence

4 Results and Discussions

A sample of the force readings recorded during a gripping trial is shown in the Fig. 6. Three sections can be visualised from the force profile. In section S1, the peduncle comes in contact with the fingers. After S1, contact force increases as the fingers continue closing/gripping until the set limit (f_d) of 2 N is reached (in section S2). Later, the end effector travels from the picking to the dropping location (from point A to C via B), and force readings during this traverse are between S2 and S3 in Fig. 6. In section S3, the grip force suddenly drops due to the opening of fingers to release the strawberry into the punnet.

From this force profile, the readings (f_m) between S2 and S3 signify how close the grip force settles around the desired grip force (f_d= 2N). The Table 4 presents the Mean Absolute Error (MAE), which quantifies the variation of f_m from f_d. It has been studied that the maximum MAE is about 0.31 N, which is negligible for this task as it doesn't damage any of the peduncles. It is to be noted that this variation in sensor reading can be for two reasons: (i). the effect of swinging the strawberry during the traverse from picking to dropping location, or (ii). the inherent calibration error of the AST skin, as discussed in Sect. 2.3.

The results of the above study provide confidence in the usability of AST skin-embedded fingers in force-controlled gripping tasks using its feedback.

Table 4. Mean Absolute Error (MAE) recorded for each gripping trial against the set grip force value of 2 N

Sample	Trials					Average MAE
	1	2	3	4	5	
1	0.079	0.167	0.143	0.102	0.089	0.116
2	0.082	0.684	0.560	0.047	0.182	0.311
3	0.219	0.307	0.401	0.178	0.358	0.293
4	0.238	0.202	0.095	0.050	0.222	0.161
5	0.106	0.059	0.142	0.031	0.115	0.090

5 Conclusion

This paper tests the usability of Acoustic Soft Tactile (AST) skin as a customisable tactile skin for enabling tactile feedback for robot manipulation tasks. This has been evaluated by building and testing an end effector finger with AST skin to perform force-controlled strawberry gripping. The AST finger facilitated the end effector to handle the strawberry by gripping its peduncle with a set target force of 2 N with a maximum mean absolute error of 0.31 N. As a future work, the AST fingers will be integrated into the custom-built end effector and tested in the field by mounting it on a strawberry harvesting robot [12]. Moreover, the entire skin area will be calibrated to measure the gripping force with the description of the peduncle location on the skin surface. This contact localisation approach helps confirming a successful grip at the target locked by the robot's integrated vision system.

References

1. Aliasgarian, S., Ghassemzadeh, H.R., Moghaddam, M., Ghaffari, H., et al.: Mechanical damage of strawberry during harvest and postharvest operations. World Appl. Sci. J. **22**(7), 969–974 (2013)
2. Chen, Z., Zhang, S., Luo, S., Sun, F., Fang, B.: Tacchi: a pluggable and low computational cost elastomer deformation simulator for optical tactile sensors. IEEE Robot. Autom. Lett. **8**(3), 1239–1246 (2023). https://doi.org/10.1109/LRA.2023.3237042
3. Deng, Z., Jonetzko, Y., Zhang, L., Zhang, J.: Grasping force control of multi-fingered robotic hands through tactile sensing for object stabilization. Sensors **20**(4), 1050 (2020)
4. Diguet, G., Froemel, J., Muroyama, M., Ohtaka, K.: Tactile sensing using magnetic foam. Polymers **14**(4), 834 (2022)
5. Fujiwara, E., de Oliveira Rosa, L.: Agar-based soft tactile transducer with embedded optical fiber specklegram sensor. Results Opt. **10**, 100345 (2023)
6. Gomes, D.F., Luo, S.: Geltip tactile sensor for dexterous manipulation in clutter. In: Tactile Sensing, Skill Learning, and Robotic Dexterous Manipulation, pp. 3–21. Elsevier (2022)

7. Gong, D., He, R., Yu, J., Zuo, G.: A pneumatic tactile sensor for co-operative robots. Sensors **17**(11), 2592 (2017)
8. Lambeta, M., et al.: Digit: a novel design for a low-cost compact high-resolution tactile sensor with application to in-hand manipulation. IEEE Robot. Autom. Lett. **5**(3), 3838–3845 (2020)
9. Li, Q., et al.: Wide-range strain sensors based on highly transparent and supremely stretchable graphene/ag-nanowires hybrid structures. Small **12**(36), 5058–5065 (2016)
10. Ono, M., Shizuki, B., Tanaka, J.: Sensing touch force using active acoustic sensing. In: Proceedings of the Ninth International Conference on Tangible, Embedded, and Embodied Interaction, pp. 355–358 (2015)
11. Park, K., Yuk, H., Yang, M., Cho, J., Lee, H., Kim, J.: A biomimetic elastomeric robot skin using electrical impedance and acoustic tomography for tactile sensing. Sci. Robot. **7**(67), eabm7187 (2022)
12. Parsa, S., Debnath, B., Khan, M.A., E, A.G.: Modular autonomous strawberry picking robotic system. J. Field Robot. (2023)
13. Rajendran, V., Mandil, W., Parsons, S.E.A.G.: Acoustic soft tactile skin (ast skin). arXiv preprint arXiv:2303.17355 (2023)
14. Rajendran, V., Parsa, S., Parsons, S.E.A.G.: Peduncle gripping and cutting force for strawberry harvesting robotic end-effector design. In: 2022 4th International Conference on Control and Robotics (ICCR), pp. 59–64 (2022). https://doi.org/10.1109/ICCR55715.2022.10053882
15. Rehan, M., Saleem, M.M., Tiwana, M.I., Shakoor, R.I., Cheung, R.: A soft multi-axis high force range magnetic tactile sensor for force feedback in robotic surgical systems. Sensors **22**(9), 3500 (2022)
16. Roberts, P., Zadan, M., Majidi, C.: Soft tactile sensing skins for robotics. Curr. Robot. Rep. **2**, 343–354 (2021)
17. for Robots, I.P.: Calculation of gripping force. https://en.iprworldwide.com/calculation-of-gripping-force/. Accessed 20 May 2022
18. Sferrazza, C., D'Andrea, R.: Design, motivation and evaluation of a full-resolution optical tactile sensor. Sensors **19**(4), 928 (2019)
19. Song, K., et al.: Pneumatic actuator and flexible piezoelectric sensor for soft virtual reality glove system. Sci. Rep. **9**(1), 8988 (2019)
20. Visentin, F., Castellini, F., Muradore, R.: A soft, sensorized gripper for delicate harvesting of small fruits. Comput. Electron. Agric. **213**, 108202 (2023)
21. Vishnu, R.S., Parsons, S.E.A.G.: Single and bi-layered 2-d acoustic soft tactile skin. In: 2024 IEEE 7th International Conference on Soft Robotics (RoboSoft), pp. 133–138 (2024). https://doi.org/10.1109/RoboSoft60065.2024.10522056
22. Wall, V., Zöller, G., Brock, O.: Passive and active acoustic sensing for soft pneumatic actuators. Int. J. Robot. Res. **42**(3), 108–122 (2023)
23. Wang, C., et al.: Tactile sensing technology in bionic skin: a review. Biosens. Bioelectron. **220**, 114882 (2023)
24. Ward-Cherrier, B., et al.: The tactip family: soft optical tactile sensors with 3d-printed biomimetic morphologies. Soft Rob. **5**(2), 216–227 (2018)
25. Wei, Y., Xu, Q.: An overview of micro-force sensing techniques. Sens. Actuators A **234**, 359–374 (2015)

26. Wu, H., Zheng, B., Wang, H., Ye, J.: New flexible tactile sensor based on electrical impedance tomography. Micromachines **13**(2), 185 (2022)
27. Zimmer, J., Hellebrekers, T., Asfour, T., Majidi, C., Kroemer, O.: Predicting grasp success with a soft sensing skin and shape-memory actuated gripper. In: 2019 IEEE/RSJ International Conference on Intelligent Robots and Systems (IROS), pp. 7120–7127. IEEE (2019)

Open Source Hardware Whisker Sensor

Robert Stevenson[✉], Dimitris Paparas, Omar Faris, Xiaoxian Xu,
Catherine Merchant, Elliot Smith, Benjamin Nicholls, and Charles Fox

School of Computer Science, University of Lincoln, Lincoln LN6 7TS, UK
28260769@students.lincoln.ac.uk
https://github.com/Ratatouille-Whiskers/Ratatouille-Whisker,
https://zenodo.org/records/11080462

Abstract. Robot whisker sensors have been researched in many studies but have been difficult or impossible to replicate across labs. An open source hardware robot whisker is presented, including easy-to-follow visual build and test instructions. The design can be used by any researcher as a standard platform, enabling replication and comparison of studies in both open source software and open source hardware. The build instructions have been validated by a complete working rebuild by a researcher independent of the design. Numerical specifications are provided and validated. A rotating multi-whisker module design is also provided, which can be used in robots with rat-like whisker arrays.

1 Introduction

Whiskers are tactile sensors used by rodents to obtain information about their surroundings, particularly in scenarios where vision-based sensing provides limited perception, such as navigating in dark underground environments [27]. Based by the structure of these whiskers and the behaviours in which these animals use them, whisker-based artificial tactile sensing has appeared in bio-inspired robotics. Researchers have developed and applied artificial whiskers for robotic tactile sensing in several tasks, such as object recognition [21] and contour reconstruction [28]. Existing publications on whisker sensors do not provide a comprehensive bill of materials, build instructions, or firmware files sufficient for other researchers to accurately replicate their manufacturing and experimental results. The research community lacks a common standard to enable collective replication, modification and extension of designs and results.

Inspired by the approaches used to Open Source Software systems, the Open Source Hardware (OSH) movement has recently gained momentum in an attempt to increase the availability of hardware systems to the community in a way that enables other researchers to easily re-reproduce, combine, and extend each others work. As a recent movement, the definitions and publications behind OSH conventions continue to evolve. Widely accepted definitions of OSH include requirements for low-cost and widely available components, detailed step-by-step build

---

This work was supported by EPSRC grant EP/S023917/1 (AgriFoRwArdS CDT).

instructions that are easy to follow by a robotics undergraduate student, integrated quantitative specifications and validation tests against them, permanent archival publication, and correct use of formal OSH licences and certifications. (Simple publication of final CAD files alone is thus insufficient for work to be considered OSH.)

We present the *Ratatouille-Whisker (RWhisker)*: an OSH artificial whisker module for robotic tactile sensing under the CERN-OHL-W license. The developed RWhisker module adopts a similar approach to the whisker sensor found in SCRATCHbot [19] and CrunchBot [10]. The whisker module mainly comprises of a whisker shaft and casing 3D printed using Fused Decomposition Modelling (FDM), off-the-shelf neodymium magnet and magnetometer, and a flexible bearing made from elastic silicone. As required by the OSH CERN-OHL-W license, we include 3D computer-aided design (CAD) files for all the components and full documentation of basic step-by-step instructions on building, assembling, and operating the whisker module in a permanent archival repository, zenodo.org/records/11080462 (Future development may occur on GitHub). An independent researcher who has not been exposed to the work beforehand has validated the build instructions and used them to rebuild a completely new and functional prototype of the design. Additionally, we demonstrate the module's functionality in an experimental setup that tests the magnetometer response to the whisker tapping on different positions along its shaft. Finally, we demonstrate how several whisker modules can be combined together to validate the tactile sensing capabilities in a texture classification task. Figure 1 demonstrates our open-source RWhisker assembled as a single module and as part of a three multi-whisker module.

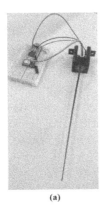 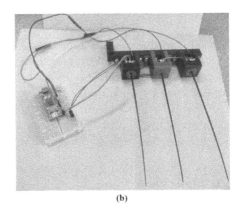

(a)   (b)

Fig. 1. (a) RWhisker single module. (b) Three-RWhisker module.

1.1 Related Work

Whisker sensors are researched in robotics due to their ability to provide detailed tactile information [4], mimicking the highly sensitive whiskers found in many mammals [1]. Whiskers can sense both contact distance [5,7], and textures of objects [6,8], enabling tactile SLAM [11,25] in dark, dusty or smoke filled environments where vision is difficult. Both rat and robot whiskers are found to be most informative when actively 'whisked', i.e. rotated back and forth, with the whisk pattern changing to reverse and retract the whiskers as soon after contact as possible [12,24]. As in rodents, the whiskers are simple passive components, with active sensing only at their base. So losing and replacing whiskers is a low cost for both rodents and robots.

Artificial whisker sensors have evolved significantly since their inception [14,16,23]. Designs such as Whiskerbot [20] focused on mimicking biological structures to enhance sensitivity to touch. SCRATCHbot [19] extended this to a robot with large arrays of whiskers, together with smaller but more precise microvibrissae. Recent advances have seen materials and technologies designed to mimic the tactile sensing mechanics found in nature, such as BioTac [9], comparable to human touch, due to significant improvements in detecting tactile microvibrations. These studies show a shift towards more sophisticated and sensitive artificial tactile systems. However, challenges of accessibility, cost, reproducibility, and extensibility remain due to their proprietary design details.

The open source hardware (OSH) movement has recently emerged [18], enabling rapid and low cost dissemination and collaborative innovation across teams and fields. A few OSH tactile sensors have been published, including the DIGIT optical fingertip [15] and capacitive multitouch sensors for human interaction interfaces [13,22]. Among these, ReSkin is a versatile tactile skin sensor designed to provide robots with a sense of touch that mimics human skin [3]. Another significant contribution in OSH is BioInTacto, a bio-inspired tactile sensor that utilises advanced materials and design principles to closely mimic the tactile sensing capabilities found in natural systems [17]. Lastly, the Allsight sensor project broadens the scope of sensory input by integrating multiple sensing modalities [2]. This comprehensive approach allows for more robust interpretations of environmental data. While these projects exemplify the principles of the OSH movement by offering open access to their designs, promoting transparency, and encouraging collaboration across various disciplines, the specific area of whisker tactile sensors has not been explored to the same extent within the OSH framework. In [29], an OSH calibration platform for whisker tactile sensors was developed based on an affordable off-the-shelf 3D printer.

2 Design Overview

Our whisker module design is inspired by the whiskers used in SCRATCHbot and CrunchBot systems, in which the whisker shaft ends in a magnet embedded in a damped elastic gel with a magnetometer underneath, as shown in Fig. 2. The whisker components are enclosed by a 3D printed rigid casing base, and the magnetometer connects to a microcontroller unit for data acquisition. The final design and prototype of our whisker utilized a different set of materials and components compared to the whiskers of SCRATCHBot and CrunchBot systems.

These materials and components were chosen several potential candidates based on the properties, cost, accessibility and availability, and ease of use.

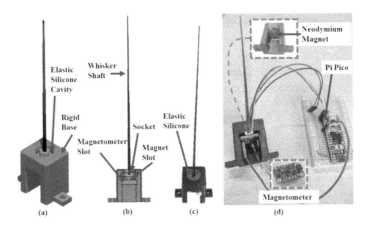

Fig. 2. Design overview of our RWhisker: (a) Isometric view of the RWhisker 3D CAD model, (b) Frontal cross-section view, (c) Prototype of the RWhisker showing the elastic silicone inside its cavity, and (d) The RWhisker prototype connected with a Raspberry Pi Pico microcontroller, along with a closer view of the magnetometer and the neodymium magnet.

2.1 Hardware Components

Shaft: The whisker shaft is manufactured from a 3D printed Polylactic Acid (PLA) material. Several other potential materials were considered: Polycaprolactone (PCL), Nylon, and polyethylene terephthalate glycol (PETG). Each offered unique properties. PCL has a low melting point, biodegradability, and ease of shaping and moulding. However, its mechanical properties and durability under varying environmental conditions posed concerns. PLA is a popular choice in 3D printing, bringing the advantage of being readily available and accessible to print. Its rigidity, though useful in some applications, limited its functionality in mimicking the flexible nature of biological whiskers. Nylon was initially a top contender. However, its higher cost and challenging printability became a significant drawback, leading us to seek an alternative that could provide similar benefits at a lower cost. PETG is a suitable substitute for Nylon, known for its flexibility, durability, and ease of printing, making it a robust material for our application. Based on this analysis, we suggest PLA and PETG as the materials for the whisker shaft. After testing both materials, we chose PLA for our RWhisker despite its lower flexibility relative to PETG. The reason for this choice is that PLA is a more common material, and the lower flexibility of the PLA resulted in better readings from the magnetometer while still offering a level of flexibility that allows significant bending of the whisker shaft without breakage or permanent deformation.

Follicle: A highly elastic silicone with 6-8A shore hardness level was chosen to manufacture the elastic follicle joint. Also considered was polyurethane rubber, specifically Polytec Poly 74–20 RTV, which was used in CrunchBot whiskers. Despite the proven effectiveness of the Poly 74–20 RTV, the potential toxicity during handling and high costs imposed concerns over its use. These factors conflicted with the objective of creating an accessible design. Therefore, we opted for elastic silicone with a low shore hardness level, characterised by its low viscosity and self-degassing properties. In particular, the BBDINO Super Elastic Silicone with a 6-8A shore hardness level proved to be highly flexible, readily available, and ideal for our assembly moulds. Nonetheless, elastic silicone materials with a similar shore hardness from other manufacturers are expected to produce the intended behaviour from the elastic joint. Figure 2a highlights the cavity in the whisker before pouring the elastic silicone, and Fig. 2c demonstrates the whisker prototype with the elastic silicone in place.

Magnetometer: We choose the MLX90393 after comparing with LIS3MDL and LIS2MDL models. All these are 3-axis, 16-bit magnetometers and helped satisfy the goal of keeping the whisker design affordable, as these options are all relatively cheap. Nonetheless, upon examining their datasheets, we found that the MLX90393 is the optimal sensor for the RWhisker due to its larger full-scale range compared to the others, making it suitable as an off-the-shelf sensor. Figure 2d shows the MLX90393 magnetometer and how it fits inside its slot.

Magnet: A 5 mm × 2 mm neodymium magnet is glued to a socket to connect it with the whisker shaft, as seen in Fig. 2 and Fig. 2d. A rigid base was also built to assemble all these components together, along with a casting jig to pour the silicone and manufacture the elastic joint. The base contains a place to slide the magnetometer in right below the magnet, with enough space to connect the wires. Additionally, the base includes holes for screws that can attach each RWhisker to external parts, which can be seen in Fig. 2c.

Microcontroller: A Raspberry Pi Pico is used as the main data acquisition and processing component for RWhisker. It was selected after a comparison with competitors including ESP32, Teensy, and Arduino. The major advantages of the Pi Pico over the other candidates are its low cost, wide availability, and dual-core architecture that allows a relatively higher speed and sampling rate. We utilised one core to receive and process data from the magnetometer through I2C and the other to transmit the results over UART to a computer for post-processing. Figure 2d shows a single RWhisker module connected to a Pi Pico, whereas Table 1 summarises the parts that have been chosen and their corresponding cost.

2.2 Software Development

After choosing the Raspberry Pi Pico, the next step is to develop the software that can be used to acquire data from the sensor. The Arduino IDE was used to develop the necessary firmware as it is considered a convenient wrapper for the

Table 1. Bill of Materials

Part	Quantity	Cost
MLX90393 Magnetometer	1	£10–£20
Raspberry Pi Pico	1	£4–£12
PLA 1kg Spool	1	£18
Elastic Silicone (6-8A Shore Hardness)	1	£28
5 mm × 2 mm Neodymium Magnet	1 (pack of 10)	£8
Demoulding Vaseline	1	£6

Raspberry Pi Pico SDK and enables rapid prototyping. An MLX90393 Arduino library [26] was forked and modified to allow defining the I2C address manually. This modification was necessary as the featured I2C addresses did not align with the respective documentation in the original library. This modification also allows the definition of the sensor boards within the Pico code, which has shown to be useful for operating the multi-whisker module.

With the MLX90393 ready, the RWhisker can be directly connected to a computer device for data acquisition. We provide simple software codes in the RWhisker official repository to read and record data from the RWhisker. The provided code can acquire data from a single magnetometer at a sampling rate of around 1 kHz when the magnetometer is activated with two axes.

2.3 Multi-Whisker Module

The multi-whisker module incorporates three RWhiskers placed next to each other using a 3D printed connecting bar. The connecting bar is also designed with a hole that allows it to attach to servo motors and perform whisking operations. Our RWhisker repository contains instructions on how to use a single microcontroller unit to obtain data from the three sensors simultaneously at a sampling rate of around 300 Hz.

3 Build Overview

Manufacturing and building the RWhisker module involves buying several components and then completing a few steps to use these components and assemble them. The full build instructions, wiring guide, and a build video are provided in the repository. Here, we highlight only key points and major steps:

1. Download and print the 3D CAD models for the whisker shaft, whisker-magnet socket, rigid base, and the elastic silicone casting jig from the repository.
2. Glue the magnet into the socket.
3. Place the socket with the magnet attached into the jig. The magnet should face upwards. The pin should slot in easily to secure it in place.

4. Place the base firmly into the jig.
5. Prepare silicone mixture as per the instructions on the packet. Then pour the silicone mixture into the hole around the socket carefully.
6. Remove the jig carefully, leaving the silicone and the socket in place.
7. Flip the base right-side-up. Place the whisker into the hole in the socket at the top. This may need a gentle push to fit firmly into place.
8. Insert the magnetometer into its slot in the whisker rigid base.
9. The RWhisker module can then be wired to the Pico microcontroller and the firmware installed.

4 User Guide

After building and assembling the system, the user can connect the Pico directly to a computer using a USB cable and start observing the data from the sensor. The repository provides a sample Python script that enables users to operate and record data in a CSV file from the RWhisker. The user must ensure the *pyserial* package is installed, and the correct serial port name is identified when using the script.

5 Characterisation

5.1 Radial Distance Estimation and Signal Repeatability Test

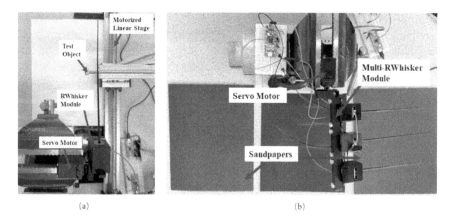

Fig. 3. (a) Radial distance estimation experimental setup. (b) Texture classification experimental setup

This experiment consists of a single RWhisker module attached to a servo motor that rotates the whisker to hit a test object. The test object is fixed on a motorised linear stage. The servo motor is programmed to rotate onto the object and to retract when a threshold angle is reached to prevent it from breaking or permanently deforming. The experiment starts from a distance of 165 mm from the whisker base (i.e., near the tip) and ends at a distance of 65 mm with a step size of 10 mm for each radial distance. At each distance, the whisking operation is performed 10 times, after which the linear stage accurately moves the object 10 mm closer to the whisker base. In total, 10 signals are collected for 11 radial distances. We use the same dataset to perform the repeatability analysis of the collected signals and develop a radial distance estimation regression model. The regression model is developed in the same approach used to predict radial distances for the CrunchBot whisker [7]. The maximum magnetometer reading is taken for each radial distance, and a line least-squares model is used to capture the relation between the data.

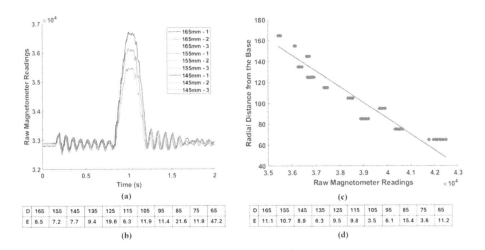

Fig. 4. (a) Raw magnetometer response for three samples of three radial distances and (b) the average absolute mean error when the first experimental run is compared against the remaining nine runs. (c) Linear regression model to predict the radial distance (mm) based on raw magnetometer readings and (d) the average absolute mean error between the regression line and the data.

Figure 4a shows the magnetometer response to three experimental samples of three radial distances to illustrate the repeatability of the signal. Clearly, signals for the same radial distance show high similarity and are visually almost identical. To quantify the difference between the signals, we compare the last nine signals against the first for each radial distance. We report the average of the absolute mean error in Fig. 4b. The error is generally extremely low relative to the range of the raw readings of the magnetometer. The largest reported average absolute mean error is 47.2 when the radial distance is 65 mm, whereas the magnetometer raw readings have a range that reaches around 10,000 units.

Figure 4c and d demonstrate the linear regression model results used to estimate the radial distance and the corresponding absolute mean error for each radial distance. The model achieves an absolute mean error lower than 10 mm in most radial distances, whereas the average absolute mean error of all radial distances is around 8.7 mm.

5.2 Texture Classification Test

The three-RWhisker module was attached to a servo motor to perform a rotational whisking operation on three different textures: P40, P120, and P240 sandpapers, as seen in Fig. 3b. The whisking was performed in clockwise and counterclockwise directions with five repetitions in each rotation at a speed of around 0.342 rad/s, and the magnetometer signal during the experiment was recorded, yielding a dataset with a total of 10 signals per whisker per texture (i.e., 90 signals for the full dataset). We process each signal such that the signal's peak is in the middle. Then, 500 recorded points are taken from each side, resulting in a signal with 1000 points that cover the entire whisking sequence for a single whisking operation in a single direction. Finally, the signal is subtracted from the first reading value, and a moving average filter of five points was applied to the signal.

We use the spectral template based classification method proposed in [8] and [6] to test the classification performance of the Multi-RWhisker module used in the experiment. This method compares a test signal's discrete Fourier transform (DFT) against the DFT of reference signals for each texture by calculating the element-wise sum of squared errors between the signals. The test signal is assigned the category of the texture with the lowest error among the three reference signals. We obtain the DFT of the signals using the fast Fourier transform (FFT) algorithm, and the first reading of the DFT is discarded from the analysis as it represents the mean signal component. Figure 5 demonstrates samples of recorded signals for each texture in each whisking direction and their corresponding DFT absolute magnitude.

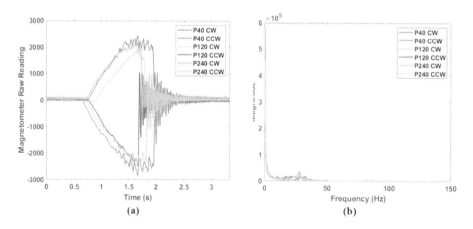

Fig. 5. (a) Raw signals for each texture at each whisking direction. (b) Corresponding DFT frequency plot for each signal.

For each texture and each whisker, we choose the reference signal as the first recorded signal in the dataset and use the remaining nine signals for testing purposes. Figure 6 reports the confusion matrix of this classifier. The classifier can classify each texture with an accuracy of 81.5%. We examined the errors and found that counter-clockwise whisking resulted in a different signal pattern than clockwise. Therefore, we repeated the analysis while having two separate reference signals for each whisker, one from each whisking direction. The remaining eight signals are compared against the reference signal with the same whisking direction. Figure 6b shows the results of this classifier, showing a boost in classification accuracy from 81.5% to 98.6%, with only one misclassification across the test set.

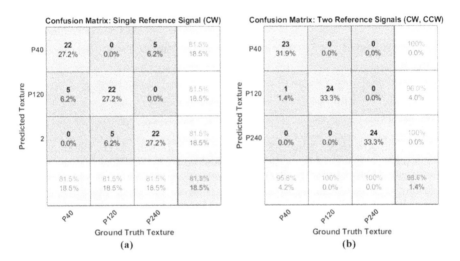

Fig. 6. (a) Confusion matrix for the classifier that uses the first clockwise (CW) signal as a reference. (b) Confusion matrix for the classifier that uses two separate reference signals, a signal for the CW whisking and a different one for the counter-clockwise (CCW) whisking.

Table 2 summarises the characterisation test results together with additional specifications for the RWhisker module. Our radial distance accuracy of 9 mm and texture accuracy of 81% across sandpapers use different experimental setups but are comparable with previous state of the art in non-open systems [6,7], which reported around 2.65 mm accuracy and 72% texture accuracy. Future work could use standardized OSH whisker test rigs [29] to directly evaluate systems against each other and help whisker design mature into a baseline- and metric-driven research field. Additionally, studying the effect of whisking speed on the texture classification accuracy and characterising the properties of the elastic joint material are required to further enhance the applicability of the whisker.

Table 2. Specifications and characterisation of the OSH RWhisker

Feature	Value
Whisker length	180 mm
Base dimensions	37 × 49 × 26 mm
Weight	19.5 g
IP rating	IP20
Sampling rate	1.1 kHz (Single)/300 Hz (Multi)
Baseline distance detection accuracy	9 mm
Baseline texture classification accuracy	81%
Max bending angle before deformation	80°
Output	Serial port
Cost	93–116 USD/19–33 USD (Per Module)
Build tools needed	Standard Fablab
Build skills needed	Undergrad robotics

References

1. Amoli, V., Kim, S.Y., Kim, J.S., Choi, H., Koo, J., Kim, D.H.: Biomimetics for high-performance flexible tactile sensors and advanced artificial sensory systems. J. Mater. Chem. C **7**(47), 14816–14844 (2019)
2. Azulay, O., et al.: Allsight: a low-cost and high-resolution round tactile sensor with zero-shot learning capability. IEEE Robot. Autom. Lett. **9**(1), 483–490 (2023)
3. Bhirangi, R., Hellebrekers, T., Majidi, C., Gupta, A.: Reskin: versatile, replaceable, lasting tactile skins. In: 5th Annual Conference on Robot Learning (2021)
4. Dahiya, R.S., Metta, G., Valle, M., Sandini, G.: Tactile sensing–from humans to humanoids. IEEE Trans. Robot. **26**(1), 1–20 (2009)
5. Evans, M., Fox, C., Pearson, M., Lepora, N., Prescott, T.: Whisker-object contact speed affects radial distance estimation. In: IEEE ROBIO (2010)
6. Evans, M., Fox, C., Pearson, M., Prescott, T.: Spectral template based classification of robotic whisker sensor signals in a floor texture discrimination task. In: TAROS (2009)
7. Evans, M., Fox, C., Lepora, N., Pearson, M., Sullivan, C., Prescott, T.: The effect of whisker movement on radial distance estimation: a case study in comparative robotics. Front. Neurorobot. **6**, 12 (2013)
8. Evans, M., Pearson, M., Lepora, N., Prescott, T., Fox, C.: Whiskered texture classification with uncertain contact pose geometry. In: IEEE IROS (2012)
9. Fishel, J., Loeb, G.E.: Sensing tactile microvibrations with the biotac—comparison with human sensitivity. In: IEEE BioRob (2012)
10. Fox, C., Evans, M., Lepora, N., Pearson, M., Ham, A., Prescott, T.: Crunchbot: a mobile whiskered robot platform. In: TAROS (2011)
11. Fox, C., Evans, M., Pearson, M., Prescott, T.: Towards hierarchical blackboard mapping on a whiskered robot. Robot. Auton. Syst. **60**(11), 1356–1366 (2012)
12. Grant, R., Mitchinson, B., Fox, C., Prescott, T.: Active touch sensing in the rat: anticipatory and regulatory control of whisker movements during surface exploration. J. Neurophysiol. **101**(2), 862–874 (2009)

13. Grosse-Puppendahl, T., Berghoefer, Y., Braun, A., Wimmer, R., Kuijper, A.: Opencapsense: a rapid prototyping toolkit for pervasive interaction using capacitive sensing. In: IEEE PerCom (2013)
14. Kim, D., Möller, R.: Biomimetic whisker experiments for tactile perception. In: International Symposium on Adaptive Motion in Animals and Machines (2005)
15. Lambeta, M., et al.: Digit: a novel design for a low-cost compact high-resolution tactile sensor. IEEE RAL **5**(3), 3838–3845 (2020)
16. Lungarella, M., Hafner, V.V., Pfeifer, R., Yokoi, H.: An artificial whisker sensor for robotics. In: IEEE IROS (2002)
17. de Oliveira, T.E.A., da Fonseca, V.P.: Bioin-tacto: a compliant multi-modal tactile sensing module for robotic tasks. HardwareX **16**, e00478 (2023)
18. Pearce, J.: Building research equipment with free, open-source hardware. Science **337**(6100), 1303–1304 (2012)
19. Pearson, M., Mitchinson, B., Welsby, J., Pipe, T., Prescott, T.: Scratchbot: active tactile sensing in a whiskered mobile robot. In: SAB, pp. 93–103 (2010)
20. Pearson, M., Pipe, A., Melhuish, C., Mitchinson, B., Prescott, T.: Whiskerbot: a robotic active touch system. Adapt. Behav. **15**(3), 223–240 (2007)
21. Pearson, M.J., Salman, M.: Active whisker placement and exploration for rapid object recognition. In: IEEE IROS (2019)
22. Pourjafarian, N., Withana, A., Paradiso, J.A., Steimle, J.: Multi-touch kit: a do-it-yourself technique for capacitive multi-touch. In: ACM UIST (2019)
23. Solomon, J., Hartmann, M.: Robotic whiskers used to sense features. Nature **443**(7111), 525–525 (2006)
24. Sullivan, J.C., et al.: Tactile discrimination using active whisker sensors. IEEE Sens. J. **12**(2), 350–362 (2011)
25. Suresh, S., Bauza, M., Yu, K.T., Mangelson, J., Rodriguez, A., Kaess, M.: Tactile slam: real-time inference of shape and pose from planar pushing. In: ICRA (2021)
26. Yapo, T.: Arduino-mlx90393 library (2016). https://github.com/tedyapo/arduino-MLX90393
27. Yu, Z., et al.: Bioinspired, multifunctional, active whisker sensors. IEEE RAL **7**(4), 9565–9572 (2022)
28. Zhang, Y., Yan, S., Wei, Z., Chen, X., Fukuda, T., Shi, Q.: A small-scale, rat-inspired whisker sensor for the perception of a biomimetic robot. IEEE Robot. Autom. Mag. **29**(4), 115–126 (2022)
29. Zhou, L., et al.: 3D printer based open source calibration platform for whisker sensors. In: TAROS (2024)

Resonant Inductive Coupling Power Transfer for Mid-Sized Inspection Robot

Mohd Norhakim Bin Hassan[(✉)][⬤], Simon Watson[⬤], and Cheng Zhang[⬤]

Department of Electrical and Electronic Engineering, University of Manchester, Manchester M13 9PL, UK
{mohdnorhakim.binhassan,simon.watson,cheng.zhang}@manchester.ac.uk

Abstract. This paper presents a wireless power transfer (WPT) for a mid-sized inspection mobile robot. The objective is to transmit 100 W of power over 1 m of distance, achieved through lightweight Litz wire coils weighing 320 g held together with a coil structure of 3.54 kg. The Wireless Power Transfer System (WPTS) is mounted onto an unmanned ground vehicle (UGV). The study addresses an investigation of coil design, accounting for misalignment and tolerance issues in resonance-coupled coils. In experimental validation, the system effectively transmits 109.7 W of power over a 1-meter distance, with obstacles present. This achievement yields a system efficiency of 47.14%, a value that is remarkably close to the maximum power transfer point (50%) when the WPTS utilises the full voltage allowance of the capacitor. The paper shows the WPTS charging speed of 5 min for 12 V, 0.8 Ah lead acid batteries.

Keywords: robotic application · resonant inductive coupling

1 Introduction

The use of robots is on the rise, and as tasks increasingly require mobility beyond the constraints of a fixed workspace, there is a growing demand for deploying mobile robots [1]. This can be observed from the vast development of various types of mobile robots at an unprecedented pace of operation in the ground, air, and underwater. Several implementations for such mobile robots include oil and gas refinery inspection [2], radiation mapping [3,4], underwater mapping [5,6], and nuclear-decommissioning [7,8].

The main constraint for long-term mobile robot deployment is the onboard battery capacity, especially in scenarios where robots operate behind sturdy concrete walls [1]. Whilst current technological advancements have increased the energy capacity and output power of batteries, this capacity is still insufficient for mobile robots. As a result, the only options are to return to the charging station for recharging or to perform manual battery replacement on-site [1]. These two options merely address the mobile robots' battery capacity limitation, albeit at the expense of increased downtime.

The other option is to use a tether to control the mobile robot. However, this approach has additional difficulties such as tether crossover, restrictions on bending around obstacles, and a decrease in the payload of the mobile robot from the tether system, all of which restrict the mobile robots' performance and mobility [9]. One feasible solution to the battery capacity limitation for mobile robots is the use of wireless power transfer (WPT) technologies.

The common implementations of WPT are mainly concentrated on either high-power applications (kW) such as electrical vehicles which generally operate with a transmission distance of less than 0.3 m [12] or low-power (less than a few Watts) applications such as consumer electronic devices and medical implants within centimetre distances and wireless sensor network (WSN) at kilometre distances [11]. Cheah et al. discussed WPT technologies limitations and the implementation for mobile robots with mid-power range and transmission distances of from 1 m up to 20 m.

This paper discusses the implementation of WPT technology into mobile robot applications. The aim is to incorporate WPT effectively while maintaining robustness without contributing a significant weight to the robot's payload. The use of WPT in this paper is limited to a mid-sized inspection robot with a nominal operating power of 100 W and transmission distances of up to 1 m. The objective is to reach a 50% efficiency target at the maximum power point, thus fully leveraging the stress-handling capabilities of the involved components. The selection of transmission distance is based on the deployment of mobile robots in remote areas, potentially encountering gaps (whether air or obstacles).

2 Mobile Robot and Wireless Power Transfer Systems

2.1 Mobile Robot

A mid-sized Unmanned Ground Vehicle (UGV) is taken into consideration to be a testing platform at the initial stage of conceptual approval apart from the other various types of mobile robots. Hence, a UGV Agile X Scout Mini is chosen. It is a compact entry-level field mobile robot for the research platform for research and development purposes with a four-wheel differential drive and independent suspension. The platform weighs 26 kg and has a maximum permissible payload of 10 kg. Scout Mini has been chosen to provide insight into the practical implementation aspects concerning the proportional relationship between the coil frame's size and that of the mobile robot. The problem statement revolves around assessing the viability of a WPT technology to supply necessary power across various transmission mediums for a moving mobile robot operating within mid-range distance. This study exclusively concentrates on an end-to-end WPT system, wherein power is directly transmitted to the robot's receiver without the need for intermediary relay systems.

2.2 Wireless Power Transfer Technologies

WPT is an umbrella term for a variety of technologies that use electromagnetic fields to transmit energy. The transmitting distance over which the system can efficiently transfer power differs between these technologies. This is determined by whether the energy-transmitting means are directional or non-directional. Instead of using conventional cables, WPT technology can transmit energy from the power source to the target through non-conducting mediums such as air, concrete or water [13].

There are two primary categories of WPT which are radiative and non-radiative. Within the realm of radiative WPT, two prominent techniques are microwave and laser. On the other hand, examples of non-radiative WPT include several technologies, including inductive coupling, magnetic resonance inductive coupling and capacitive coupling. Each of these technologies can be further classified into either direct or background energy harvesting. The distinction lies in their energy source: direct energy harvesting retrieves power from a specifically established transmitter. Meanwhile, background energy harvesting utilizes ambient energy incidental to other processes, like heat from a cooling system or radio waves from wireless communication.

Far-field WPT is usually used to accomplish a longer range of power transmission. This often includes an extended distance to be covered in one transmission. Laser beams and microwaves are the two best-suited forms of electromagnetic radiation techniques for power transmission. The transmission path for a high-efficiency microwave system must be unbroken, and it must be extremely directed [23].

A near-field magnetic solution will enable transmission through numerous barriers, including walls, obstructions, and even humans. The magnetic field remains unaffected by these factors, offering superior tolerance when compared to the far-field WPT. The near-field technique makes use of the inductive coupling effect of non-radiative electromagnetic fields, including the inductive and capacitive mechanisms [24].

2.3 Analysis

Table 1 summarised the main characteristics of different WPT technologies through non-conducting mediums. Parameters of interest are the transmission range (T_r), transmitter-to-receiver diameter ratio (R_{Tx-Rx}) and the maximum efficiency (η_{max}). Laser and microwave WPT can be eliminated due to higher potential hazardous effects on humans within its radiative field and line-of-sight requirement which requires an accurate tracking system. The Maximum Permissible Exposure (MPE) can be determined using IEEE C95.1-2005 regulations [25]. Capacitive coupling is constrained to short-range operations, limited to a maximum transmission distance of 0.3 m. This restriction comes from the minimal capacitance generated by the permittivity of space, leading to a small transmitter-to-receiver ratio. Enhancing capacitance involves enlarging capacitive plate dimensions and reducing transmission distance, but these adjustments pose challenges given the limited dimensions of the receiver on a mobile robot.

The inductive coupling power transfer is accomplished by employing an elongated ferrite core to ensure magnetic flux lines connect between the transmitter and receiver. Nonetheless, the transmitter utilized is substantial and incon-

Table 1. Summary of different performances of WPT technology based on the range of transmission distance with respect to transmitter-to-receiver ratio [1].

WPT Technology	T_r		R_{Tx-Rx}	η_{max}
Laser	$d \geq 20\,\text{m}$		916 [14]	14% [15]
Microwaves			222 [16]	62% [17]
Capacitive	$0.1\,\text{m} \leq d \leq 0.3\,\text{m}$		0.5 [18]	90% [18]
Inductive	$0.1\,\text{m} \leq d \leq 20\,\text{m}$		3.5 [19]	98% [20]
Resonance Inductive	$0.1\,\text{m} \leq d \leq 5\,\text{m}$		6 [21]	85% [22]

venient, leading to an increased transmitter-to-receiver ratio. While inductive coupling has demonstrated a higher transmission range compared to resonance inductive coupling, the transmitter-to-receiver ratio is lower, suggesting that resonance inductive coupling can indeed achieve a greater transmission range. As discussed in the previous subsections, a combination of a UGV which is Scout Mini and the resonance inductive coupling power transfer are the most feasible for the context of this paper.

2.4 Parameters of Resonant Inductive Power Transfer System

Resonant Inductive Power Transfer (RIPT) is regarded as a unique instance of inductive coupling power transfer, where strong electromagnetic coupling is attained by operating at the resonance frequency of the coils. This operational principle can be realized using two or more coils. Two operational principles exist; power delivered to load (PDL) and power transfer efficiency (PTE), where a trade-off between power delivered and system efficiency is observed. High system efficiency is influenced by the subsystems such as the power supply, coil configuration, and coil material.

Figure 1 depicts the simplified circuit topology of the WPT for a mid-sized inspection mobile robot. Equation 1 shows the relationship of angular frequency between self-inductance and compensation capacitors at the primary and secondary coils [17]. Ideally, the system operates at the nominal frequency (f_s) where both sides of the resonators shall be tuned as in Eq. 1 where L_p and L_s are the self-inductance, while C_p and C_s are the compensation capacitors at the primary and secondary sides respectively.

$$f_s = \frac{1}{2\pi\sqrt{L_p C_p}} = \frac{1}{2\pi\sqrt{L_s C_s}} \tag{1}$$

The efficiency of energy transmission is primarily influenced by the load impedance, along with other operational factors including frequency, tolerances of inductive and capacitive components, and alignment. Equation 2 can be used to obtain the coupling coefficient (k) from the self- and mutual inductances (M):

$$k = \frac{M}{\sqrt{L_p L_s}}, \tag{2}$$

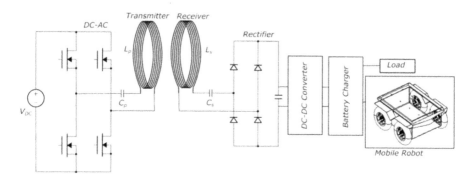

Fig. 1. Circuit topology of resonant inductive coupling for a mid-sized inspection mobile robot.

By employing rms values for voltages and currents, the output power across the load (P_{R_L}) can be determined using the expression where V_L is the load voltage, R_L is the resistance across the load, and I_s is the ac secondary current:

$$P_{R_L} = \frac{|V_L|^2}{R_L} = |I_s|^2 R_L \tag{3}$$

I_p and I_s can be expressed using matrix calculation in Eq. 5 using the component of Z-matrix in Eq. 4, where R_p denotes the resistance on the primary side, R_s the resistance on the secondary side, and V_p the external source voltage.

$$[Z] = \begin{bmatrix} R_p + j\omega L_p + \frac{1}{j\omega C_p} & j\omega M \\ j\omega M & R_s + R_L + j\omega L_s + \frac{1}{j\omega C_s} \end{bmatrix} \tag{4}$$

The relationship between Z_{1_2} and Z_{2_1} can be expressed as $Z_{1_2}=Z_{2_1}$. During resonance condition, it simplifies as:

$$\begin{bmatrix} R_p & j\omega M \\ j\omega M & R_s + R_L \end{bmatrix} \begin{bmatrix} I_p \\ I_s \end{bmatrix} = \begin{bmatrix} V_p \\ 0 \end{bmatrix} \tag{5}$$

The current at the primary side of the coil (I_p) is given in Eq. 6 whilst V_p is given in Eq. 7 as the following:

$$I_p = \frac{j(R_s + R_L) \cdot I_s}{\omega M} \tag{6}$$

$$R_p \cdot I_p + j\omega M \cdot I_s = V_p \tag{7}$$

Substituting Eq. 6 into Eq. 7:

$$R_p \cdot \left(\frac{j(R_s + R_L) \cdot I_s}{\omega M} \right) + j\omega M \cdot I_s = V_p \tag{8}$$

Hence, I_s is obtained as follows:

$$I_s = -V_p \left(\frac{j\omega M}{R_p(R_s + R_L) + (\omega M)^2} \right) \quad (9)$$

Substituting Eq. 9 into Eq. 3 to obtain power across the load:

$$P_{R_L} = \frac{(\omega M)^2 V_p^2 R_L}{(R_p(R_s + R_L) + (\omega M)^2)^2} \quad (10)$$

Whilst the power transfer efficiency, defined as the ratio of the output power across the load P_{R_L} to the input power delivered to the primary coil ($P_{in,AC}$), can be expressed as follows:

$$\eta = \frac{P_{R_L}}{P_{in,AC}} = \frac{(\omega M)^2 R_L}{(R_s + R_L)(R_p(R_s + R_L) + (\omega M)^2)} \quad (11)$$

The self and mutual inductances of the coils and their equivalent series resistances have been discovered to have a direct impact on coupling effectiveness and power transfer efficiency.

3 Constraints

3.1 Dimensional

The resonator is formed by the connection of external compensation capacitors with each coil. Practically the winding also has its parasitic resistance and capacitance. X_n is the total series reactance of resonator circuit n and R_n is the total equivalent series resistance. The transmitter circuits are connected to a voltage or current source and the receiver circuits are connected to loads [26].

The mutual and self-inductance values are determined by the physical dimensions and positions of the winding, unlike other values which are appointed externally. The inductance values can be calculated from the vectors of winding segments using Neumann's formula [27].

$$M_{ij} = \frac{\mu_0}{4\pi} \oint \oint \frac{dl_i \cdot dl_j}{|rdl_i \cdot rdl_j|} \quad (12)$$

where l_i and l_j are contours of two coils, dl_i and dl_j are infinitesimal segment vectors on the two coils and r vectors are the displacement vectors refer to the reference origin. In the case that $i = j$, M_{ij} becomes the self-inductance of the coil and the integral term cannot be computed if dl_i and dl_j are the same segments. An alternative closed-form formula is used to replace the integral term. For wires with a circular cross-section, the following formula can be used [28].

$$L_{part} = \frac{\mu_0}{2\pi} \left[l \log \left(\frac{l + \sqrt{l^2 + \rho^2}}{\rho} \right) - \sqrt{l^2 + \rho^2} + \frac{l}{4} + \rho \right] \quad (13)$$

where the cross-section's radius is ρ and the segment's length is l. The predicted self-inductance is the best predictor of the measurement value when the segment's length is equal to the wire's diameter.

3.2 Power Losses, Interference and Translational Offsets

Additional elements like fine-tuning coil designs, effectively managing heat, implementing shielding, and mitigating electromagnetic interference (EMI) ultimately play a pivotal role in assessing the efficiency of WPT, depending on the specific approaches employed. Higher transmission frequencies (f $\geq$ 50 kHz), resonant switching, and optimised coil components like HF-litz wire and ferrite cores can all help achieve this [29].

Special consideration is needed for the interactions between transmission and reception information and the AC charging flux in electronic circuit loads. The high charging flux density can induce eddy currents in unintended metallic components within these loads, leading to internal temperature increases and circuit damage. Additionally, power losses occur in the secondary and primary circuits, coils, and magnetics. Unusual entities like ferromagnetic or metallic materials near the flux routes can also absorb radiated power. When such materials are positioned in the AC magnetic flux, they generate induced eddy currents, resulting in temperature rise and conduction losses. Significant conduction loss may lead to safety concerns and potential system damage or failure [30]. For instance, a power loss of 0.5–1 W in metallic materials can elevate their temperature above 80 °C [31].

RIPT typically involves tuning the primary and secondary coils to the same resonant frequency, facilitating efficient power transfer across a specific distance. However, achieving perfect alignment between the coils may be impractical in real-world scenarios. Translational offsets arise from this scenario, enabling the coils to transfer power effectively even when not perfectly aligned, albeit possibly with reduced efficiency. It presents both advantages and challenges. On one hand, it permits power transfer across various orientations and distances, thereby enhancing system flexibility and robustness. On the other hand, it complicates system design and optimization, as engineers need to consider factors like coil geometry, alignment tolerances, and potential interference from nearby objects [32]. This fits perfectly with the implementation of RIPT for the mobile robot which provides charging flexibility when movement and space are limited.

4 Simulations and Experimental Setup

The optimization method for designing the coil shape and structure was conducted to identify a collection of solutions that effectively balance multiple competing objectives. In this specific context, there are two key objectives: the quantity of coil turns and the transmitted power. The simulation was conducted using IPTVisual [27] to model coils based on the parameters of interest. The transmitter (TX) and receiver (RX) coils are depicted in Fig. 2a in magenta (left) and green (right) respectively to show the simulation setup of the circular coils with 0.75 mm cross-sectional radius wires and five-turn-per layer helical structure each. The total length of each coil is 15.363 m. This is used as the base of the simulation for which this will be the specification of the wire aimed to be used in the experimental setup.

The self-inductance for both transmitter and receiver coils at resonance condition evaluated from IPTVisual is 63.15 μH ($L_p = L_s$) and is compensated with a capacitance of 1 nF at each side. A 10 Ω load is applied to the receiver circuit. Simulation results show that five coil turns are the optimal amount of winding which provides a balance between the coil resistance, coil length, WPT efficiency and the amount of power to be transmitted. For practicality, octagonal frames are employed to secure Litz wire coils. They closely resemble circular coils. Each frame is octagonal with a 1-meter opening, and the coil has five turns of Litz wire.

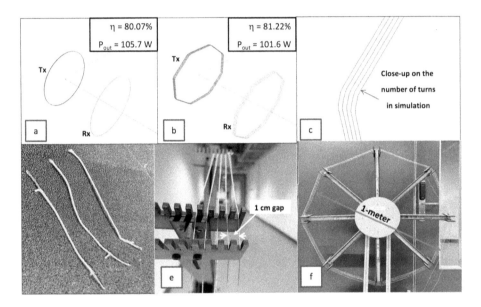

Fig. 2. a) Simulated circle coils, b) simulated octagon-shaped coils, c) a close-up view of the number of turns for each coil in the simulation, d) A sample of lightweight litz wire to construct the transmitter and receiver coil, e) a 1 cm gap for each individual turn, f) an aperture of 1 m for the transmitter and receiver coils.

Figure 2a, 2b and 2c depict the simulations for two geometrical shapes (circular and octagon coils). Simulation results from the circle diameter of 1 m and inscribed octagon of the same diameter show a close approximation of efficiency for the ideal case (no external disturbances in 3D space) in simulation (80.07% for circular, 81.22% for octagon). Both have been simulated with the same transmission distance of 1 m and approximation results of 100 W to match the common operating power of a mid-sized inspection mobile robot (105.7 W for circular, 101.6 W for octagon) with 5 coil turns of identical self-inductance and mutual coupling. The variation in transmitted power between circular and octagonal coils can be attributed to the distinct allocation of coil length when shaping them according to their respective geometries. This disparity in coil length distribution results in differing coil resistances, which, in turn, impacts both the resonance frequency and overall output power. Thus, octagon coils were selected for the experimental setup as the results approached the system efficiency of circular coils and matched the objective of achieving 100 W of transferred power.

Figure 2d shows the sample for the lightweight litz wires used for the coils. Figure 2e shows the 1 cm gap between each coil turn designed to mitigate proximity effects. Figure 2f shows the diameter of the structure frame to hold the litz wire in place forming the transmitter and receiver coils. The experiment will utilise the implementation of RIPT and both coil frames are designed to hold the coil turns through a wall or obstacles of 1 m distance separating these two coils. The observation of power transmission efficiency is expected to achieve a minimum of 50%.

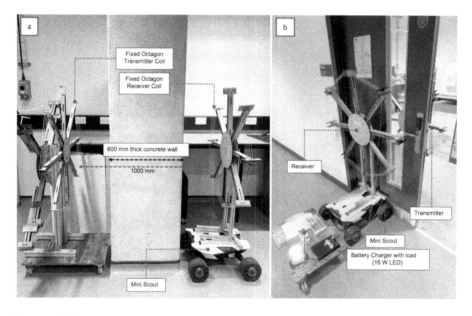

Fig. 3. a) Transmitter and receiver octagon coil frames in between 600 m thick concrete wall, b) Powering up 16 W LED with WPT with obstacles in between transmitter and receiver coils.

The transmitter coil frame is separately mounted at a height of approximately 0.9 m from the ground, aligned coaxially with the receiver coil frame. A receiver coil is installed onto the Agile X Mini Scout (refer Fig. 3a). The WPT was powered by a GS61004B-EVBCD evaluation board. Table 2 shows the key parameters of the experimental RIPT. The receiver coil was linked to a rectifier circuit, a DC-DC power converter, a UPS battery charger, and a 16 W LED with a 12 V, 0.8 Ah lead-acid battery.

Table 2. Component parameter for octagon coil WPT (OC-WPT)

Parameter	Symbol	Value	Unit
DC supply voltage	V_{DC}	43	V
Operating frequency	fs	615	kHz
Capacitance at primary coil	C_p	1	nF
Capacitance at secondary coil	C_s	1	nF
Self-inductance at primary coil	L_p	63.15	μH
Self-inductance at secondary coil	L_s	65.73	μH
Mutual inductance	M	1.4525	μH

Experiment 1 was conducted to test WPT's capability to transmit power through a 600 mm concrete pillar over a 1 m distance (as shown in Fig. 3a. Experiment 2 was conducted aimed to test WPT for power transmission through metallic obstacles at a 1 m distance. A fire door and metallic bin were placed between the transmitter and receiver coils (as shown in Fig. 3b). The results for both experiments showed that WPT transmitted 100 W of power without significant power losses.

5 Results and Observations

5.1 Power Transmission

The measured results are based on the experimental setups as discussed in the previous section, with an input voltage of 43 V and peak input current (7.284 A), the output voltage (33.12 V) measured across a 10 Ω load resistance and the receiving power of (109.7 W). The system's resonance operating frequency, determined by inductance and capacitance, was operated at 615 kHz. The input power was half of the output transmitted at the receiver with 1 nF 2500 V C0G (NP0) ceramic capacitance at each side. This targets a 50% energy efficiency for achieving the maximum power transfer point, ensuring the full utilization of component stress capabilities. A current probe was used to monitor the transmitting circuit's current. Meanwhile, the receiver circuit measures output voltage across the load resistance.

5.2 Effect of Translational Offsets

Figure 4a illustrates the WPTS efficiency, output power and the transmission distance for coaxially aligned coils and the efficiency of y-axis offset coils. It is observed that the estimated efficiency can be achieved when these two coils are positioned coaxially whilst it reduces when the receiver coil is positioned farther away from the coaxial points. In Fig. 4b, the offset positions along y-axis proved the reduction of WPTS efficiency with transmitted power as low as

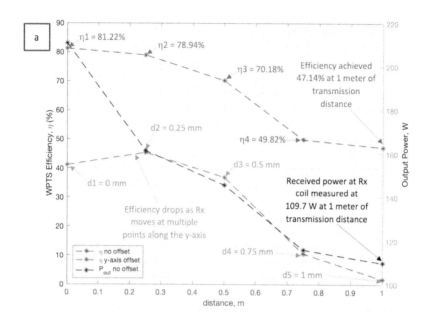

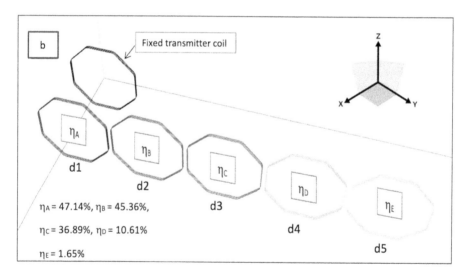

Fig. 4. a) WPTS efficiency and output power vs. transmission distance for coaxially aligned coils and efficiency of y-axis offset coils, b) WPTS efficiency reduces when the receiver coil is positioned at multiple points along y-axis for operational space offset testing (with d5 showing the maximum distance of 1 m)

0.466 W when at d5 position. The objective of this experiment is to obtain 100 W receiving power through WPTS and it is proven to be capable with satisfactory efficiency.

5.3 Battery Charging Using WPT

Smaller 12 V, 0.8 Ah lead-acid batteries were chosen for a demonstration of the WPTS's battery charging capability for easier observation. Figure 5 displays the I-V curve of the lead-acid battery being charged using the WPT. During the charging process, it was observed that the battery charger alternated between charging and discharging modes, as indicated by LED indicators. The voltage input to the battery charger exhibited fluctuations and instability, primarily due to the presence of built-in pulsating diagnostic and maintenance routines for lead-acid batteries. Given the small size of the batteries, these fluctuations were significant. Nonetheless, the WPT successfully charged these two series-connected 12 V, 0.8 Ah lead-acid batteries to their full capacity within a 5-min charging duration, confirming the charging capability.

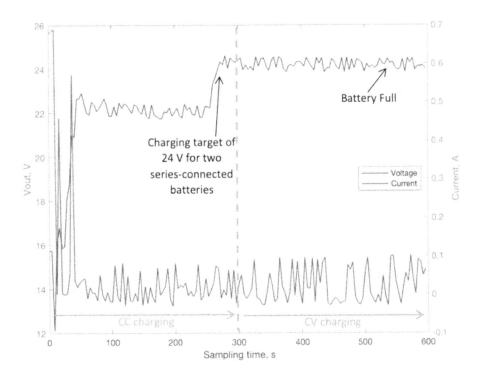

Fig. 5. I-V curve of 12 V, 0.8 Ah lead-acid battery charging with 615 kHz using the octagon-shaped transmitter and receiver coil frames.

6 Conclusions and Future Works

This paper investigated WPT performance for a mid-sized inspection robot, analyzing the impact of various parameters. The paper also presented the implementation of resonant inductive coupling WPT, demonstrating the potential to wirelessly charge mobile robot batteries. The system's efficiency, at 47.14%, is affected by heat loss and reduced coupling due to coil stretching. The efficiency of the WPT stage is close to the maximum output power point of 50% where the stresses of passive components have been well utilised in consideration of lightweight Litz wire, with each coil weighing 320 g held together with the octagon-shaped coil frames of 3.54 kg at each side, thereby fully exploiting the stress capabilities of the components. Future work will propose a collapsible coil frame design for enhanced robustness and flexibility.

References

1. Cheah, W., Watson, S., Lennox, B.: Limitations of wireless power transfer technologies for mobile robots. Wirel. Power Transf. **6**(2), 175–189 (2019)
2. Hutter, M., et al.: Anymal-toward legged robots for harsh environments. Adv. Robot. **31**(17), 918–931 (2017)
3. Li, B., Zhu, Y., Wang, Z., Li, C., Peng, Z., Ge, L.: Use of multi-rotor unmanned aerial vehicles for radioactive source search. Remote Sens. **10**(5), 728 (2018)
4. Connor, D.T., et al.: Application of airborne photogrammetry for the visualisation and assessment of contamination migration arising from a Fukushima waste storage facility. Environ. Pollut. **234**, 610–619 (2018)
5. Nancekievill, M., et al.: Development of a radiological characterization submersible rov for use at Fukushima Daiichi. IEEE Trans. Nucl. Sci. **65**(9), 2565–2572 (2018)
6. Meurer, C., Fuentes-Pérez, J.F., Palomeras, N., Carreras, M., Kruusmaa, M.: Differential pressure sensor speedometer for autonomous underwater vehicle velocity estimation. IEEE J. Oceanic Eng. **45**(3), 946–978 (2019)
7. Bird, B., et al.: Radiological monitoring of nuclear facilities: using the continuous autonomous radiation monitoring assistance robot. IEEE Robot. Autom. Mag. **26**(1), 35–43 (2018)
8. West, C., et al.: Development of a debris clearance vehicle for limited access environments, Poster Papers, p. 20, 2019
9. West, C., et al.: A debris clearance robot for extreme environments. In: Annual Conference Towards Autonomous Robotic Systems, pp. 148–159. Springer, 2019
10. Park, C., Lee, S., Cho, G.H., Choi, S.Y., Rim, C.T.: Two-dimensional inductive power transfer system for mobile robots using evenly displaced multiple pickups. IEEE Trans. Ind. Appl. **50**(1), 558–565 (2013)
11. Xie, L., Shi, Y., Thomas Hou, Y., Lou, A.: Wireless power transfer and applications to sensor networks. IEEE Wirel. Commun. **20**(4), 140–145 (2013)
12. Choi, S.Y., Gu, B.W., Jeong, S.Y., Rim, C.T.: Advances in wireless power transfer systems for roadway-powered electric vehicles. IEEE J. Emerg. Sel. Top. Power Electron. **3**(1), 18–36 (2014)
13. Shinohara, N.: Wireless Power Transfer Via Radiowaves. John Wiley and Sons, Hoboken (2014)

14. Becker, D.E., Chiang, R., Keys, C.C., Lyjak, A.W., Nees, J.A., Starch, M.D.: Photovoltaic-concentrator based power beaming for space eleva- tor application. In: AIP Conference Proceedings, pp. 271–281
15. Ortabasi, U., Friedman, H.: A photovoltaic cavity converter for wireless power transmission using high power lasers. In: IEEE 4th World Conference on Photovoltaic Energy Conference, Waikoloa, HI, USA (2006)
16. Dickinson, R.M.: Evaluation of a Microwave High-Power Reception- Conversion Array for Wireless Power Transmission, Technical report (1975)
17. Wan, S., Huang, K.: Methods for improving the transmission- conversion efficiency from transmitting antenna to rectenna array in microwave power transmission. IEEE Antennas Wirel. Propag. Lett. **17**, 538–542 (2018)
18. Lu, F., Zhang, H., Hofmann, H., Mi, C.: A double-sided LCLC-compensated capacitive power transfer system for electric vehicle charging. IEEE Trans. Power Electron. **30**, 6011–6014 (2015)
19. Choi, B.H., Thai, V.X., Lee, E.S., Kim, J.H., Rim, C.T.: Dipole-coil-based widerange inductive power transfer systems for wire- less sensors. IEEE Trans. Ind. Electron. **63**, 3158–3167 (2016)
20. Bati, A., Luk, P.C., Aldhaher, S., See, C.H., Abd-Alhameed, R.A., Excell, P.S.: Dynamic analysis model of a class E2 converter for low power wireless charging links. IET Circuits Devices Syst. **13**, 399–405 (2019)
21. Chen, W.X., Chen, Z.P.: Optimization on the transmission distance and efficiency of magnetic resonant WPT system. In: CSAA/IET International Conference on Aircraft Utility Systems, Guiyang, China, pp. 148–154 (2018)
22. Akuzawa, Y., Tsuji, K., Matsumori, H., Ito, Y., Ezoe, T., Sakai, K.: A 95% efficient inverter with 300-W power output for 6.78 MHz magnetic resonant wireless power transfer system. In: IEEE MTT-S International Microwave Symposium (IMS), Phoenix, AZ, USA (2015)
23. Haihong, M.: Research on microwave wireless power transmission technology in aerospace. Space Electron. Technol. **10**(05), 590–594 (2014)
24. Zhang, Z., Pang, H., Georgiadis, A., Cecati, C.: Wireless power transfer-an overview. IEEE Trans. Ind. Electron. **66**(2), 1044–1058 (2019)
25. IEEE Standard for Safety Levels With Respect to Human Exposure to Radio Frequency Electromagnetic Fields, 3 kHz to 300 GHz. IEEE Std. C95.1-2005, p. 27, October 2000
26. Imura, T.: Basic Circuit for Magnetic Resonance Coupling (S–S Type). In: Wireless Power Transfer. Springer, Singapore (2020). https://doi.org/10.1007/978-981-15-4580-1_4
27. Wang, H., Zhang, C., Hui, S.Y.R.: Visualization of energy flow in wireless power transfer systems. In: IEEE Wireless Power Transfer Conference (WPTC) 2019, pp. 536–541 (2019). https://doi.org/10.1109/WPTC45513.2019.9055544
28. Ruehli, A.E.: Equivalent circuit models for three-dimensional multiconductor systems. IEEE Trans. Microw. Theory Tech. **MTT-22**, 216–221 (1974)
29. Rosa, E.B., Cohen, L.: Formulae and tables for the calculation of mutual and self-inductance. Bull. Bureau Stand. **5**, 35–50 (1907)
30. Mecke, R., Rathge, C., Fischer, W., Andonovski, B.: Analysis of inductive energy transmission systems with large air gap at high frequencies. In: European Conference on Power Electronics and Applications, 2003, Toulouse
31. Kuyvenhoven, N., Dean, C., Melton, J., Schwannecke, J., Umenei, A.E.: Development of a foreign object detection and analysis method for wireless power systems. In: Proceedings of the IEEE Symposium on Product Compliance Engineering (PSES), San Diego, CA, USA, pp. 1–6, 10–12 October 2011

32. Hui, S.: Planar wireless charging technology for portable electronic products and qi. Proc. IEEE **101**, 1290–1301 (2013)
33. Fu, M., Zhang, T., Zhu, X., Luk, P.C.K., Ma, C.: Compensation of cross coupling in multiple-receiver wireless power transfer systems. IEEE Trans. Ind. Inform. **12**(2), 474–482 (2016). https://doi.org/10.1109/TII.2016.2516906

3D Printer Based Open Source Calibration Platform for Whisker Sensors

Liyou Zhou(✉), Omar Ali, Soumo Emmanuel Arnaud, Eden Attenborough, Jacob Swindell, George Davies, and Charles Fox

School of Computer Science, University of Lincoln, Lincoln, UK
liyou.zhou@outlook.com
https://zenodo.org/records/11081338,
https://github.com/FoR-Group1/OpenWhisker

Abstract. Whisker sensors have been an area of active research in recent years for their interesting use cases in tactile robotics and mammal physiology research. Several attempts have been made to develop open-source versions of the sensor to promote wider adoption. However, the existing calibration solutions for these sensors are highly proprietary, cost-prohibitive and error-prone. In this paper, we present a low-cost open-source calibration and testing platform for whisker sensors based on an off-the-shelf 3D printer. We demonstrate its effectiveness by calibrating a whisker sensor for radial contact distance inference. All artefacts of the design are open-sourced and are fully reproducible.

Keywords: Whisker Sensor · Calibration · Open-Source

1 Introduction

Whisker sensor is a type of tactile sensor that mimics the whiskers of rodents. It has been widely produced to study whiskered mammal physiology [8]. It has also garnered interest in the field of robotics in recent years as a low-cost and effective tactile sensor. [3] shows its use in material defect identification. [10] and [4] used whisker sensors in tactile SLAM tasks.

Existing research initiatives endeavour to build the whisker sensor and testing platform from scratch. This is time-consuming and often involves expensive proprietary hardware. For calibration, [11] and [3] used industrial robot arms to make the controlled movements required. [2] relied on a Yamaha-PXYX closed-source hardware Cartesian robot and a Yamaha RCX 222 controller, together costing several thousand pounds.

The cheap, accurate and repeatable calibration of whisker sensors is crucial for its wider adoption. While open-source sensor designs have been made available to researchers and hobbyists [9], no attempt has been made to provide a

---

This work was partly supported by the Engineering and Physical Sciences Research Council Grant [EP/S023917/1].

solution for sensor calibration. Our project sets out to fill this gap by developing a low-effort open-source whisker calibration and testing platform. We demonstrate the use of an off-the-shelf open-source 3D printer for automated and repeatable calibration.

2 Solution Overview

2.1 Hardware Components

The solution consists of the following hardware components:

1. **3D Printer** A Prusa i3 (MK2) [1] was used as the base for the platform.
2. **Whisker Sensor Mount** A mount is 3D printed to secure the whisker sensor onto the printer bed. The base clamps onto the bed using grub screws. An adaptor is designed to hold the specific shape of the whisker sensor in place and allow for easy removal and replacement. The base features through-holes where the adaptor can be secured via zip ties.
3. **Whisker Sensor** The whisker sensor used in our experimental setup is of a similar design to the one presented in [9]. It consists of a 3d printed whisker shaft with a gel material at the base acting as a hinge. The whisker is $150mm$ long and is tapered from base to the tip. A magnet is attached to the bottom of the shaft and a digital magnetometer is used to sense the displacement of the magnet.
4. **End Effector** Different end effectors can be mounted onto the printer head to make contact with the whisker sensor to carry out various calibration routines. For radial contact distance calibration, a metal ruler is mounted to make point contact with the whisker.

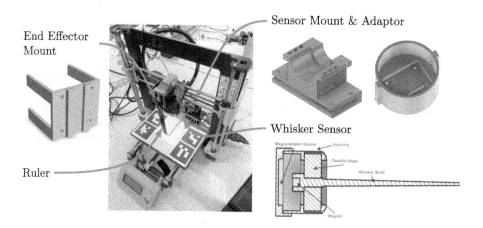

Fig. 1. Calibration Platform Overview

2.2 Software Components

Printer Control. The printer natively supports a GCode [5] interface via a serial connection to a host computer. A Python API is written to interface with the printer and provide easy access to calibration routines. It provides functions to reset the printer to a known state and drive the printer head through a parameterized list of locations. The 3D coordinates of the printer head is also exposed via the API.

Whisker Sensor Interface. The whisker sensor uses a digital magnetometer to sense displacement. The magnetometer is read by a microcontroller and is in turn connected to the host computer via a serial port. The microcontroller prints the x, y and z readings to the serial port in hex format. The magnetometer has 16 bits per channel and is read at 800 Hz.

Different sensors will have different interfaces, hence a ROS 2 abstraction is provided so that the calibration routine can be easily adapted to different sensor designs.

ROS Integration. The sensor and the calibration process, including the 3D printer, are integrated into the ROS 2 framework. The integration has the following main components:

- **whisker_driver_node** Interfaces with the whisker sensor micro-controller via a serial port. Publishes data on a ROS topic.
- **printer_driver_node** Interfaces with the 3D printer, and drives it to go through a calibration sequence upon a ROS service call.
- **whisker_interfaces** Message and service definitions for the drivers.

The nodes utilize standard ROS 2 messages and service definitions where possible. Interoperability is ensured through careful control of the interface messages. As a result, the project benefits from a wide range of open-source ROS 2 tools for data collection and analysis. The `ros2_bag` utility is used to record data for calibration. `foxglove` is used to visualize the data (Fig. 2) during development.

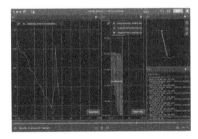

Fig. 2. Live visualization of the sensor system via Foxglove

3 Evaluation

To demonstrate the effectiveness of the calibration platform, a whisker sensor is calibrated for radial contact distance inference. This is a crucial step in using the sensor for tactile SLAM [6,7]. In this type of task, an array of whisker sensors is actuated to whisk back and forth. When the whisker makes contact with an obstacle, location of the contact point is estimated to build a map of the environment as the robot moves around.

Using the ROS 2 **printer_driver_node**, the printer head is driven to make contact with the whisker shaft at a series of known locations. The contact is made in a swift back-and-forth motion to mimic the whisking action. Data from the sensor as well as the 3D coordinates of the printer head is recorded via the ROS 2 **rosbag** utility. Figure 3 shows the raw data collected from 3 consecutive calibration runs. The following analysis focuses on the y channel only as it contains the most significant signals, but the same process can be applied to the x and z channels.

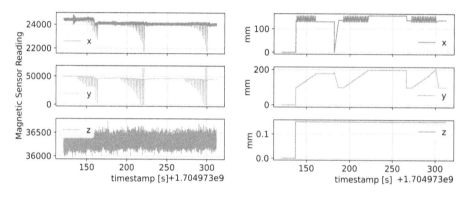

Fig. 3. Raw Magnetometer (left) and 3D Printer Data (right) from 3 consecutive Calibration Routines

3.1 Calibration Data Processing

A low-pass Butterworth filter is first applied to the sensor readings to reduce noise. A comparison of raw and filtered readings is shown in Fig. 4.

Each episode of contact is isolated and extracted from the time series. It is done by detecting a negative gradient in the sensor reading at the start of contact and a positive gradient at the end.

During the routine, the printer head is always moving at a constant speed. In a realistic whisking motion, the shaft rotates around the base and the orthogonal speed $\dot{y}$ is proportional to the distance from the base x and the angular speed $\dot{\theta}$.

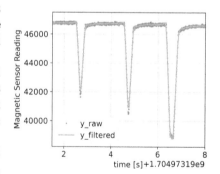

Fig. 4. Raw and Filtered Magnetometer Reading in the Y-Axis

$$\dot{y} = \dot{\theta}x \tag{1}$$

To correct this difference, we scale the derivative of the sensor ($\dot{y}$) reading by a factor of x to simulate a constant angular velocity in all data. The corrected $\dot{y}$ is plotted in Fig. 5a.

There is a clear correlation between $\dot{y}$ and the radial distance of the contact location x. Figure 5b plots the Nth data point in each episode against the contact distance from the base of the whisker x. Note that the relationship is non-linear as suggested by Eq. 1. This is because the deflection in y at the contact point has 2 components: $y_{contact} = y_{twist} + y_{bend}$. y_{twist} is due to twisting of the gel material and y_{bend} is driven by bending of the whisker shaft.

While y_{twist} creates a directly proportional reading at the magnetometer sensor, the bending of the whisker shaft y_{bend} causes non-linearity. Hence, a non-linear model is required to model the relationship between x and $\dot{y}$.

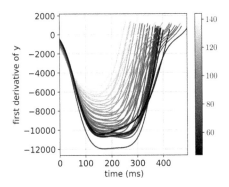 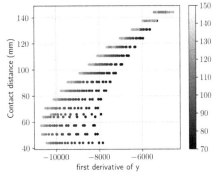

(a) Time series of $\dot{y}$ of each Episode. The colour corresponds to the radial contact distance in mm.

(b) Nth data point of $\dot{y}$ in each episode against the radial contact distance. The colour corresponds to the value of N

Fig. 5. First derivative of y for each episode of contact

Picking the 120th sample in each episode, the relationship between x and $\dot{y}$ is regressed using a third-order polynomial (Fig. 6). The root mean squared error is 5.31 mm for the training set and 4.36 mm for the test set. The largest error (17.4 mm) occurs towards the base of the whisker where the deflection is dominated by y_{twist} and the correlation is weak. If a contact is made within 70 mm from the tip, the contact location can be accurately inferred to within 2 mm by the model.

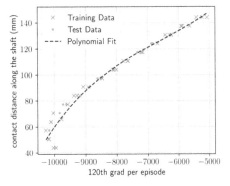

Fig. 6. Polynomial Regression of the relationship between 120th sample of $\dot{y}$ in each episode and the radial contact distance x

The parameters of the polynomial are saved and used in **whisker driver node** to infer the contact location in real time. A polynomial model benefits from low compute resource requirement and is particularly suited for resource-constrained robotic applications with many whisker sensors.

4 Conclusion

As a result of the project, a fully re-producible cost-effective whisker sensor calibration and testing platform is developed and made available via open-source software and hardware. The utilization of a 3D printer as the main component drastically reduces the cost and barrier to reproduction. The integration with ROS 2 provides a convenient interface for data collection and analysis. We demonstrate the accurate and repeatable calibration of the whisker sensor in estimating the radial contact location. There is huge potential for further development based on the project's work, and we hope to see more people joining the effort. All project artefacts are available at https://github.com/FoR-Group1/OpenWhisker.

References

1. Original Prusa i3 MK3S+ | Original Prusa 3D printers directly from Josef Prusa. https://www.prusa3d.com/category/original-prusa-i3-mk3s/
2. Evans, M., Fox, C.W., Pearson, M.J., Lepora, N.F., Prescott, T.J.: Whisker-object contact speed affects radial distance estimation. In: 2010 IEEE International Conference on Robotics and Biomimetics, pp. 720–725, December 2010. https://doi.org/10.1109/ROBIO.2010.5723415
3. Fotouhi, S., Khayatzadeh, S., Pui, W.X., Damghani, M., Bodaghi, M., Fotouhi, M.: Detection of barely visible impact damage in polymeric laminated composites using a biomimetic tactile whisker. Polymers **13**(20), 3587 (2021). https://doi.org/10.3390/polym13203587
4. Fox, C., Evans, M., Pearson, M., Prescott, T.: Tactile SLAM with a biomimetic whiskered robot. In: 2012 IEEE International Conference on Robotics and Automation, pp. 4925–4930, May 2012. https://doi.org/10.1109/ICRA.2012.6224813
5. Kramer, T.R., Proctor, F.M., Messina, E.R.: The NIST RS274NGC Interpreter - Version 3. NIST (August 2000)
6. Lepora, N.F., et al.: Naive Bayes novelty detection for a moving robot with whiskers. In: 2010 IEEE International Conference on Robotics and Biomimetics, pp. 131–136 (December 2010). https://doi.org/10.1109/ROBIO.2010.5723315
7. Pearson, M.J., Fox, C., Sullivan, J.C., Prescott, T.J., Pipe, T., Mitchinson, B.: Simultaneous localisation and mapping on a multi-degree of freedom biomimetic whiskered robot. In: 2013 IEEE International Conference on Robotics and Automation, pp. 586–592 (May 2013). https://doi.org/10.1109/ICRA.2013.6630633
8. Prescott, T., Lepora, N., Mitchinson, B., Pearson, M., Martinez-Hernandez, U., Grant, R.: Active touch sensing in mammals and robots. In: Reference Module in Neuroscience and Biobehavioral Psychology (January 2020). https://doi.org/10.1016/B978-0-12-805408-6.00031-2
9. Stevenson, R., et al.: Open source hardware whisker sensor. In: TAROS (2024)

10. Struckmeier, O., Tiwari, K., Salman, M., Pearson, M.J., Kyrki, V.: ViTa-SLAM: a bio-inspired visuo-tactile SLAM for navigation while interacting with aliased environments. In: 2019 IEEE International Conference on Cyborg and Bionic Systems (CBS), pp. 97–103 (September 2019). https://doi.org/10.1109/CBS46900.2019.9114526
11. Sullivan, J.C., et al.: Tactile discrimination using active whisker sensors. IEEE Sens. J. **12**(2), 350–362 (2012). https://doi.org/10.1109/JSEN.2011.2148114

Mitigating the Time Delay and Parameter Perturbation by a Predictive Extended State Observer-Based Active Disturbance Rejection Control

Syeda Nadiah Fatima Nahri(✉), Shengzhi Du, Barend J. van Wyk, and Tawanda Denzel Nyasulu

Department of Electrical Engineering, Tshwane University of Technology, Pretoria 0001, South Africa
{nahrisnf,dus,vanwykb}@tut.ac.za, 217334420@tut4life.ac.za

Abstract. This paper engages the predictive extended state observer (ESO)-based active disturbance rejection control (ADRC) method to estimate and compensate for parameter perturbations in control systems with time delay. This is important because such variations are certain in real-world control plants (systems). In most systems, time delays and disturbances further challenge the controller system's operation, necessitating a balance between controller performance (transient response) and disturbance rejection operations. Thus, this paper focuses on performance analysis in a specific area of disturbance rejection pertaining to parameter perturbation present in a predictive ESO-based ADRC (PESO-ADRC) control system. Experimental analysis is performed on different systems using the PESO-ADRC method and compared with state-of-the-art methods, such as the delay-based ADRC and the extended state predictor observer (ESPO)-based controller methods. It is observed that the PESO-based ADRC outperformed the delay-based ADRC and ESPO-based controllers by showing tolerance to disturbances, faster estimation, and compensation of parameter perturbation under time-delay control. This predictive method can be used to control the future operations of robotic applications.

Keywords: Predictive ESO-based ADRC (PESO-ADRC) · Parameter Perturbation · Delay-based ADRC · Extended State Predictor Observer (ESPO)-based Controller · Disturbance Compensation

1 Introduction

In control systems theory, a system to be controlled is affected by various uncertainties, such as unmodelled dynamics, nonlinearities, varying time delay, internal parameter perturbation, system disturbances and sensor noise. These uncertainties are broadly characterised as internal and external disturbances [1]. The former disturbance constitutes parametric and dynamic model uncertainties in the controlled object (system), which includes unmodelled system dynamics and perturbation that causes a change in the modelled system parameters. The latter includes outward disturbances, such as sensor noise, input and output system disturbances, and hysteresis with time delay.

Consequently, studying the system's performance depends on how efficiently a control system can overcome the undesirable effects of the aforementioned disturbances, as their encounter with a system is inevitable in the real world, thus forming an important fundamental research topic of disturbance rejection in control systems theory. The systems that are widely controlled cater to the fields of robotics, for example, teleoperation systems [2] and smart automotive industries [3].

The controllers that perform disturbance rejection are classified as model-based and model-free control methods. The model-based control method includes, for example, disturbance observer-based control (DOBC) [4], proportional integral derivative (PID) [5], adaptive and robust control methods [6], and sliding mode controller (SMC) [7]. However, to overcome the model-dependency problem, some model-free or intelligent control methods have been proposed, such as neural network (NN)-based [8], fuzzy logic-based [9], and active disturbance rejection control (ADRC) [10].

Han first recognised the ADRC controller in the mid-1990s [11], followed by further realisation by Gao [12]. This paper focuses on the model-free ADRC that has evolved to efficiently control systems by compensating for the system's non-linearities and delay-related impacts in real-time [13]. Thus, ADRC is known for its effective disturbance rejection of external disturbances. However, in most situations, parameter perturbation under time-delay control is still one of the most significant factors, as process control industries are susceptible to outside-world uncertainties that affect the process model.

1.1 Motivation and Paper Contribution

Various controllers can mitigate the detrimental effects of parametric perturbations and other exogenous disturbances under time-delay control to a certain degree. However, under non-ideal conditions, the system model state experiences modelling errors when subjected to time-varying parameters owing to the effect of environment-changing factors. In light of the aforementioned, some researchers consider the major parameter perturbations an element of the total disturbance to determine the stability requirement in the event that the total disturbance is bounded. Since external disturbances and parameter perturbations are regarded as "total disturbances", it is challenging to investigate how each affects the system independently.

Keeping the above in mind, the major contribution of this paper serves to experimentally test the performance of one of the recently published works in [14] with regard to the system's parameter perturbation (variation). Previously, the proposed predictive extended state observer-based ADRC (PESO-ADRC) in [14] displayed stable transient response, robustness, and quick recovery to various external disturbances (input disturbance, output disturbance, including white noise) in the presence of time delay. This is achieved by using the controller design's predictive feature to predict unknown system dynamics that were caused during the delay, followed by their compensation through disturbance rejection under time-delay control. However, performance analysis of the system's parameter perturbations was not part of the study. Thus, this paper focuses on studying the behaviour of the PESO-ADRC [14] for systems subjected to parameter perturbations, system output delay and an external disturbance added in progressive stages.

The remainder of the paper is organised as follows: Sect. 2 presents related works in literature, followed by problem formulation and preliminary concepts addressed in Sect. 3. Section 4 discusses the results obtained from experiments performed on perturbed System 1 and System 2 using the PESO-ADRC, and other state-of-the-art methods. Section 5 concludes this paper by proposing future recommendation.

2 Related Work

A model-based method presented in [4] overviews the various disturbance observer control methods by addressing different internal and external disturbance constraints using the disturbance estimation techniques applied to distinct electrical and mechanical systems applications. Further, an SMC-based tracker is generated for a perturbed robotic manipulator [15], and perturbation is overcome without prior knowledge of the uncertainty upper bound using a parameter-tuning adaptive control law. Furthermore, a robust H_∞-fuzzy logic control approach is proposed to improve the tracking performance of a mobile robot controlled by DC motors exposed to model parameter uncertainties [9]; however, limitations on the input signal and structure's model need further work. In addition, a linear adaptive controller was developed for tracking control of robotic manipulators with uncertain input parameters and disturbances by converting the perturbed robotic system to a double integrator system with bounded disturbance [16]; however, further work is needed to deal with non-minimum phase systems.

Moreover, ADRC shows robust behaviour to unmodelled dynamics and small parameter variations; however, large parameter variations in second-order systems significantly impact system stability. In [7], a combination of predictive-based ADRC with the SMC approach attained good position tracking for a perturbed servo motor model under time delays. However, the presence of high mechanical resonances was ignored, thus limiting the proposed method's gain tuning factor. In addition, in [17], strong robustness was shown under parameter perturbation scenarios for an autonomous land vehicle based on ADRC and differential flatness theory. However, further analysis of time-delayed systems was recommended.

In addition, ADRC control methods that deal with overcoming the system parameter perturbation problem include, for example, the active disturbance rejection and modified twice optimal control (ADRMTOC) [18], robust-based ADRC [19], and automatic current-constrained double loop ADRC for the electro-hydrostatic actuator (EHA) [20]. These techniques offered different approaches to handling disturbances, including system internal disturbance. In [18], the ADRMTOC design showed robustness and quick recovery when system parameters were subjected to random perturbation within a certain range. However, further algorithm improvement was encouraged to improve the initial response curve.

Robust rejection of uncertain system parameters was attained by introducing two gain blocks, one in series and the other in parallel with the plant output [19]. However, a challenge due to uncertainty in a non-minimum phase system was noted. Further, a cascade double-loop ADRC design, including the position and speed-current control loops, helped improve the performance of EHA using the single perturbation theory, thus enabling the usage of a few controller-tuned parameters [20]; however, performance based on the time-delay factor demands future study. Thus, it has been seen that for efficient disturbance rejection, the control system design faces a trade-off between reference tracking (control performance) and the disturbance rejection property. Hence, this paper puts more effort into this regard by investigating the parameter perturbation effect on systems controlled by the PESO-ADRC method proposed in [14].

3 Problem Formulation and Preliminaries

3.1 System Description

In control theory literature, higher-order systems are analysed using their second-order approximation; thus, most controllers focus on simpler or lower-order systems for better accessible analysis and their application to a real controlled plant system. Hence, this paper uses similar linear systems subjected to parameter perturbation as a benchmark.

Consider the following unperturbed ($G(s)$) and perturbed system ($\tilde{G}(s)$) represented using the transfer function and state-space representation:

$$G(s) = \frac{k}{as^2 + bs + c}; \quad \begin{cases} \dot{x} = Ax + Bu \\ y = Cx \end{cases} \quad (1)$$

$$\tilde{G}(s) = \frac{\tilde{k}}{\tilde{a}s^2 + \tilde{b}s + \tilde{c}}; \quad \begin{cases} \dot{\tilde{x}} = \tilde{A}\tilde{x} + \tilde{B}u \\ \tilde{y} = \tilde{C}\tilde{x} \end{cases} \quad (2)$$

In the transfer function $\tilde{G}(s)$, $\tilde{k} = k + \Delta k$, $\tilde{a} = a + \Delta a$, $\tilde{b} = b + \Delta b$ and $\tilde{c} = c + \Delta c$. 's' denotes the Laplace transform. Terms k, a, b and c represent the original system dynamics, followed by the second terms Δk, Δa, Δb and Δc that indicate parameter changes from k, a, b and c. Therefore $\tilde{k}$, $\tilde{a}$, $\tilde{b}$ and $\tilde{c}$ are the perturbed terms of a given perturbation system. Further, Δk, Δa, Δb and Δc. These are also called the perturbation coefficients, which indicate a shift in the controlled system's model parameters after a certain time period.

In the state-space representation, matrices $\tilde{A}$, $\tilde{B}$ and $\tilde{C}$ are the perturbed system, input and output matrices, respectively. In the observer canonical form of state space, matrices A', B' and C' are represented by the following equation,

$$\begin{cases} \tilde{A} = \begin{bmatrix} \tilde{A}_{11} & 1 \\ \tilde{A}_{12} & 0 \end{bmatrix}, \tilde{B} = \begin{bmatrix} 0 \\ \tilde{B}_{12} \end{bmatrix} \\ \tilde{C} = \begin{bmatrix} 1 & 0 \end{bmatrix} \end{cases} \quad (3)$$

where $\tilde{A}_{11} = A_{11} + \Delta A_{11}$, $\tilde{A}_{12} = A_{12} + \Delta A_{12}$ and $\tilde{B}_{12} = B_{12} + \Delta B_{12}$. The matrix elements A_{11}, A_{12}, and B_{12} are the real system elements, and ΔA_{11}, ΔA_{12} and ΔB_{12} indicate the deviation from the real system elements due to the effect of system parameter perturbations.

3.2 Delay-Based ADRC Method

An essential component of the ADRC is the extended state observer (ESO), which provides real-time estimation of system states and disturbances (nonlinear dynamics of the manipulator, internal and external disturbances). This is followed by active compensation of total disturbance using the feedback loop. The other functional blocks include the Tracking differentiator (TD) and nonlinear state error feedback controller (NLSEF) [12]. TD provides a softened signal with an estimated derivative signal, and the NLSEF generates the control signal u_0. Figure 1 is a modified structure of the standard ADRC, and is called the delay-based ADRC, wherein the delay present in the input path of ESO allows for time-delay compensation when the system's output is also delayed [21]. In this study, the Fig. 1 control method is used for experimental analysis as one of the methods to control the parameter-perturbed system.

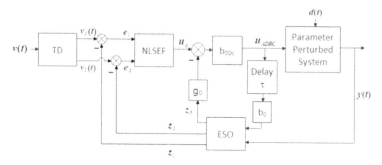

Fig. 1. Delay-based ADRC structure

The TD, NLSEF, ESO, and the control law (u_{ADRC}) are represented by the following equations, respectively [14],

$$\begin{cases} \dot{v}_1 = v_1 + hv_1 \\ \dot{v}_2 = v_2 + h\text{fhan}(v_1 - r, v_2, r_0, h) \end{cases} \quad (4)$$

$$\begin{cases} e_1 = v_1 - z_1 \\ e_2 = v_2 - z_2 \\ u_o = -\text{fhan}(e_1, c \cdot e_2, r, h_1) \end{cases} \quad (5)$$

$$\begin{cases} e = z_1 - y \\ \dot{z}_1 = z_2 - \beta_{01}e \\ \dot{z}_2 = z_3 - \beta_{02}\text{fe} + b_0 u_{ADRC}(t - \tau) \\ \dot{z}_3 = -\beta_{03}\text{fe}_1 \end{cases} \quad (6)$$

$$u_{ADRC} = (u_0 - z_3 g_0) b_{0Dc} \quad (7)$$

β_{01}, β_{02} and β_{03} refer to the ESO gains. The parameters r_0 and h are the transition process speed and the simulation step of the TD, respectively. fhan($v_1 - r, v_2, r_0, h$) is the nonlinear function that enables quick convergence of v_1 (tracking signal) to v

(input signal) with minimal overshoot, which is achieved by adjusting r_0 to r. The stepwise relations of the 'fhan' function is provided in [12]. e_1 and e_2 are fed as inputs to NLSEF. From ESO equation, the nonlinear feedback functions, fe = fal(e, 0.25, h_{ESO}) and fe$_1$ = fal(e, 0.5, h_{ESO}), where error (e) will reach zero soon for α = 0.25 and 0.5 ($\alpha < 1$), with the 'fal' function relations provided in [12, 14]. The simulation step size of ESO is given by h_{ESO}. z_1 and z_2 are the estimated state variables, and z_3 is the estimated signal of total disturbances. The controller parameters to be tuned are c (damping coefficient), r (control gain) and h_1 (simulation step). Gain factors such as g_0, b_0 and b_{0Dc} are also tuned to obtain the desired transient response. The parameter-perturbed system used is a second-order system as an example to show the ADRC operation.

3.3 ESPO-Based Control Method

The full-state feedback pole assignment technique is used to compute the controller parameter K for an ESPO-based controller method. Hence, it is important to note this controller's dependency on the model of the system under control [22]. In this study, experimental tests are conducted on second-order parameter perturbed systems used as a template to study the ESPO-based controller operation (in Fig. 2). Hence, from Fig. 2, the extended state '$\hat{d}$ (t)' performs the estimation and compensation of the total external disturbances.

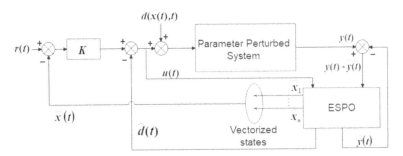

Fig. 2. ESPO-based mechanism

The following set of equations mathematically represents the ESPO [22],

$$\begin{bmatrix} \dot{\hat{x}}(t) \\ \dot{\hat{d}}(t) \end{bmatrix} = A_e \begin{bmatrix} \hat{x}(t) \\ \hat{d}(t) \end{bmatrix} + e^{A_e \tau} L_e(y(t) - \hat{y}(t)) + B_e u(t) \tag{8}$$

$$\hat{y}(t) = C_e^T \begin{bmatrix} \hat{x}(t-\tau) \\ \hat{d}(t-\tau) \end{bmatrix} + C_e^T \int_{t-\tau}^{t} e^{A_e(t-s)} L_e(y(s) - \hat{y}(s)) ds \tag{9}$$

$$A_e = \begin{bmatrix} A & B \\ 0 & 0 \end{bmatrix}, B_e = \begin{bmatrix} B \\ 0 \end{bmatrix}, C_e = \begin{bmatrix} C \\ 0 \end{bmatrix}, L_e = \begin{bmatrix} 3\omega_e \\ 3\omega_e^2 \\ \omega_e^3 \end{bmatrix} \tag{10}$$

The ESPO's control law is given in the following equation,

$$u(t) = -K^T \left(\hat{x}(t) - r(t) \right) - \hat{d}(t) \quad (11)$$

$r(t)$ is the reference signal, ω_e is the ESPO bandwidth, τ is the time delay, $e^{A_e \tau}$ and L_e are the state transition and parameter matrices, respectively. $\hat{x}(t)$, $\hat{d}(t)$ and $\hat{y}(t)$ are the estimated system, disturbances, and output states, respectively. The difference $(y(t) - \hat{y}(t))$ provides the estimation error.

3.4 Predictive ESO-Based ADRC Design

The predictive ESO-based ADRC mechanism shown in Fig. 3 is developed by cascading features of the model-free delay-based ADRC controller (in Fig. 1) with the model-based ESPO controller (in Fig. 2) [14]. This controller method uses its predictive feature to conquer the unknown system dynamics caused during the delay, followed by their rejection as a disturbance ($u_{ADRC}(t) - \hat{d}(t)$), by using the advantages of ADRC.

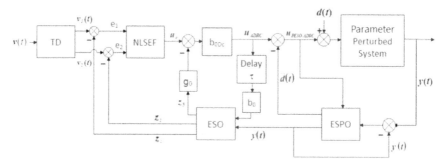

Fig. 3. Predictive ESO-based ADRC (PESO-ADRC) structure

The PESO-ADRC's control law ($u_{PESO-ADRC}(t)$) and its ESO equations [14], given by,

$$\begin{cases} u_{PESO-ADRC}(t) = u_{ADRC}(t) - \hat{d}(t) \\ u_{PESO-ADRC}(t) = (u_0(t) - z_3 g_0)b_{0Dc} - \hat{d}(t) \\ u_{PESO-ADRC}(t) = (-\text{fhan}(e_1, c \cdot e_2, r, h_1) - z_3 g_0)b_{0Dc} - \hat{d}(t) \end{cases} \quad (12)$$

$$\begin{cases} e = z_1 - \hat{y} \\ \dot{z}_1 = z_2 - \beta_{01} e \\ \dot{z}_2 = z_3 - \beta_{02}\text{fe} + b_0 u_{ADRC}(t - \tau) \\ \dot{z}_3 = -\beta_{03}\text{fe}_1 \end{cases} \quad (13)$$

In Sect. 4, the controller design algorithm in Fig. 3 is studied on second-order parameter perturbed systems to compare with controller algorithms in Fig. 1 and Fig. 2.

4 Experiments and Results

In this section, two different perturbed second-order systems are used as examples to present their response under the control of different controller algorithms. The systems are characterised as System 1 (Type 0 system) and System 2 (Type 1 system) systems. Each of the systems is controlled by the delay-based ADRC, ESPO-based control, and PESO-ADRC method, followed by their experimental analysis under different scenarios: with system parameter perturbation, with system parameter perturbation and delay, and with system parameter perturbation along with system input disturbance (also called external disturbance) and delay.

The transfer functions of System 1 and System 2, without and with parameter perturbations $G(s)$ and $\tilde{G}(s)$ respectively, are given in the following equation,

$$\begin{cases} G_1(s) = \frac{2}{s^2+3s+2} = \frac{1}{(0.5s+1)(s+1)} \\ G_2(s) = \frac{1}{s^2+s} = \frac{1}{s(s+1)} \end{cases} \rightarrow \begin{cases} \tilde{G}_1(s) = \frac{1.5123}{s^2+2.6087s+1.5123} \\ \tilde{G}_2(s) = \frac{0.7561}{s^2+0.8696s} \end{cases} \quad (14)$$

Table 1 gives the values of the unperturbed and perturbed matrix elements for both System 1 and System 2 in the case of a 15% incrementation of their respective time constant values. The reference input signal is a step signal of final value 1 for all the experiments. Moreover, parameter perturbation is applied at 20 s. Hence, System 1 and System 2 behave unperturbed for 19 s, followed by the system's response deviation from its normal behaviour starting from 20 s.

Table 1. Parameter values for systems under experiment

System	System Matrix	Input Matrix	Output Matrix
$G_1(s)$	$A = \begin{bmatrix} -3 & 1 \\ -2 & 0 \end{bmatrix}$	$B = \begin{bmatrix} 0 \\ 2 \end{bmatrix}$	$C = [1\ 0]$
$G_2(s)$	$A = \begin{bmatrix} -1 & 1 \\ 0 & 0 \end{bmatrix}$	$B = \begin{bmatrix} 0 \\ 1 \end{bmatrix}$	$C = [1\ 0]$
$\tilde{G}_1(s)$	$\tilde{A} = \begin{bmatrix} -2.6087 & 1 \\ -1.5123 & 0 \end{bmatrix}$	$\tilde{B} = \begin{bmatrix} 0 \\ 1.5123 \end{bmatrix}$	$\tilde{C} = [1\ 0]$
$\tilde{G}_2(s)$	$\tilde{A} = \begin{bmatrix} -0.8696 & 1 \\ 0 & 0 \end{bmatrix}$	$\tilde{B} = \begin{bmatrix} 0 \\ 0.7561 \end{bmatrix}$	$\tilde{C} = [1\ 0]$

Parameters values of the delay-based ADRC (Fig. 1) include $\beta_{01} = 100$, $\beta_{02} = 300$, $\beta_{03} = 1000$, $h_{ESO} = 0.01$, and its remaining values for the two system types were obtained from the optimization tool using the Genetic Algorithm in MATLAB R2020a version with the stopping criteria specified in [14]. The obtained optimal set of parameter values used in both delay-based ADRC and the PESO-ADRC methods are

$[c, h_1, r, g_0, b_0, b_{0DC}] = [0.7645, 1.0831, 56.0350, 0.3097, 3.1303, 1.0314]$ for System 1; and $[1.0710, 0.9340, 41.5480, 1.0000, 1.0000, 1.0000]$ for System 2. Further, for the ESPO-based controller, the parameter $K = [49/72 \ -1/3]$ and $[1/4 \ 0]$ for System 1 and System 2, respectively [14]. The observer bandwidth (ω_e) is tuned as per the system and controller used to reach a steady state.

In this section, ADRC indicated in the experimental response plots from Fig. 4 to Fig. 6 corresponds to the delay-based ADRC method seen in Fig. 1. The comparison is based on transient response analysis and system parameter perturbation compensation observed from their response curves. Further, the analysis also uses the integral of time-weighted absolute error (ITAE) as a performance index measure. The smaller the ITAE value, the less the controlled system error.

4.1 Effect of Parameter Perturbation

This subsection analyses the control of perturbed System 1 and System 2 using the delay-based ADRC, ESPO, and PESO-ADRC-based controllers. Their respective response curves are shown in Fig. 4. For PESO-ADRC and ESPO methods, $\omega_e = 12$ and 7 for System 1, and $\omega_e = 10$ and 5 for System 2, respectively.

As seen in Fig. 4(a) and 4(b), for System 1, the PESO-ADRC method at startup showed a slightly higher overshoot than the delay-based ADRC method. However, the PESO-ADRC method showed a decent decrease in overshoot when perturbation was applied at 20 s. Under the impact of parameter perturbation, the percentage of improvement relating to a decrease in overshoot for the PESO-ADRC method for System 1 is around 29% and 55% and around 15% and 73% for System 2 relative to the delay-based ADRC and ESPO-based methods, respectively. These values mentioned earlier were obtained from the following equation,

$$P_i(\%) = \frac{OS_0 - OS_1}{OS_0} \quad (15)$$

P_i is the percentage of improvement by the decrease in overshoot. Due to the application of parameter perturbation at 20 s, the overshoot arising is expressed as an overshoot (%) value for delay-based ADRC or the ESPO method given by OS_0, and for the PESO-ADRC method is given by OS_1.

From both System 1 and System 2, it was observed that the PESO-ADRC method showed more robustness to the parameter perturbation effect. The transient response curve showed stability by attaining quick recovery and reaching a steady state faster than delay-based ADRC and ESPO-based methods. Further, for the PESO-ADRC, the time width of the overshoot due to perturbation at 20 s was reduced by around 50% (Fig. 4(a) and 4(b)). Moreover, after the parameter perturbation was applied, for System 1, the ESPO control attained zero-steady state error (SSE) later at around 26 s, compared to the PESO-ADRC control method that attained zero SSE at around 23 s (Fig. 4(a)). However, for System 2, the ESPO method did not attain zero SSE (Fig. 4(b)). Table 2 shows ITAE values for the response curves of System 1 and System 2.

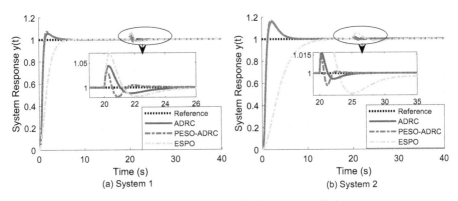

Fig. 4. Effect of Parameter Perturbation on controlled systems

Table 2. ITAE performance index for systems with Parameter perturbation

ITAE	ADRC	PESO-ADRC	ESPO
System 1	1.7070	1.2400	5.0410
System 2	2.0390	1.9140	17.3600

4.2 Effect of Parameter Perturbation in the Presence of System Time-Delay

In this experiment, the PESO-ADRC method is contrasted with the delay-based ADRC and ESPO-based control methods to study the control of perturbed systems in the presence of a system time delay of $\tau = 0.1$ s. Figure 5 illustrates the respective system responses for this scenario. For PESO-ADRC and ESPO methods, $\omega_e = 20$ and 6 for System 1, and $\omega_e = 26$ and 5 for System 2, respectively.

Under time-delay control, the predictive feature of the PESO-ADRC method reduced the overshoot at the initial period of the response curve in Fig. 5(a); this slight overshoot was observed in Fig. 4(a). Further, the time width of the overshoot under the influence of parameter perturbation indicates almost a 50% decrease for the PESO-ADRC method compared to the delay-based ADRC method. Furthermore, the zoomed-in plots of Fig. 5(a) and 5(b) show a decrease in overshoot using the PESO-ADRC method, i.e., the P_i (%) relating to a decrease in overshoot (from Eq. (15)) for System 1 is around 10% and 39%, and for System 2 is around 10% and 65% compared to the delay-based ADRC and ESPO control methods, respectively. Thus, this indicated faster recovery and better compensation performance using the PESO-ADRC method.

Further, Fig. 5(a) and 5(b) show that the perturbed systems under PESO-ADRC control reached a steady state earlier, at around 22 s, compared to the delay-based ADRC and the ESPO controller methods. Furthermore, the response curve of the perturbed system under ESPO control could not attain zero SSE after the application of parameter perturbation (refer to Fig. 5(b)). Table 3 gives the ITAE values of the response curves of System 1 and System 2 for the three different control methods under analysis. Therefore,

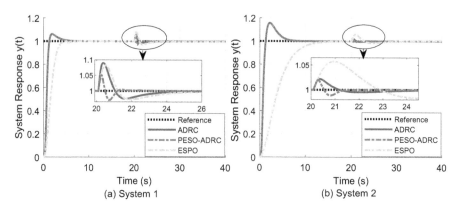

Fig. 5. Effect of Parameter Perturbation on controlled systems with time-delay control

the PESO-ADRC method demonstrated better performance response under the effect of parameter perturbation with system time delay.

Table 3. ITAE performance index for systems with Parameter perturbation and time-delay

ITAE	ADRC	PESO-ADRC	ESPO
System 1	1.7940	1.1790	5.7470
System 2	2.0460	1.7640	17.1000

4.3 Effect of Parameter Perturbation in the Presence of External Disturbance and System Time-Delay

In this experiment, the control of the perturbed systems under a time-delay case was further challenged by adding an external disturbance $d(x(t), t)$, given by the equation,

$$d(x(t), t) = -0.2e^{\cos x_2} + 0.1u + 0.2\text{sign}(\sin\frac{\pi}{8t}) \tag{16}$$

For PESO-ADRC and ESPO methods, $\omega_e = 37$ and 5 for System 1, and $\omega_e = 36$ and 12 for System 2, respectively. Figure 6 provides the system responses when the controlled perturbed systems are subjected to both time delay and external disturbance.

In Fig. 6(a) and 6(b), the perturbed system, in the presence of external disturbance and system time delay, shows an oscillatory response under the control of delay-based ADRC and ESPO alone, which can cause harm to the system when the parameters stay within a certain range of perturbation, unlike the PESO-ADRC method that shows an acceptable stable behaviour by attaining zero SSE. Further, PESO-ADRC showed an overshoot decrease at the start of the response curve by about 44% (System 1) and 21% (System 2) compared to the delay-based ADRC method. Furthermore, the key observation is that under the effect of parameter perturbation, the P_i (%) relating to

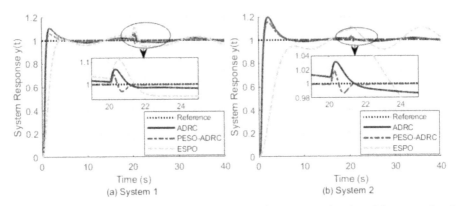

Fig. 6. Effect of Parameter Perturbation on controlled systems under time-delay control and external disturbance

the decrease in overshoot (from Eq. (15)) is around 34% for System 1 (Fig. 6(a)) and 41% for System 2 (Fig. 6(b)) when compared to the delay-based ADRC method. Hence, PESO-ADRC showed stability and decent resilience to parameter perturbation, external disturbance, and system time delay. Table 4 compares ITAE values for perturbed systems under different control methods.

Table 4. ITAE performance index for systems with Parameter perturbation, time delay and external disturbance

ITAE	ADRC	PESO-ADRC	ESPO
System 1	10.8500	1.5520	25.5000
System 2	8.7370	2.1230	54.8900

5 Conclusion

This study focuses on a specific target area of the parameter perturbation phenomenon. By incorporating this experiment into the formerly proposed predictive ESO-based ADRC control method (PESO-ADRC), we provided empirical evidence supporting the robustness and effectiveness of the PESO-ADRC algorithm in mitigating the effects of disturbances caused by parameter perturbations in systems with time delay.

Three different controller algorithms, the delay-based ADRC method, the extended state predictor observer method (ESPO), and the PESO-ADRC method were used to control the specified systems (System 1 and System 2). The PESO-based ADRC control method outperformed the other state-of-the-art methods by exhibiting stability in transient response, efficient robustness, estimating and compensating for the existing disturbance and parameter perturbation impacts under time-delay control. Hence, this controller shows an improved disturbance rejection mechanism. Furthermore, this work offered realisable transfer functions to analyse the performance of predictive ADRC-based systems with significant parameter perturbations.

Future work should include the experimental analysis of higher-order systems, such as a robotic manipulator subjected to parameter perturbation and time delay effects.

Acknowledgements. This work is based on the research supported in part by the National Research Foundation of South Africa (grant number PSTD2204143453).

References

1. Raja, G.L., Mohammad, S.: Introductory chapter: introduction to disturbance rejection control. In: Disturbance Rejection Control. IntechOpen (2023). https://doi.org/10.5772/intechopen.112020
2. Nahri, S.N.F., Du, S., Van Wyk, B.J.: A review on haptic bilateral teleoperation systems. J. Intell. Robot. Syst. **104**(1), 1–23 (2022). https://doi.org/10.1007/s10846-021-01523-x
3. Darvish, K., et al.: Teleoperation of humanoid robots: a survey. IEEE Trans. Robot. **39**(3), 1706–1727 (2023). https://doi.org/10.1109/TRO.2023.3236952
4. Chen, W.-H., Yang, J., Guo, L., Li, S.: Disturbance-observer-based control and related methods—an overview. IEEE Trans. Ind. Electron. **63**(2), 1083–1095 (2015). https://doi.org/10.1109/TIE.2015.2478397
5. Rawa, M., et al.: Disturbance rejection-based optimal PID controllers for new 6ISO AVR systems. Fractal Fractional **7**(10), 765 (2023). https://doi.org/10.3390/fractalfract7100765
6. Xia, J., Li, Z., Yu, D., Guo, Y., Zhang, X.: Robust speed and current control with parametric adaptation for surface-mounted PMSM considering system perturbations. IEEE J. Emerg. Sel. Top. Power Electron. **9**(3), 2807–2817 (2020). https://doi.org/10.1109/JESTPE.2020.3015288
7. Li, P., Wang, L., Zhu, G., Zhang, M.: Predictive active disturbance rejection control for servo systems with communication delays via sliding mode approach. IEEE Trans. Ind. Electron. **68**(12), 12679–12688 (2020). https://doi.org/10.1109/TIE.2020.3039203
8. Yao, W., Wang, C., Sun, Y., Zhou, C.: Robust multimode function synchronization of memristive neural networks with parameter perturbations and time-varying delays. IEEE Trans. Syst. Man Cybern. Syst. **52**(1), 260–274 (2020). https://doi.org/10.1109/TSMC.2020.2997930
9. Ahmad, N.S.: Robust h∞-fuzzy logic control for enhanced tracking performance of a wheeled mobile robot in the presence of uncertain nonlinear perturbations. Sensors **20**(13), 3673 (2020). https://doi.org/10.3390/s20133673
10. Fareh, R., Khadraoui, S., Abdallah, M.Y., Baziyad, M., Bettayeb, M.: Active disturbance rejection control for robotic systems: a review. Mechatronics **80**, 102671 (2021). https://doi.org/10.1016/j.mechatronics.2021.102671
11. Han, J.Q.: Auto disturbance rejection controller and it's applications. Control Decis. **13**(1), 19–23 (1998)
12. Han, J.: From PID to active disturbance rejection control. IEEE Trans. Ind. Electron. **56**(3), 900–906 (2009). https://doi.org/10.1109/TIE.2008.2011621
13. Feng, H., Guo, B.-Z.: Active disturbance rejection control: old and new results. Annu. Rev. Control. **44**, 238–248 (2017). https://doi.org/10.1016/j.arcontrol.2017.05.003
14. Nahri, S.N.F., Du, S., van Wyk, B.J.: Predictive extended state observer-based active disturbance rejection control for systems with time delay. Machines **11**(2), 144 (2023). https://doi.org/10.3390/machines11020144

15. Mobayen, S., Mofid, O., Din, S.U., Bartoszewicz, A.: Finite-time tracking controller design of perturbed robotic manipulator based on adaptive second-order sliding mode control method. IEEE Access **9**, 71159–71169 (2021). https://doi.org/10.1109/ACCESS.2021.3078760
16. Nguyen, P.D., Nguyen, N.H., Nguyen, H.T.: Adaptive control for manipulators with model uncertainty and input disturbance. Int. J. Dyn. Control **11**(5), 2285–2294 (2023). https://doi.org/10.1007/s40435-023-01115-7
17. Xia, Y., Pu, F., Li, S., Gao, Y.: Lateral path tracking control of autonomous land vehicle based on ADRC and differential flatness. IEEE Trans. Ind. Electron. **63**(5), 3091–3099 (2016). https://doi.org/10.1109/TIE.2016.2531021
18. Wang, X., Zhou, Y., Zhao, Z., Wei, W., Li, W.: Time-delay system control based on an integration of active disturbance rejection and modified twice optimal control. IEEE Access **7**, 130734–130744 (2019). https://doi.org/10.1109/ACCESS.2019.2939905
19. Zachi, A.R., Correia, C.A.M., Filho, J.L.A., Gouvêa, J.A.: Robust disturbance rejection controller for systems with uncertain parameters. IET Control Theory Appl. **13**(13), 1995–2007 (2019). https://doi.org/10.1049/iet-cta.2018.5291
20. Yang, R., Ma, Y., Zhao, J., Zhang, L., Huang, H.: Automatic current-constrained double loop ADRC for electro-hydrostatic actuator based on singular perturbation theory. In: Actuators, vol. 12, p. 381. MDPI (2022). https://doi.org/10.3390/act11120381
21. Zhao, S., Gao, Z.: Modified active disturbance rejection control for time-delay systems. ISA Trans. **53**(4), 882–888 (2014). https://doi.org/10.1016/j.isatra.2013.09.013
22. Xue, W., Liu, P., Chen, S., Huang, Y.: On extended state predictor observer based active disturbance rejection control for uncertain systems with sensor delay. In: 2016 16th International Conference on Control, Automation and Systems (ICCAS), pp. 1267–1271. IEEE (2016). https://doi.org/10.1109/ICCAS.2016.7832475

Accessibility Framework for Determining Collisions and Coverage for Radiation Scanning

Joshua Bettles[1]([✉])[iD], Andrew West[1][iD], Jeremy Andrew[2][iD], Iain Darby[3,4][iD], and Barry Lennox[1][iD]

[1] Manchester Centre for Robotics and AI, The University of Manchester, Manchester M13 9PL, UK
`joshua.bettles@manchester.ac.uk`
[2] Nuclear Restoration Services Ltd. Dounreay, Thurso, Caithness KW14 7TZ, UK
[3] National Nuclear Laboratory, Chadwick House, Birchwood Park, Warrington, Cheshire WA3 6AE, UK
[4] School of Physics and Astronomy, University of Glasgow, Glasgow G12 8QQ, UK

Abstract. Radioactive contamination monitoring is an important part of radiological protection. Automation of surface monitoring poses difficulties, with a major challenge being determining the coverage of a radiation probe over an object in close proximity without collision. We propose a new accessibility framework to determine if radiation probes, modelled as convex hulls, collide with 3D point clouds representing the objects. We explain how to structure and analyze point clouds to extract properties such as the surface normal for each point. Our method for approximating radiation probes is demonstrated using the BP4 probe. This approach models both the probe and the sensor's effective scan volume with geometric primitives, providing a computationally efficient way to detect collisions with flat surfaces. The accessibility assessment builds on common methods within computer science for determining intersections. A laptop in various positions was used to demonstrate that the framework can efficiently categorise points as accessible or inaccessible, identifying unscannable regions. The output of this framework can then be used to plan collision-free paths over objects and will be the foundation of a robotic survey assistant.

Keywords: Accessibility Assessment · Contamination Monitoring · Discrete Collision Detection · Health Physics · Point Cloud · Nuclear

1 Introduction

Within the nuclear industry, many roles involve possible contact with radioactive materials, from laboratory experimentation to decommissioning work; these activities take place in controlled or supervised areas. The scientists and engineers who work in these areas require the use of Personal Protective Equipment

(PPE). This PPE protects individuals and reduces the possibility of spreading contamination from high-level controlled zones to lower zones. To limit the spread of contamination, the removal of PPE must be carried out in buffer zones and at barriers between controlled areas. All sites have contamination monitoring equipment and health physics specialists situated at each egress point between buffer zones, barriers and controlled areas to perform examinations of individuals to ensure that no residual contamination is present [9]. Several reports [1,14] have highlighted the declining availability of the radiological protection workforce. Robotics could alleviate the demand for roles such as health physics specialists by automating these mundane tasks.

Small and large article transfer monitors [4] and full body contamination monitors [18] are frequently used for contamination monitoring at these barriers but are limited in their application. Many systems are single purpose, i.e. only used for scanning humans or objects of a certain size, and for use detecting β/γ contamination due to the short mean free path of α. Automation of contamination monitoring poses a challenging application for robotics. For example, robotic manipulators equipped with radiation probes could perform autonomous monitoring operations, and advances in perception and control could allow for the system to be used on any object, material, or personnel presented.

Radioactive contamination monitoring using robotics is a multi-faceted problem which is a growing field in the literature. Recent work by White et al. [21] used an industrial robotic manipulator to map and profile radioactive sources on a flat surface. Whilst work by Mauer and Kawa [8] used a manipulator and forklift to characterise large building surfaces, performing at a distance of 1 cm whilst avoiding unknown obstacles. These systems are limited by their application, with [8] constrained to flat surfaces and paths. Monk et al. developed a low-cost system which was able to scan over the tops of objects and identify source locations, this was able to perform over various shaped objects however only performed these tasks in a line, not completely covering the object [10]. The work presented differs from the previous approaches as it aims to be used over complex geometries, where accessibility is considered to maximise coverage by a given probe. It will be tested on various complex geometries that will likely be found at barriers.

Contamination monitoring using manipulators is analogous to many surface following approaches. Many industries are concerned with the topic of the surface following from spray-painting, polishing, laser scanning, and more recently medical applications. Surface following can be divided into two broad categories: in-contact or at-distance. Each of these follows a similar approach of scanning the surface to create a point cloud, generating a path, and then executing a task along that path over the Volume of Interest (VoI), differing only in their control strategies, overlap, collision avoidance and stand-off distance. In-contact is common in the field of medical ultrasound, Graumann et al. utilised a depth camera to capture a point cloud which was used to generate trajectories over a convex hull representing the VoI [2]. A review of robotic medical ultrasound imaging and their levels of autonomy [20] concluded that there is a current limitation in robust and reliable navigation as well as the safety strategies used.

At-distance systems need to work in close proximity to the surface whilst avoiding collisions with the object. It has been demonstrated that high levels of coverage can be achieved through projecting 3D meshes created from point clouds to 2D for coverage planning [7] and was demonstrated on flat and curved surfaces. A mode-switching motion controller [12] for vehicle inspection can swap between surfaces following approaches. Obstacle avoidance was performed by moving to a safe distance away during the scan path [11]. Collision avoidance during the motion was considered. In [11] the motion did not follow the surface of the obstacle resulting in decreased coverage. This cannot occur in contamination monitoring, both works considered the size of the scanner to assess the coverage of the surfaces. However, with the complex geometries of radiation probes and the objects they will scan, the framework will place accessibility and coverage at the forefront.

Radiation scanning requires the detector to be in close proximity, 10–20 mm to the surface. Collisions are strictly prohibited as they could result in contamination or damage to the sensor, invalidating the sensor readings. Furthermore, contact with the object or personnel could spread contamination or potentially cause damage to the object or cause injury to the individual.

None of the previous work looked at the geometry of the tools and their accessibility to cover areas of a surface. This framework aims to fill this gap by considering the geometry of the probes and the effective scan volumes to determine the accessibility and coverage of objects whilst ensuring no collisions. The framework is able to analyse and identify areas of residual contamination which would require further examination.

Automating the monitoring process can be broken down into 3 stages: Data Processing, Motion Planning and Control, and Human Robot Interaction (HRI). The focus of this paper will be to develop the data processing framework. The main contribution of this paper is a new framework which:

- Enables modelling of the radiation probe and effective scan volume.
- Expands on current discrete collision detection approaches with application specific parameters.
- Determines accessibility of point clouds and regions of residual contamination.

The remainder of this paper is structured as follows. Section 2 will introduce the framework used for processing and determining the accessibility of a point cloud. The results of this approach and discussion of the outcomes will be highlighted in Sect. 3. Concluding remarks and future work will be discussed in Sect. 4.

2 Framework

This framework draws on work done by [3,16] to combine Point Cloud (PC) analysis techniques with those of spatial partitioning and normal estimation [5].

The framework extends the Discrete Collision Detection (DCD) approach [15] and describes a method to model probe geometries and their Effective Scan Volume (ESV), that is the volume of space in front of the probe in which it can detect radiation, as convex hulls. This results in a novel accessibility assessment which can identify areas of inaccessibility on a PC, subsequently highlighting areas which pose a risk of containing residual contamination.

2.1 Spacial Partitioning and Surface Normal Estimation

When working with unorganised PC data, the initial task is to spatially partition the data in such a way that it allows for efficient searching of the data. Two common approaches for spacial subdivision are k-d trees and octrees; each allows for efficient and quick searching for the Nearest Neighbours (NN) in a given region. Octrees are shallow data structures which have 8 child nodes, splitting around a point rather than an axis like k-d trees, and do not suffer from becoming unbalanced with the insertion of new data.

Octrees generate a bounding box around the point data with equal edge lengths. Each bounding box which encapsulates a series of points is known as an octant. When each of these octants exceeds the number of allowable points within (bin number), the space is subdivided into 8 child octants with edge lengths equal to half that of the parent octant. This continues recursively until either the number of points within an octant is less than or equal to the bin number or the minimum allowable edge length of the octant is reached. This allows for control over the resolution of the generated Octree.

It has been shown by [3] that utilising Principal Component Analysis (PCA), it is possible to discern certain properties of a PC given the set of NN. The set of points forming the point cloud can be described as $P = \{\mathbf{p}_1, \cdots, \mathbf{p}_K\}$, with each point representing a location in three dimensions $\mathbf{p} \in \mathbb{R}^3$, with the set of Nearest Neighbours as

$$N_i = \{\mathbf{p}_i \in P : ||\mathbf{p}_i - \mathbf{q}_j|| < r\}, \quad \forall j \in K \tag{1}$$

where $K = |P|$ is the cardinal of the point set, $\mathbf{q}_j$ is our query point and r is a specified radius. This forms a family of sets $\{N_i\}_{\mathbf{p}_i \in P}$ where each element of N_i is a set of nearest neighbours indexed by P.

From this, it is possible to build a 3×3 covariance matrix $\mathbf{\Sigma}$, $\forall \mathbf{p} \in P$ which is given by

$$\Sigma_i = \frac{1}{k} \sum_{j \in N_{\mathbf{p}_i}} (\mathbf{p}_j - \overline{\mathbf{p}_i})(\mathbf{p}_j - \overline{\mathbf{p}_i})^T \tag{2}$$

where k is the number of nearest neighbours and $\mathbf{p}_j$ is each point in $N_{\mathbf{p}_i}$, $\overline{\mathbf{p}_i}$ is the centroid calculated at $\mathbf{p}_i$ from the nearest neighbours. The eigendecomposition of this covariance matrix can be expressed as $\mathbf{\Sigma} \cdot \mathbf{X} = \lambda \cdot \mathbf{X}$, where $\mathbf{X}$ is the matrix of eigenvectors and λ is a diagonal matrix of the eigenvalues. The eigenvectors form an orthogonal frame at the point, $\overline{\mathbf{p}_i}$, with each vector corresponding to a principal component of the spread of points [16]. Given that $\lambda_3 \geq \lambda_2 \geq \lambda_1 \geq 0$, λ_1 is the eigenvalue which corresponds to the normal and therefore the first column of the $\mathbf{X}$ matrix corresponds to the normal vector, $\mathbf{n}_i$, at the point $\mathbf{p}_i$. The estimated normals are aligned utilising a Minimum Spanning Tree (MST) as described in [5].

2.2 Modelling of A Radiation Probe and Effective Scan Volume

The nuclear industry uses radiation probes for a variety of scanning applications, one of which is radiological protection. These devices have become ingrained in the daily operation of many facilities, and a surplus of these devices typically exists. Generating Computer-Aided Design (CAD) models for these probes is time-consuming, and requires knowledge of 3D modelling software. However, most devices can be simplified and represented by a combination of geometric primitives such as planes, cones, spheres, and cylinders. Two examples of common probes are shown in Fig. 1.

Fig. 1. DP6 (left) and BP4 (right) radiation probes supplied by Thermo Fisher Scientific [19].

Consider the BP4 probe shown in Fig. 1, it can be approximated as a closed cylinder. The Effective Scan Volume (ESV), can be approximated as a second cylinder extending from the end of the probe. This representation of a bounding volume can be achieved by measuring a series of parameters of the probe such as length, radius, and radius of the ESV (equal to the radius of the detector film) as seen in Fig. 2. The depth of the ESV is determined by the Mean Free Path (MFP) of the radiation which is to be detected.

The top and face of the probe can be described as two hyperplanes separated by the length of the probe. Let a cylinder of radius, r, connect the 2 hyperplanes such that the intersection of these forms a convex hull, representing the probe, see Fig. 2 left. The ESV can be thought of as another cylinder with a smaller radius protruding from the front of the probe, see Fig. 2 right.

A hyperplane splits space in 2 which for 2 half-spaces, these can be represented in the form $B = \{\mathbf{x} \in \mathbb{R}^3 : \mathbf{n}^T(\mathbf{x} - \mathbf{x_0}) \leq 0\}$ for a given normal n and point on the hyperplane $\mathbf{x_0}$. For each $\mathbf{p}_i$, we can define a new point $\mathbf{s}_i = \mathbf{p}_i + d \cdot \mathbf{n}_i$, where d is equal to the desired stand-off distance of the probe and $\mathbf{n}_i$ is the normal at that point. This will act as the base point for the convex hull representing

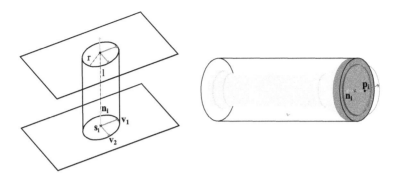

Fig. 2. Representation of two hyperplanes and cylinder forming a convex hull (left) and overlay of the convex hull representing the probe and ESV (purple) on top of the probe (right). (Color figure online)

the probe and is the centre of the detector film. By defining 2 support vectors $\mathbf{v}_1 = \mathbf{n}_i \times [0,0,1]$, and $\mathbf{v}_2 = \mathbf{n}_i \times \mathbf{v}_1$, the set of half-spaces can be described by substituting the different values for normals and points into B to generate

$$\mathcal{C}_{b_i} = \{\mathbf{x} \in \mathbb{R}^3 : \mathbf{n}_i^T(\mathbf{x} - \mathbf{s}_i) \geq 0\}$$
$$\mathcal{C}_{t_i} = \{\mathbf{x} \in \mathbb{R}^3 : \mathbf{n}_i^T(\mathbf{x} - (\mathbf{s}_i + l \cdot \mathbf{n}_i)) \leq 0\}$$
$$\mathcal{C}_{c_i} = \{\mathbf{x} \in \mathbb{R}^3 : \mathbf{v}_1^T(\mathbf{x} - \mathbf{p}_i)(\mathbf{x} - \mathbf{p}_i)^T \mathbf{v}_1 + \mathbf{v}_2^T(\mathbf{x} - \mathbf{p}_i)(\mathbf{x} - \mathbf{p}_i)^T \mathbf{v}_2 - r^2 \leq 0\}$$
$$\mathcal{C}_{\text{probe}} = \{\mathcal{C}_{t_i}, \mathcal{C}_{b_i}, \mathcal{C}_{c_i}\} \tag{3}$$

where $\mathcal{C}_{\text{probe}}$ is the convex hull of the probe at each point $\mathbf{p}_i$, l is the length of the probe and r is the radius of the probe. The same approach can be taken for the modelling of the ESV

$$\mathcal{C}_{b_i} = \{\mathbf{x} \in \mathbb{R}^3 : \mathbf{n}_i^T(\mathbf{x} - \mathbf{s}_i) \leq 0\}$$
$$\mathcal{C}_{p_i} = \{\mathbf{x} \in \mathbb{R}^3 : \mathbf{n}_i^T(\mathbf{x} - (\mathbf{p}_i - (a - d) \cdot \mathbf{n}_i)) \geq 0\}$$
$$\mathcal{C}_{c_i} = \{\mathbf{x} \in \mathbb{R}^3 : \mathbf{v}_1^T(\mathbf{x} - \mathbf{p}_i)(\mathbf{x} - \mathbf{p}_i)^T \mathbf{v}_1 + \mathbf{v}_2^T(\mathbf{x} - \mathbf{p}_i)(\mathbf{x} - \mathbf{p}_i)^T \mathbf{v}_2 - r_s^2 \leq 0\}$$
$$\mathcal{C}_{\text{ESV}} = \{\mathcal{C}_{p_i}, \mathcal{C}_{b_i}, \mathcal{C}_{c_i}\} \tag{4}$$

with $\mathcal{C}_{\text{ESV}}$ being the convex hull representing the ESV at $\mathbf{p}_i$, a is the range over which the probe can reliably detect radiation determined by the MFP of radiation, and r_s is the radius of the scan zone.

2.3 Accessibility Assessment

The accessibility assessment determines the points within the set P representing the point cloud, which a given probe can both reach and effectively scan. For a point $\mathbf{p}_i$ to be valid, that is reachable without colliding, then none of the points within its neighbourhood can satisfy $\mathcal{C}_{\text{probe}}$. If valid, then points it can effectively scan at $\mathbf{p}_i$ become any point within its neighbourhood which satisfies C_{ESV}. This is visually represented in Fig. 4 This approach builds upon methods used within computer science for determining collisions between meshes or other objects [6]. The assessment considers collisions between the probe and the surface on which the object is placed, between the probe and the PC, as well as between the probe and other constraints modelled as planes.

It expands upon Discrete Collision detection (DCD) [15], which returns a boolean value to whether two objects have collided. This can be generalised to a function which assesses the intersections of a family of sets. Let I be an indexing set of the family of sets $\{A_i\}_{i \in I}$ then

$$f(A) = \begin{cases} 0, & \text{if } \bigcap_{i \in I} A_i = \emptyset \\ 1, & \text{otherwise} \end{cases} \tag{5}$$

that is to say, if the intersection of the family of sets is empty, it is classed as a disjoint set. If we define N_{r_i} using Eq. 1 with $r = l + 2 \cdot d$ and $A_i = \{N_{r_i}, C_{\text{probe}}\}$, we can use the function to evaluate whether the probe collides with any neighbouring points. If the function returns 0, the sets do not intersect and the point is accessible, and if it returns 1, the point is inaccessible. Equation 5 can also be used to evaluate whether the probe will collide with the surface it's placed on, assumed to be a table, or whether the probe will collide with other constraints. Collisions with other constraints will not be considered. The table can be modelled as $C_{\text{table}} = \{\mathbf{x} \in \mathbb{R}^3 : [0, 0, 1] \cdot \mathbf{x} \leq b\}$. While the intersection can be solved using linear programming, a more efficient method is posed in Sect. 2.4.

2.4 Collision with Table Surface

All objects to be scanned must be placed on a surface, whether that is the floor or a table, which adds a challenge in determining point accessibility. This issue can be simplified to evaluate if the vertices or the minimum point of C_{probe} intersects with C_{table}. There are many approaches for solving this [6]. For convex hulls with vertices, detecting intersections with a half-space is straightforward. However, for C_{probe} without vertices, the minimal point of the convex hull is used to evaluate the collision with the flat surface. This can be posed as an optimisation problem, though it is computationally expensive.

Alternatively, leveraging the definition of C_{probe} allows reducing this to a constant time evaluation $\mathcal{O}(1)$. This is possible due to the definition of $\mathbf{v}_1$ and $\mathbf{v}_2$. Consider the intersection of the cylinder with the half-space C_{b_i} which forms a circle $Q \in \mathbb{R}^3$. If $\mathbf{n}_i$ has a positive z component, then the minimal point of the circle will also be the minimum point of C_{probe}. Let Q be expressed as

$$Q = \mathbf{s}_i + r \cdot \cos t \cdot \mathbf{v}_1 + r \cdot \sin t \cdot \mathbf{v}_2, \quad t = [0, 2\pi) \tag{6}$$

where r is the radius of the cylinder. Due to the properties of a vector product, $\mathbf{v}_1$ will always be in the xy-plane and orthogonal to $\mathbf{n}_i$. As a result of the right-hand rule, the second vector product, $\mathbf{v}_2$, will always be collinear with and in the direction of the minimum point. This minimum point will lie on the intersections of the circle Q with the line $\mathbf{r} = \mathbf{s}_i + t \cdot \mathbf{v_2}$. It can be shown that this intersection occurs at: $t = \frac{\pi}{2}, \frac{3\pi}{2}$, with the minimum point at $\frac{\pi}{2}$. Thus, it is only necessary to evaluate the Eq. 6 at $t = \frac{\pi}{2}$ to determine whether the probe intersects with $\mathcal{C}_{\text{table}}$. If $\mathbf{n}_i$ has a negative z component, then $\mathbf{s}_i$ is replaced by $\mathbf{s}_i + l \cdot \mathbf{n}_i$ in Eq. 6.

3 Results and Discussions

To demonstrate this approach several test models were selected. A flat box, a laptop in two positions: Laptop (1) - half closed, and Laptop (2) - open, along with a model of a torse [13] were used. PCs were generated from vertices taken from the .STL files.

The framework was implemented within MATLAB R2023a. The Octree library [17] was modified with additional functions to enable nearest neighbour searches, normal estimation and normal alignment using an MST. The framework can be described by Algorithm 1. The parameters used for the algorithm were taken from the radiation probe with radius 0.0335 m, length 0.222 m, stand-off distance 0.015 m, scan radius and range of 0.03 m and 0.02 m respectively. The points which represented the base of the objects were removed as these would not be captured during a 3D scan.

Algorithm 1. Accessibility Assessment

Require: r, l, d, r_s, a ▷ Radius, Length, Stand-off, Scan Radius, Scan Range
    function ACCESS(P, N_i)
        for $\forall \mathbf{p}_i \in P$ **do**
            $\mathbf{n}_i \leftarrow NormalEstimation(N_{p_i})$
            $\mathbf{v}_1 \leftarrow \mathbf{n}_i \times [0, 0, 1]$
            $\mathbf{v}_2 \leftarrow \mathbf{n}_i \times \mathbf{v}_1$
            if !$CollideWithTable(\mathbf{p}_i, \mathbf{v}_1, \mathbf{v}_2, r, l, d)$ **then**
                if !$f(\{N_{p_i}, \mathcal{C}_{\text{probe}}\})$ **then**
                    $A_i \leftarrow True$ ▷ Set the accessibility of the point to True
                    $DetermineScannedPoint(\mathbf{p}_i, \mathbf{v}_1, \mathbf{v}_2, r_s, a)$
                else
                    $A_i \leftarrow False$ ▷ Set the accessibility of the point to False
                end if
            end if
        end for
        return A
    end function

Figure 3 shows the output of the framework, the blue points represent poses within the PC where the probe will not collide with the surface if it approaches that point. The red points represent poses where the probe would collide with the table or PC. Consider Fig. 3c, the points which are considered inaccessible are those around the base and underside of the laptop, due to colliding with the half-space representing the table. Not as obvious are the points along the hinge which are also flagged as inaccessible, this is expected due to the radius of the sensor and the small stand-off distance causing a collision between the probe and the screen/keyboard. Figure 3a and Fig. 3d show successful coverage of the models with only the armpits and those along the base being inaccessible, The results of the framework showing accessible points and total coverage is presented in Table 1.

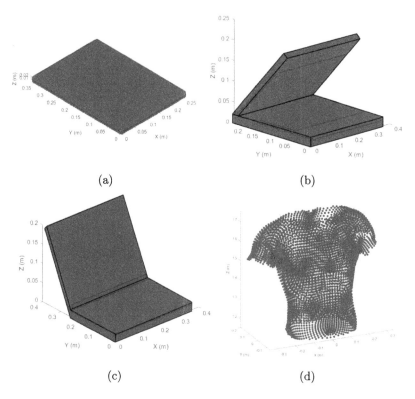

Fig. 3. Plots of the accessible points (blue) and inaccessible points (red) of both (a) flat box, (b) Laptop (1), (c) Laptop (2) and (d) Torso. (Color figure online)

Figure 3b has been used in Fig. 4 to illustrate several example cases, where the probe has been overlaid onto the PC to show how the framework simulates the probe and determines if a point will be accessible. Points around the base are inaccessible due to collisions with the table, and points at the innermost part of the screen are inaccessible as the probe collides with the points of the screen/keyboard. Those on the flat exterior surfaces are accessible and as such shown in blue.

This assessment does not consider the workspace of the manipulator or human conducting the assessment and is to identify areas of potential risk and

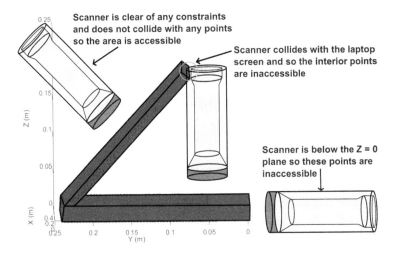

Fig. 4. Overlap of the probe onto the PC to illustrate the assessment process carried out in determining accessibility.

points which can be used to plan safe trajectories. When assessing the performance of the approach, it is assumed that the octree, associated normals and clusters for each point have already been calculated via Sect. 2.1 with a minimum octant size of 0.005 m and maximum bin capacity of 8, which achieved a good compromise between the performance and resolution.

The performance of the algorithm can be assessed using several metrics such as execution time and time complexity, coverage, and visual inspection. From inspecting the generated plots, it is intuitive to see that the framework is performing as expected. However, there are some discrepancies in the normal estimation caused by the surface geometry. These points can be seen in Fig. 4 close to the hinge (e.g. $y = 0.2$ m, $z = 0.05$ m), this causes an uneven cut-off line instead of a straight cut-off. Although this inaccuracy is a limitation of the approach, it is representative of the process when using real-world data. Furthermore, being conservative in this way minimises the risk of collision.

Table 1. Results of the accessibility assessment for each generated point cloud.

Model	Points	Leafs	Accessible (%)	Coverage (%)	Assessment Time (s)
Flat Box	33932	7339	89.33	94.96	0.62
Laptop (1)	69653	18722	43.11	53.90	3.86
Laptop (2)	49646	17460	61.23	75.84	3.76
Torso	32424	5398	70.84	90.08	0.50

The evaluation of the execution time and time complexity, given the stated assumption, requires assessing the performance of Algorithm 1. Assuming all calculations and conditional statements execute in constant time, the dominant term in determining time complexity is the loop over each leaf octant of the octree. It is possible to conclude that this algorithm will execute in linear time $\mathcal{O}(n)$. In Table 1, accessible refers the percentage of points within P which do not cause a collision with the probe if it approach that pose. However, some points may not be directly accessible but fall within the ESV of the probe at another pose and therefore can be monitors, this is referred to as coverage.

The assessment achieving high accessibility and coverage over all models in reasonable execution time. With the flat box and torso achieving over 90% coverage, the results of the torso model are significant as the final intended system will perform radiological surveys over humans.

4 Conclusion and Future Work

The results presented in this paper demonstrate the effectiveness of the framework in distinguishing between accessible and inaccessible areas within point cloud data for radioactive contamination monitoring. By modelling the probe with geometric primitives and expanding on conventional discrete collision detection, the framework quickly assesses point clouds. Constraining the probe definition reduces the time complexity for detecting collisions between the probe and a surface to $\mathcal{O}(1)$. With minimal modification, this framework can be applied to other fields by adjusting parameters or defining new end-effector geometries.

This work could be included in existing manipulation frameworks and expanded to include non-convex end effector geometries that could be used outside of surface following, such as grasping. Future work will develop this framework to include various probe geometries in a dedicated library and use the output of the framework to investigate methods for discerning a near-optimal set of scan points to be used for constructing viable trajectories over the objects which will be used to perform surface contamination monitoring as part of a robotic radiological survey assistant.

Acknowledgements. This work is supported by funding from an Industrial Cooperative Awards in Science & Engineering (iCASE) by Nuclear Restoration Services Dounreay, the Nuclear Decommissioning Authority and EPSRC. AW acknowledges support from UK Research and Innovation via project EP/V026941/1.

References

1. Bryant, P.A.: The role of radiation protection societies in tackling the skills shortage and development of young professionals and researchers. J. Radiol. Prot. **41**(3), S79–S88 (2021). https://doi.org/10.1088/1361-6498/abf815
2. Graumann, C., Fuerst, B., Hennersperger, C., Bork, F., Navab, N.: Robotic ultrasound trajectory planning for volume of interest coverage. In: 2016 IEEE International Conference on Robotics and Automation (ICRA). IEEE (May 2016). https://doi.org/10.1109/icra.2016.7487201

3. Hackel, T., Wegner, J.D., Schindler, K.: Contour detection in unstructured 3d point clouds. In: 2016 IEEE Conference on Computer Vision and Pattern Recognition (CVPR). IEEE (June 2016). https://doi.org/10.1109/cvpr.2016.178
4. Hasegawa, T., Hashimoto, T., Hashimoto, M.: Radioactive contamination monitors. Fuji Jiho **36**, 373–379 (2004)
5. Hoppe, H., DeRose, T., Duchamp, T., McDonald, J., Stuetzle, W.: Surface reconstruction from unorganized points. In: Proceedings of the 19th Annual conference on Computer Graphics and Interactive Techniques. SIGGRAPH92, ACM (July 1992). https://doi.org/10.1145/133994.134011
6. Lin, M., Manocha, D.: Overview on Collision and Proximity Queries, pp. 181–203. A K Peters/CRC Press, Boca Raton (July 2008). https://doi.org/10.1201/b10636-12
7. Ly, K.T., Munks, M., Merkt, W., Havoutis, I.: Asymptotically optimized multi-surface coverage path planning for loco-manipulation in inspection and monitoring. In: 2023 IEEE 19th International Conference on Automation Science and Engineering (CASE). IEEE (August 2023). https://doi.org/10.1109/case56687.2023.10260625
8. Mauer, G.F., Kawa, C.: Accuracy Analysis of a Robotic Radionuclide Inspection and Mapping System for Surface Contamination. American Nuclear Society, United States (2008)
9. Management of Radiation Protection in Defence: part 2 guidance (jsp 392) (1 2023)
10. Monk, S.D., West, C., Bandala, M., Dixon, N., Montazeri, A., Taylor, C.J., Cheneler, D.: A low-cost and semi-autonomous robotic scanning system for characterising radiological waste. Robotics **10**(4), 119 (2021). https://doi.org/10.3390/robotics10040119
11. Nakhaeinia, D., Payeur, P., Chávez-Aragón, A., Cretu, A.M., Laganière, R., Macknojia, R.: Surface following with an rgb-d vision-guided robotic system for automated and rapid vehicle inspection. Int. J. Smart Sens. Intell. Syst. **9**(2), 419–447 (2016). https://doi.org/10.21307/ijssis-2017-877
12. Nakhaeinia, D., Payeur, P., Laganiere, R.: A mode-switching motion control system for reactive interaction and surface following using industrial robots. IEEE/CAA J. Autom. Sin. **5**(3), 670–682 (2018). https://doi.org/10.1109/jas.2018.7511069
13. nixor: Male base mesh 3d model. https://free3d.com/3d-model/male-base-mesh-6682.html. Accessed 24 June 2024
14. Nuclear Skills Strategy Group: Nuclear Workforce Assessment (December 2023)
15. Pan, J., Chitta, S., Manocha, D.: FCL: a general purpose library for collision and proximity queries. In: 2012 IEEE International Conference on Robotics and Automation. IEEE (May 2012). https://doi.org/10.1109/icra.2012.6225337
16. Pauly, M., Gross, M., Kobbelt, L.: Efficient simplification of point-sampled surfaces. In: IEEE Visualization, 2002. VIS 2002. VISUAL-02, IEEE (2002). https://doi.org/10.1109/visual.2002.1183771
17. Sven: octree - partitioning 3d points into spatial subvolumes
18. Thakur, V.M., et al.: Design and development of a plastic scintillator based whole body β/γ contamination monitoring system. Nucl. Sci. Tech. **32**(5), 1–10 (2021). https://doi.org/10.1007/s41365-021-00883-1
19. Thermo Fisher Scientific: Non-Intelligent and Intelligent Scintillation Probes Instruction Manual (August 2006)

20. von Haxthausen, F., Böttger, S., Wulff, D., Hagenah, J., García-Vázquez, V., Ipsen, S.: Medical robotics for ultrasound imaging: current systems and future trends. Curr. Robot. Rep. **2**(1), 55–71 (2021). https://doi.org/10.1007/s43154-020-00037-y
21. White, S.R., et al.: Radiation mapping and laser profiling using a robotic manipulator. Front. Robot. AI **7** (2020). https://doi.org/10.3389/frobt.2020.499056

A New Hybrid Teleoperation Control Scheme for Holonomic Mobile Manipulator Robots Using a Ground-Based Haptic Device

Bandar Aldhafeeri[✉][iD], Joaquin Carrasco[iD], Bruno V. Adorno[iD], and Erwin Jose Lopez Pulgarin[iD]

Manchester Centre for Robotics and AI, The University of Manchester, Manchester, UK
bandar.aldhafeeri@postgrad.manchester.ac.uk,
{joaquin.carrasco,bruno.adorno}@manchester.ac.uk

Abstract. Teleoperating mobile manipulator (MM) robots using a single ground-based haptic (GBH) device is challenging due to differences in workspaces and mechanical structures. A hybrid control scheme combines navigation and manipulation modes to operate a robot using manual or automatic mode switching. Although an automatic scheme tends to reduce the user's mental workload, in some cases, the bilateral teleoperation system loses transparency when interacting with the environment; in others, the interaction with the environment may happen in inappropriate operation mode without the user's awareness. To overcome those challenges, this paper proposes a new hybrid control scheme for bilateral teleoperation of holonomic MM robots. In this scheme, while navigating, the user's hand motion is restricted by an artificial force to prevent reaching the boundary of the GBH device's workspace. When the end-effector touches the environment, a method not requiring any exteroceptive sensor automatically switches the operation mode to manipulation mode. After completing the remote task, the user must switch to navigation mode manually. The simulation results indicate that the baseline bubble technique, a widely used automatic scheme, creates over 50% of forces on the haptic interface unrelated to environmental interaction, resulting in transparency deterioration. On the other hand, with our proposed technique, the bilateral system becomes transparent in under 0.3 s.

Keywords: Bilateral Teleoperation · Switching · Mobile Manipulator

1 Introduction

Mobile manipulator (MM) robots have significant potential in bilateral teleoperation tasks thanks to their manipulation and locomotion capabilities. They

---

This work was funded by Majmaah University, Riyadh, Saudi Arabia.

© The Author(s), under exclusive license to Springer Nature Switzerland AG 2025
M. N. Huda et al. (Eds.): TAROS 2024, LNAI 15051, pp. 283–295, 2025.
https://doi.org/10.1007/978-3-031-72059-8_24

are usually redundant [1] for most manipulation tasks, and since tasks are typically formulated in the task space, operating these types of robots is carried out via whole-body control (WBC) schemes. For instance, Jin et al. [2] propose a haptic teleoperation framework for a legged manipulator controlled by a whole-body model predictive control scheme, whereas Quiroz-Omaña and Adorno [3] propose WBC for nonholonomic mobile manipulators using linear programming.

When teleoperating MM robots, switching between manipulation and navigation modes (i.e., operation modes) is common practice [4]. Typically, in the manipulation mode, the remote robot's base remains stationary while the robotic arm is teleoperated (i.e., position-position (PP) control). In the navigation mode, the local robot's end-effector movements determine the velocity of either the remote robot's end-effector or its base (i.e., position-velocity (PV) control). Integrating PP and PV control schemes to operate a robotic system is called a hybrid control scheme [5]. The process of switching can be carried out either manually [6,7] or automatically [8,9]. The latter aims to reduce the user's mental workload by automatically managing operation modes.

The bubble technique is a common automatic hybrid control scheme, originally designed to help users navigate large virtual environments (VEs) with haptic feedback [10]. The ground-based haptic (GBH) device's physical workspace is divided virtually into manipulation and navigation regions. The operation mode is automatically determined based on the location of the GBH device's end-effector within the virtual volume. If the probe is inside the navigation area, an elastic force is induced to restrict the user's motion; therefore, the user is aware of this mode through the haptic feedback. The technique has been extended to applications in underactuated aerial robots [11], augmented reality [12], and mobile manipulator robots [5]. Although this approach's automatic mechanism is appealing due to its simplicity and local implementation, force overlapping compromises the transparency of the bilateral system. This overlap happens when the force felt by the human operator is a result of remote interaction with the environment combined with an elastic force generated due to the probe being outside the virtual sphere.

To the best of our knowledge, the transparency deterioration of the force feedback in the bubble technique has not been addressed before in applications related to MM robots. In [5], although a simple grasping task has been evaluated, force reflection has been left for future studies. This limitation is an inherent flaw in this automatic hybrid control scheme.

To overcome this limitation, this paper proposes an alternative hybrid control scheme for holonomic MM robots. The transition from the navigation to the manipulation mode is automated, whereas the converse requires a push-button mechanism for simplicity. The proposed methodology is independent of force sensors and simple to implement. We use a whole-body control scheme to reduce the user's burden, so controlling the MM robot using a single GBH device becomes potentially more intuitive. The contribution of this paper can be summarised as follows:

1. A new hybrid control scheme for bilateral teleoperation of holonomic MM robots using a single GBH device is designed to provide the user with haptic feedback without deteriorating the transparency.
2. Using only proprioceptive sensors, we exploit the robot's control inputs to detect possible end-effector collisions, enabling the automatic transition from navigation to manipulation modes.

2 Constrained Kinematic Control

Given the task vector $x \in \mathcal{T} \subseteq \mathbb{R}^m$ representing the end-effector pose, position, orientation, etc., the goal is to drive it to a constant desired task vector $x_d \in \mathcal{T}$ using the coordinated motion of all elements in the kinematic chain, including the mobile base and the robot manipulator joints, while considering constraints in the control inputs.

To this aim, let $q \in \mathcal{Q} \subseteq \mathbb{R}^n$ be the generalized configuration vector comprising the base and the manipulator robot. The whole-body forward kinematics is given by $x = f(q)$ and the whole-body differential kinematics is given by $\dot{x} = J(q)\dot{q}$, where $J(q) = \partial f(q)/\partial q \in \mathbb{R}^{m \times n}$ is the analytical Jacobian matrix of the holonomic mobile manipulator [13]. We define the task error $\tilde{x} \triangleq x - x_d$ and the desired closed-loop task-space error dynamics as

$$\dot{\tilde{x}} + \beta\tilde{x} = 0 \implies J(q)\dot{q} + \beta\tilde{x} = 0$$

aiming at an exponential convergence, where $\beta \in (0, \infty)$ determines the convergence rate. Following [14], a kinematic control law ensuring closed-loop stability is given by

$$\begin{aligned} u \in \arg\min_{\dot{q}} \; & \|J(q)\dot{q} + \beta\tilde{x}\|_2^2 + \lambda^2\|\dot{q}\|_2^2 \\ \text{subject to } & W\dot{q} \preceq w, \end{aligned} \quad (1)$$

where $W \triangleq W(q) \in \mathbb{R}^{s \times n}$, $w \triangleq w(q) \in \mathbb{R}^s$ are used to define s linear constraints in the control inputs (i.e., configuration velocities), and $\lambda \in (0, \infty)$ is a damping factor penalizing solutions yielding large configuration velocities.

Although constrained control law (1) admits general constraints using vector field inequalities [15], in this paper, we only constrain the whole-body configuration velocities and include joint limits for the robot manipulator. Let $q_{\text{vel}_u}, q_{\text{vel}_\ell} \in \mathbb{R}^n$ represent the upper and lower limits for the configuration velocities, respectively; therefore, the constraint in the configuration velocities is given by

$$q_{\text{vel}_\ell} \preceq \dot{q} \preceq q_{\text{vel}_u} \implies \begin{bmatrix} I_n \\ -I_n \end{bmatrix} \dot{q} \preceq \begin{bmatrix} q_{\text{vel}_u} \\ -q_{\text{vel}_\ell} \end{bmatrix},$$

where $I_n \in \mathbb{R}^{n \times n}$ is the identity matrix.

Since the base can move in unlimited space, we constrain only the robot manipulator configurations. For that, consider the partitioned configuration vector $q = \begin{bmatrix} q_b^T & q_a^T \end{bmatrix}^T$, where $q_b \in \mathbb{R}^{n_b}$ is the mobile base configuration and $q_a \in \mathbb{R}^{n_a}$ is the robot arm configuration, such that $n = n_a + n_b$. Given the vectors $q_u, q_\ell \in \mathbb{R}^{n_a}$ of constant upper and lower joint limits, we define $\tilde{q}_u \triangleq q_a - q_u$ and $\tilde{q}_\ell \triangleq q_a - q_\ell$. It is possible to show that enforcing the inequalities $\dot{\tilde{q}}_u + \eta \tilde{q}_u \preceq 0$ and $\dot{\tilde{q}}_l + \eta \tilde{q}_l \succeq 0$ with $\eta \in (0, \infty)$ ensures that, if $\tilde{q}_u(0) \preceq 0$ and $\tilde{q}_l(0) \succeq 0$, then $\tilde{q}_u(t) \preceq 0$ and $\tilde{q}_l(t) \succeq 0$ for all $t > 0$ [15]. Hence, the inequalities enforcing joint limits are rewritten as

$$\begin{bmatrix} B \\ -B \end{bmatrix} \dot{q} \preceq \eta \begin{bmatrix} -\tilde{q}_u \\ \tilde{q}_l \end{bmatrix}$$

where $B = \begin{bmatrix} 0_{n_a \times n_b} & I_{n_a} \end{bmatrix}$ such that $\dot{q}_a = B\dot{q}$.

3 Bilateral Teleoperation Control Scheme

3.1 Overview of the Bilateral Teleoperation Architecture

A typical bilateral system comprises five components: the human operator, a haptic device, a communication channel, a remote robot, and a remote environment. In this paper, both Position-Force and PP architectures [16] are adopted, depending on the availability of a force sensor, in which the local device's displacement is transmitted to the distant site as desired commands. When the robot contacts the remote environment, the force data, measured or induced by the position error, is fed back to the human operator.

Because MMs can navigate and manipulate, the desired commands are executed differently. To achieve this, we use an operation mode index, $\zeta \in \{0, 1\}$, such that $\zeta = 0$ represents navigation, whereas $\zeta = 1$ represents manipulation. Besides defining the strategy for each mode and the force rendering mechanism depending on the availability of external sensors, the transition between modes adds a new layer of complexity that must be dealt with appropriately to ensure transparency and low cognitive load.

3.2 Position-Position Manipulation Scheme ($\zeta = 1$)

Let $\mathcal{F}_H$, $\mathcal{F}_W$, and $\mathcal{F}_O$ represent the frames of the haptic device, the remote world, and an offset frame, respectively. Let $p_{m_d}^W, p_m^W \in \mathbb{R}^3$ represent the desired and current positions of the MM's end-effector in $\mathcal{F}_W$, and $p_h^H \in \mathbb{R}^3$ represent the position of the haptic device's end-effector in $\mathcal{F}_H$. Thus, the desired position of the MM robot's end-effector is

$$p_{m_d}^W = \alpha p_h^W + p_O^W, \qquad (2)$$

where $\alpha \in (0, \infty)$ is a scaling factor, $p_O^W \in \mathbb{R}^3$ is the origin of $\mathcal{F}_O$ in $\mathcal{F}_W$, and p_h^W is the haptic device's end-effector position expressed in the world frame.

Therefore, whenever the human operator moves the haptic device's end-effector, the MM's end-effector replicates the motion using the offset p_O^W as the starting point according to the scale factor. If $\alpha > 0$, the motion is amplified at the expense of less precision on the MM's end-effector motion; conversely, if $\alpha < 0$, the MM's end-effector motion is reduced but precision is increased.

3.3 Force Rendering and Collision Detection

Depending on the availability of a force sensor in the end-effector, we propose two approaches to providing the operator with haptic feedback.

Method 1. In this method, we assume a force sensor is available. Let $f_e^H \in \mathbb{R}^3$ be the force measurement at the MM's end-effector expressed in $\mathcal{F}_H$. The rendered force f_*^H in the haptic device is

$$f_*^H = \begin{cases} \kappa f_e^H, & \text{if } \|f_e^H\| > \gamma, \\ 0, & \text{otherwise,} \end{cases} \quad (3)$$

where $\kappa \in (0, \infty)$ is the force feedback constant and $\gamma \in (0, \infty)$ is a small threshold to avoid generating spurious force feedback due to sensor noise. In addition, the sensor enables straightforward automatic switching from the navigation to the manipulation mode because the force magnitude can be used for collision detection. For instance, suppose $\zeta = 0$ and consider a threshold $\delta \in (0, \infty)$ to avoid incorrectly detecting collisions due to noise. A simple rule can be defined as

$$\|f_e^W\| > \delta \implies \zeta = 1,$$

which means that whenever the measured force is beyond the predefined threshold due to contact with the environment, the operation mode changes to manipulation.

Method 2. In this method, we assume that a force sensor is unavailable. In the spirit of classic bilateral teleoperation [16], we use the position error to infer the interaction with the environment. The rendering of the force can be carried out without the use of a force sensor by defining $\tilde{p}^W \triangleq p_{m_d}^W - p_m^W$, and using

$$f_*^W = -k_s \tilde{p}^W - k_d \dot{\tilde{p}}^W, \quad (4)$$

where $k_s \in (0, \infty), k_d \in [0, \infty)$ and f_*^W is then expressed in $\mathcal{F}_H$ to provide the rendered force f_*^H in the haptic interface. Similarly, the control input (1) triggers the automatic switching from the navigation to the manipulation mode as follows. Let $\dot{p}_{m_u}^W = J_p(q)u$, where $\dot{p}_{m_u}^W$ is the MM's end-effector velocity due to control input (1), expressed in $\mathcal{F}_W$, and $J_p(q) \in \mathbb{R}^{3 \times n}$ is the position Jacobian [13]. We define the following velocity error

$$\dot{\tilde{p}}_{m_u}^W = \dot{p}_m^W - \dot{p}_{m_u}^W. \quad (5)$$

Suppose $\zeta = 0$; if $\|\dot{\tilde{p}}_{m_u}^W\| > \sigma$, then $\zeta = 1$, where $\sigma \in (0, \infty)$ must be designed to differentiate between the typical tracking error during navigation and actual contact with the environment. See Sect. 4.4 for a discussion on how to select an appropriate threshold.

3.4 Hybrid Control Schemes

Two hybrid control schemes for bilateral teleoperation are presented. The first is the standard bubble technique [10], which we use as the baseline for comparisons with our method. The second is entirely novel and constitutes the main contribution of this paper.

Bubble Technique. In the navigation mode (i.e., $\zeta = 0$), the user "moves" the bubble, the visual representation of the virtual sphere with radius $\alpha \cdot r$, by changing the desired MM's end-effector position whenever the haptic interface's end-effector position is outside a virtual sphere of radius r (i.e., $\|p_h^H\| > r$) according to

$$\|p_h^H\| > r \implies p_{m_d}^W = \alpha r \bar{p}_h^W + p_O^W(t), \qquad (6)$$

where $\bar{p}_h^W = p_h^W / \|p_h^W\|$ and $p_O^W(t)$ is the time-dependent position of the bubble's centre. This position changes with speed proportional to the distance to the sphere; that is,

$$\dot{p}_O^W(t) = \rho(d - r)\bar{p}_h^W, \qquad (7)$$

where $\rho \in (0, \infty)$ and $d \triangleq \|p_h^H\|$. Following [10], for the user to "feel" the change in the operation mode whenever the probe is moved outside the virtual sphere, a restoring force is defined as

$$f_{\text{rest}}^H = -k_r(d - r)\bar{p}_h^H, \qquad (8)$$

where $k_r \in (0, \infty)$. Thus, the automatic switching process is given by

$$\zeta = \begin{cases} 0, & \text{if } d > r, \\ 1, & \text{otherwise.} \end{cases}$$

In bilateral systems, the human operator receives haptic feedback when the remote robot interacts with the environment. In the bubble technique, the local force $f_\ell^H \in \mathbb{R}^3$ that the user feels is

$$f_\ell^H = \begin{cases} f_{\text{rest}}^H + f_*^H, & \text{if } \zeta = 0, \\ f_*^H, & \text{if } \zeta = 1, \end{cases} \qquad (9)$$

where $f_*^H \in \mathbb{R}^3$ is given by either (3) or (4) depending on the adopted method.

To illustrate the effects of the local force (9), Fig. 1 shows the remote end-effector touching a wall in the right image, whereas the local end-effector is outside the virtual sphere in the left image. Haptic feedback no longer solely comes from the restoration force f_{rest}^H, causing ambiguity in force feedback due to the additional reaction force arising from the MM's end-effector contact with the environment and potentially disturbing how the human experiences the navigation mode. In VEs, the user can view the entire bubble and move the haptic interface's end effector to avoid interaction with the environment during the navigation mode. However, this is practically difficult in real-world applications as the user's inability to differentiate between operation modes may cause interaction with the environment in the navigation mode. This can deteriorate the transparency of the bilateral system due to force overlap, and the bubble's centre may be located where the MM's end effector cannot reach.

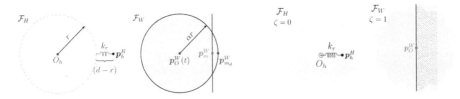

Fig. 1. Bubble technique: on the left, the haptic interface's end-effector is outside the virtual sphere (i.e., $\zeta = 0$), whereas on the right, the teleoperator interacts with a wall.

Fig. 2. Our technique: The left figure depicts the moment before the collision, whereas the right figure shows the mode switching after the collision.

Proposed Technique. This section presents a novel hybrid control scheme for bilateral teleoperation of holonomic MM robots. The navigation mode (i.e., $\zeta = 0$) is the default operation mode. The desired position that maps the position of the GBH device's end effector from limited to unlimited workspace is given by

$$\boldsymbol{p}^W_{m_d} = \alpha \boldsymbol{p}^W_h + \boldsymbol{p}^W_m. \tag{10}$$

A virtual restoration force proportional to the haptic interface's end-effector displacement counteracts the human operator's movement during navigation, ensuring that $\boldsymbol{p}^W_h = \boldsymbol{0}$ when the human stops the actuation. In the manipulation mode, the repulsive virtual force is set to zero. Hence,

$$\boldsymbol{f}^H_{rest} = \begin{cases} -k_r \boldsymbol{p}^H_h, & \text{if } \zeta = 0, \\ \boldsymbol{0}, & \text{if } \zeta = 1. \end{cases}$$

When a collision is detected either from a force sensor or inferred from monitoring the robot's internal state, the operation mode switches automatically to manipulation mode ($\zeta = 1$). To avoid a drastic change in the force at the collision time, we use a time-varying restoring variable $k_r(t)$ as follows

$$k_r(t) = \begin{cases} k_{nom}, & \text{if } t < t_c, \\ k_{nom} \cdot \left(1 - \frac{t - t_c}{t_t}\right), & \text{if } t_c \leq t \leq t_c + t_t, \\ 0, & \text{if } t > t_c + t_t, \end{cases}$$

where t_c is the time when the collision is detected, k_{nom} is the nominal value of the restoration coefficient, and $t_t \in (0, \infty)$ is a predefined transition time between modes. Finally, the position of the end effector at the collision time is selected as offset in (2) by making $\boldsymbol{p}_O^W \triangleq \boldsymbol{p}_m^W(t_c)$.

To avoid the overlap between the restoring and the rendered force, we define the local force as

$$\boldsymbol{f}_\ell^H = \begin{cases} \boldsymbol{f}_{\text{rest}}^H, & \text{if } \zeta = 0, \\ \boldsymbol{f}_*^H, & \text{if } \zeta = 1. \end{cases}$$

Figure 2 illustrates our proposed technique before and after a collision. The grey shape on the left represents the actual workspace of the GBH device, whereas the red shape on the right is its scaled version centred at the MM's end-effector. In the left figure, the user pushes the haptic interface's end-effector, causing the platform to move without stopping ($\zeta = 0$). The position of the MM robot's end-effector is used to determine the origin of $\mathcal{F}_O$; therefore, the red shape moves with the end-effector. After the collision, the right figure shows that the red shape is attached to the environment, and the system switches to the manipulation mode.

In our current formulation, users must manually switch back to navigation mode. This is easy to implement as haptic devices typically include a stylus with buttons explicitly designed for predefined actions.

4 Simulations and Discussions

We have implemented a semi-physical human-in-loop experiment to assess the effectiveness of the proposed hybrid control scheme. The GBH device, Phantom Omni, sends and receives forces via Haptic Device API (HDAPI). The CoppeliaSim 4.5.1 software has been used to carry out the simulations. The kinematic modelling and control of the MM robot was done using the C++ API of the DQ Robotics library [17]. Solving the quadratic optimization problem is achieved numerically via the DQ Robotics solver interface, which internally uses the IBM CPLEX library. The experiments were conducted in Linux OS with Intel Core i5 CPU 1.60 Hz and Intel UHD Graphics 620.

4.1 Simulation Setup

A semi-physical bilateral teleoperation system that includes a human operator, a GBH device, a communication channel, a remote robot, and a remote environment has been prepared. The remote site contains a wall 2 m from the origin of $\mathcal{F}_W$ along the x-axis, a holonomic MM robot with an 8-DOF ($n_b = 3$ and $n_a = 5$). The user manipulates a haptic device with three actuated DOFs to control the MM robot's end-effector position with a fixed orientation. The objective is to move the robot toward the wall for interaction. The simulation is advanced one step at each iteration with the help of an external C++ application that runs the controller at 20 Hz. The communication between the controller and the scene is implemented through ZeroMQ remote API. No artificial time delay is injected into the channel.

4.2 Transparency Evaluation

A performance index $\mu \in [0, 1]$ has been devised to measure the transparency of the bubble and proposed techniques. Specifically, this index indicates how much of the actual force the human operator perceives is due to the environmental force only.

To quantitatively measure the transparency level, the performance index is defined as $\mu = \|\boldsymbol{f}_*^H\|/\|\boldsymbol{f}_\ell^H\|$. Therefore, the bilateral system achieves full transparency when $\mu = 1$ and loses all transparency when $\mu = 0$. If $0 < \mu < 1$, the transparency level of the haptic feedback deteriorates proportionally.

In this evaluation, only Method 1 in Sect. 3.3 has been used to provide haptic feedback by measuring the environmental forces using an external sensor at the MM's end-effector. Method 2 was not used because it is not straightforward to determine the transparency level from the performance index, as the rendered force depends on the closed-loop position error dynamics. For the bubble technique, the parameters used are $r = 25$ mm, $\alpha = 0.01, \rho = 0.005$, $k_r = 0.3$ N/mm. For the proposed technique, the parameters are $\alpha = 0.01$ and $k_{\text{nom}} = 0.3$ N/mm. The constrained controller (1) is used for both techniques, with $\beta = 5$ and $\lambda = 0.01$.

Figure 3 shows that the bubble technique has severely impacted the transparency level, as over 50% of the force felt by the user was not caused by the remote robot's interaction with the environment. In contrast, in the proposed method, while the performance index shows low transparency at the moment of collision, the bilateral system became transparent under 0.3 s. Figure 4 shows a wall obstructing the forward movement of the MM end-effector's position along the x-axis.

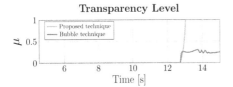

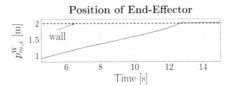

Fig. 3. The transparency deterioration level measured by the performance index μ in both techniques.

Fig. 4. The positions of the MM robot's end-effector along the x-axis and the wall in the world frame.

4.3 Haptic Feedback

We conducted two other experiments to evaluate the force rendering methods in our proposed technique described in Sect. 3.4.

First, to provide insight into the transparency level when using the sensorless Method 2 in Sect. 3.3, we ran a second experiment to compare the magnitudes of the virtual force to the force obtained from an external force sensor. Figure 5 shows that when using Method 2 (virtual force) with parameters given by $\alpha = 0.01, \beta = 5, \lambda = 0.01, k_s = 300$ N/m and $k_d = 20$ Ns/m, the human operator receives haptic feedback from 19.3 to 26.9 s that is not caused by environmental interaction, decreasing transparency.

A third experiment using parameters $\alpha = 0.009, \beta = 20, \lambda = 0.01, k_s = 200$ N/m, and $k_d = 10$ Ns/m was run to compare Methods 1 and 2 when the user quickly moves the haptic interface's end-effector during manipulation mode. Figure 6, the top figure, shows that the robot did not follow the quick desired commands, whereas when Method 2 was utilised, the middle figure, the haptic interface's end-effector was moved relatively slowly due to the virtual force being large, as shown in the bottom figure. In this regard, Method 2 prevents abrupt motions compared to Method 1, improving trajectory tracking, albeit less transparently, as the user feels a force unrelated to the robot's contact with the environment.

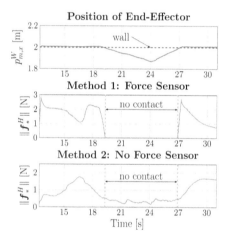

Fig. 5. Difference between the force rendering methods in the manipulation mode ($\zeta = 1$) when using our proposed method.

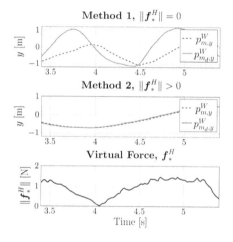

Fig. 6. MM's capability in following the desired commands when $\zeta = 1$ under Methods 1 and 2.

4.4 Collision Detection Threshold

Several trials have been conducted to determine the threshold σ for automatic switching. The teleoperator was commanded to hit the wall at different speeds to mimic different human operators' behaviours. The parameters used are $\beta = 5, \lambda = 0.01$ and $\alpha = 0.01$. Table 1 shows data before and after collision for several trials. Through trial and error and inspection of spikes in the data plots during collisions under different speeds along the x-axis, as shown in Table 1, we found a suitable threshold of $\sigma = 0.065$. The mean of the absolute errors (i.e., $|p_{e,x}^W - p_{O,x}^W|$) for the collected data is 1.4 cm with a standard deviation of 0.0033, where $p_{e,x}^W$ and $p_{O,x}^W$ are the wall's location and MM's end-effector position, respectively, along the x-axis during the detected collision. The positions $p_{O,x}^W$, which can be used as an estimate of the wall's locations and reported in Table 1, are close to the ground-truth value obtained from the simulator.

Table 1. Collision detection data.

Before Collision				After Collision				
t[s]	$\dot{p}_{m,x}^W$[m/s]	$p_{h,x}^W$[mm]	$\|\tilde{\dot{p}}_{m_u}^W\|$	t[s]	$\dot{p}_{m,x}^W$[m/s]	$p_{h,x}^W$[mm]	$\|\tilde{\dot{p}}_{m_u}^W\|$	$p_{O,x}^W$[m]
22.55	0.075	3.31	0.062	22.6	0.0251	3.31	0.0651	2.0104
15.1	0.099	5.11	0.056	15.15	-0.0055	5.11	0.1034	2.0113
10.3	0.197	8.29	0.032	10.35	0.0665	8.29	0.1474	2.0129
5.8	0.376	15.29	0.015	5.85	0.2925	15.29	0.3299	2.0133
6	0.518	21.55	0.014	6.05	0.2636	21.60	0.2459	2.0181
4.5	0.774	33.20	0.024	4.55	0.678	33.20	0.3798	2.013
4	0.791	37.30	0.026	4.05	0.463	37.15	0.3915	2.019

5 Conclusions

This paper has proposed a novel hybrid control scheme for bilateral teleoperation of holonomic MM robots using a single haptic device. Simulation results show that our method outperforms the bubble method, an automatic scheme, in transparency level, as measured by a classic transparency index. The WBC paradigm has been adopted to define teleoperation in the task space, providing a straightforward way of bidirectionally controlling the robot end-effector while all DOFs are used simultaneously. In addition, implementing the scheme is simple as it only requires a haptic device with three actuated DOFs and works with or without a force sensor. Using the proposed technique, simulation results suggest that estimating the interaction force slightly affected transparency, but it decreased the human operator's abrupt motions compared to using an external force sensor during the manipulation mode, which improved trajectory tracking. Moreover, we have developed a method that employs the control inputs to detect collisions without relying on exteroceptive sensors.

Future work will focus on addressing collision avoidance with the surrounding environment by using vector field inequalities in the constrained controller [15], and the automatic switching from manipulation to navigation modes.

References

1. Siciliano, B., Khatib, O., Kröger, T.: Springer handbook of robotics. Springer, vol. 200, 2008
2. Cheng, J., Abi-Farraj, F., Farshidian, F., Hutter, M.: Haptic teleoperation of high-dimensional robotic systems using a feedback MPC framework. In: 2022 IEEE/RSJ International Conference on Intelligent Robots and Systems (IROS), pp. 6197–6204. IEEE (2022)
3. Quiroz-Omaña, J.J., Adorno, B.V.: Whole-body kinematic control of nonholonomic mobile manipulators using linear programming. J. Intell. Robot. Syst. **91**, 263–278 (2018)
4. Pham, C.D., From, P.J.: Control allocation for mobile manipulators with on-board cameras. In: 2013 IEEE/RSJ International Conference on Intelligent Robots and Systems, pp. 5002–5008. IEEE (2013)
5. Pepe, A., Chiaravalli, D., Melchiorri, C.: A hybrid teleoperation control scheme for a single-arm mobile manipulator with omnidirectional wheels In: 2016 IEEE/RSJ International Conference on Intelligent Robots and Systems (IROS). IEEE, 2016, pp. 1450–1455 (2016)
6. Andaluz, V.H., Salinas, L., Roberti, F., Toibero, J.M., Carelli, R.: Switching control signal for bilateral tele-operation of a mobile manipulator. In: 9th IEEE International Conference on Control and Automation (ICCA). IEEE 2011, pp. 778–783 (2011)
7. Lasnier, A., Murakami, T.: Hybrid sensorless bilateral teleoperation of two-wheel mobile manipulator with underactuated joint. In: 2010 IEEE/ASME International Conference on Advanced Intelligent Mechatronics. IEEE, 2010, pp. 347–352 (2010)
8. Frejek, M.C., Nokleby, S.B.: Simplified tele-operation of mobile-manipulator systems using knowledge of their singular configurations. In: International Design Engineering Technical Conferences and Computers and Information in Engineering Conference, vol. 49002, pp. 411–418 (2009)
9. Wrock, M., Nokleby, S.B.: Decoupled teleoperation of a holonomic mobile-manipulator system using automatic switching. In: 2011 24th Canadian Conference on Electrical and Computer Engineering (CCECE). IEEE, 2011, pp. 001 164–001 168 (2011)
10. Dominjon, L., Lécuyer, A., Burkhardt, J.-M., Andrade-Barroso, G., Richir, S.: The bubble technique: interacting with large virtual environments using haptic devices with limited workspace. In: First Joint Eurohaptics Conference and Symposium on Haptic Interfaces for Virtual Environment and Teleoperator Systems. World Haptics Conference. IEEE, 2005, pp. 639–640 (2005)
11. Mersha, A.Y., Stramigioli, S., Carloni, R.: Switching-based mapping and control for haptic teleoperation of aerial robots. In: 2012 IEEE/RSJ International Conference on Intelligent Robots and Systems. IEEE, 2012, pp. 2629–2634 (2012)
12. Satriadi, K.A., Ens, B., Cordeil, M., Jenny, B., Czauderna, T., Willett, W.: Augmented reality map navigation with freehand gestures. In: IEEE Conference on Virtual Reality and 3D User Interfaces (VR). IEEE 2019, pp. 593–603 (2019)
13. Adorno, B.V.: Two-arm manipulation: From manipulators to enhanced human-robot collaboration, Ph.D. dissertation, Université Montpellier II-Sciences et Techniques du Languedoc, 2011
14. Marinho, M.M., Adorno, B.V.: Adaptive constrained kinematic control using partial or complete task-space measurements. IEEE Trans. Rob. **38**(6), 3498–3513 (2022)

15. Marinho, M.M., Adorno, B.V., Harada, K., Mitsuishi, M.: Dynamic active constraints for surgical robots using vector-field inequalities. IEEE Trans. Rob. **35**(5), 1166–1185 (2019)
16. Hokayem, P.F., Spong, M.W.: Bilateral teleoperation: an historical survey. Automatica **42**(12), 2035–2057 (2006)
17. Adorno, B.V., Marinho, M.M.: Dq robotics: a library for robot modeling and control. IEEE Robot. Autom. Mag. **28**(3), 102–116 (2020)

Variable Stiffness & Dynamic Force Sensor for Tissue Palpation

Abu Bakar Dawood[1](✉), Zhenyu Zhang[2], Martin Angelmahr[2], Alberto Arezzo[3], and Kaspar Althoefer[1]

[1] Centre for Advanced Robotics @ Queen Mary, Queen Mary University of London, London, UK
a.dawood@qmul.ac.uk
[2] Fraunhofer Institute for Telecommunications, Heinrich Hertz Institute, Goslar, Germany
[3] Department of Surgical Sciences, University of Torino, Turin, Italy

Abstract. Palpation of human tissue during Minimally Invasive Surgery is hampered due to restricted access. In this extended abstract, we present a variable stiffness and dynamic force range sensor that has the potential to address this challenge. The sensor utilises light reflection to estimate sensor deformation, and from this, the force applied. Experimental testing at different pressures (0, 0.5 and 1 PSI) shows that stiffness and force range increases with pressure. The force calibration results when compared with measured forces produced an average RMSE of 0.016, 0.0715 and 0.1284 N respectively, for these pressures.

Keywords: Palpation · Soft Force Sensor · Optical Force Sensor

1 Introduction

Palpation is a key technique in which clinicians use their hands and fingers to localise and identify tissue abnormalities. By exerting force upon human tissue, clinicians can glean information about tissue condition during open surgery [1]. However, in Minimally Invasive Surgery (MIS) conventional palpation cannot be used, making tissue examination difficult [2]. Force sensing probes, adapted for minimally invasive surgery, can be used to detect tissue abnormalities such as tumors, which are significantly stiffer than healthy tissue by a factor of ten) [3].

Numerous solutions have been proposed to compensate for the lack of haptic feedback in Minimally Invasive Surgery. These solutions leverage sensing technologies including optical [4,5], vibro-acoustic [6] and pneumatic [7]. Typically, force sensors have a predetermined force sensing range and sensitivity - indeed only a few attempts have been made to develop sensors with adjustable force range and sensitivity. Raitt et al. [8] developed a stiffness controllable sensing tip that uses a camera to observe the deformation of a silicone membrane. The membrane stiffness can be controlled by pneumatic pressure, resulting in an adjustable force range. Another dynamic force range sensor ESPRESS.0 developed by Jenkinson et al. also uses pneumatic pressure to adjust the stiffness of the membrane, although in this instance, the force is measured with a camera, by tracking the fluid, coupled with the membrane, inside a tube [9].

This extended abstract introduces a soft force sensor whose stiffness can be adjusted, that employs light reflection to estimate deformation and, from that, the force. Its adjustable stiffness, which is controlled by pneumatic input, means that it can function over a variable force range. Experiments were conducted at different internal pressures and the calculated force was compared to the measured force.

2 Materials and Methods

To fabricate the sensor, moulds were designed using Solidworks and 3D printed using Ultimaker S3. Polylactic Acid (PLA) was used to 3D print the moulds. EcoFlex 00-50 parts A & B were mixed in equal quantities along with black dye. This black silicone mixture was then degassed using a vacuum chamber. A plastic fibre was placed inside the mould for integration into the silicone before closing the mould. This fibre would restrict the ballooning of the silicone dome, preventing it from rupture. EcoFlex was poured into the mould and degassed again to remove any air pockets within the mould. The EcoFlex was then cured at room temperature.

The Silicone dome, 40mm in diameter, was then taken out of the mould and Aluminum powder was brushed onto its inner side to enhance its reflectivity. A holder for the dome was 3D printed with a pneumatic channel and holes for both emitting and receiving optical fibres, as well as for the integrated plastic fibre. The dome was glued to the holder, ensuring no air leakage. The cross-sectional view of the sensor is shown in Fig. 1.

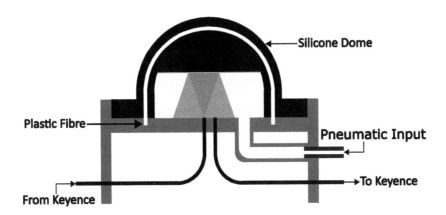

Fig. 1. Our proposed palpation sensor - a black silicone dome with an integrated plastic fibre to restrict ballooning.

The sensor was pressurised using a syringe pump, which was connected to a stepper motor and controlled by an Arduino controller. The air pressure inside the dome was measured using a pressure sensing IC (NPA-500B-005D by Amphenol Advanced Sensors).

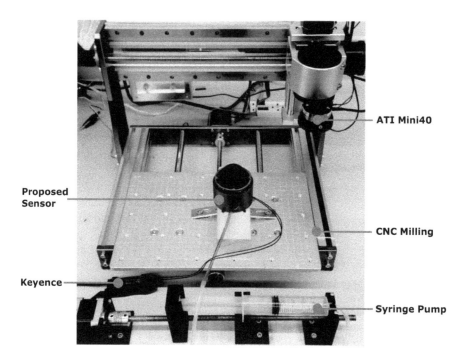

Fig. 2. The experimental setup showing a modified CNC Milling machine with an ATI Mini40, Syringe pump and Keyence.

The emitting and receiving optical fibres were connected to an optoelectronic system (Keyence). A Keyence emits light and measures the intensity of the light it then receives, converting it to a voltage signal which in our set-up was measured by the Arduino. A modified CNC milling machine (Genmitsu 3018-PRO) with an ATI Mini40 force and torque sensor was used to test the sensor. A python program was developed to control the movement of the CNC machine, communicate with the Arduino, and for data acquisition and data recording. The experimental setup is shown in Fig. 2.

The sensor was tested at 3 different pressures 0 PSI (0 Pa), 0.5 PSI (3.447 kPa) and 1 PSI (6.894 kPa). The sensor was indented using a flat indenter which was displaced by 4mm, and each experiment was repeated 3 times. For each pressure, the stiffness of the sensor was calculated by using Hooke's law,

$$F = kx \qquad (1)$$

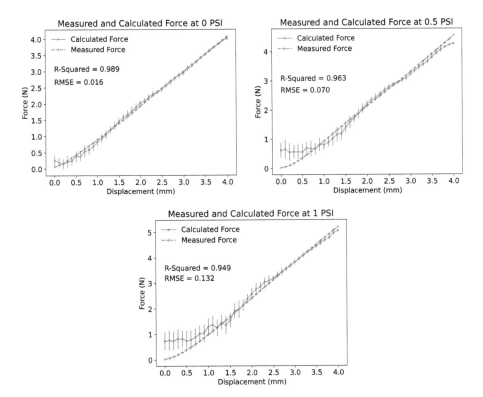

Fig. 3. Calculated and measured forces by force torque sensor at 0, 0.5 and 1 PSI, respectively. Error bars show the standard deviation. R-squared and RMSE between the calculated values and measured values are also shown.

where F is the applied force, k is the stiffness of the spring and x is the displacement. To find the stiffness of our sensor, the force measured by the ATI Mini40 sensor under a displacement of 4mm was used.

For one sample of each internal pressure, the data from the optical proximity sensor was then calibrated for distance, by using a 4th order polynomial. This calculated distance was then multiplied by the stiffness, to estimate the force value.

3 Results

Sensor stiffness values, calculated by using ground truth values of displacement and force, for 0, 0.5 and 1 PSI were $1016.13 \pm 1.29 N/m$, $1134.34 \pm 7.13 N/m$ and $1301.633 \pm 1.21 N/m$, respectively. The equations obtained after the polynomial fitting, for the calculation of displacement (y) from the optical data (x), are:

At 0 PSI:

$$y = -43.328x^4 + 4.367 \times 10^2 x^3 - 1.6605 \times 10^3 x^2 + 2.8294 \times 10^3 x - 1.82325 \times 10^3 \quad (2)$$

At 0.5 PSI:

$$y = -6.352\times 10^2 x^4 + 6.0185\times 10^3 x^3 - 2.1382\times 10^4 x^2 + 3.3765\times 10^4 x - 1.99978\times 10^4 \quad (3)$$

At 1 PSI:

$$y = -3.2323\times 10^2 x^4 + 3.0106\times 10^3 x^3 - 1.0519\times 10^4 x^2 + 1.6346\times 10^4 x - 9.53143\times 10^3 \quad (4)$$

The calculated value of displacement in meters when multiplied by the stiffness, resulted in the calculated force. The two metrics used for the comparison of calculated and measured force are R-squared and Root Mean Squared Error, shown in the Table 1. We note that curve fitting was performed only on Dataset 3 for each of the internal pressures.

Table 1. R-Squared and Root Mean Squared Error (RMSE) of all the three datasets at three pressures.

	Data 1		Data 2		Data 3	
	RMSE	R^2	RMSE	R^2	RMSE	R^2
0 PSI	0.016	0.989	0.016	0.989	0.016	0.989
0.5 PSI	0.0697	0.9637	0.0746	0.9604	0.0702	0.9633
1 PSI	0.1316	0.949	0.1286	0.9506	0.125	0.9523

4 Discussion

The stiffness of the sensor, calculated by using the ATI Mini40 and the displacement of the machine, shows a direct correlation with the applied pressure. It appears that by increasing the applied internal pressure, the force range of the sensor increases while the sensitivity decreases. This can also be observed in Fig. 3 - as the pressure increases, the standard deviation of the calculated force, especially for lower forces, also increases.

Table 1 shows the R-squared and Root Mean Squared Error of measured force and calculated force, for all three datasets. Both the metrics are very consistent for force estimation at 0 PSI and Fig. 3 confirms this. The metrics start showing deviation from previous behaviour with lower R-squared values and higher RMSE, as the pressure increases. The least favorable metrics are associated with dataset 1 at 1PSI, with an RMSE of 0.1316 N and an R-squared value of 0.949.

We plan to employ data driven methods to estimate interaction forces and the stiffness of palpated soft tissue. Future work will also focus on miniaturising the sensor for use in MIS, which will mean aiming for maximum diameters of 10mm so as to enable them to fit through trocar ports.

Acknowledgments. This work was funded by UK Research and Innovation (UKRI) under the UK government's Horizon Europe funding guarantee [grant # N°101092518] and funded by the European Union.

The design files, dataset, program and the instructions are uploaded to a Github repository *Variable Stiffness Sensor*.

References

1. Konstantinova, J., Jiang, A., Althoefer, K., Dasgupta, P., Nanayakkara, T.: Implementation of tactile sensing for palpation in robot-assisted minimally invasive surgery: a review. IEEE Sens. J. **14**(8), 2490–2501 (2014)
2. Konstantinova, J., Cotugno, G., Dasgupta, P., Althoefer, K., Nanayakkara, T.: Palpation force modulation strategies to identify hard regions in soft tissue organs. PLoS ONE **12**(2), e0171706 (2017)
3. Ahn, B.-M., Kim, J., Ian, L., Rha, K.-H., Kim, H.-J.: Mechanical property characterization of prostate cancer using a minimally motorized indenter in an ex vivo indentation experiment. Urology **76**(4), 1007–1011 (2010)
4. Ahmadi, R., Dargahi, J., Packirisamy, M., Cecere, R.: A new hybrid catheter-tip tactile sensor with relative hardness measuring capability for use in catheter-based heart surgery. In: SENSORS, 2010 IEEE, pp. 1592–1595 (2010)
5. Xie, H., Liu, H., Luo, S., Seneviratne, L.D., Althoefer, K.: Fiber optics tactile array probe for tissue palpation during minimally invasive surgery. In: 2013 IEEE/RSJ International Conference on Intelligent Robots and Systems, pp. 2539–2544 (2013)
6. Sühn, T., et al.: Vibro-acoustic sensing of tissue-instrument-interactions allows a differentiation of biological tissue in computerised palpation. Comput. Biol. Med. **164** (2023)
7. Wanninayake, I.B., Dasgupta, P., Seneviratne, L.D., Althoefer, K.: Air-float palpation probe for tissue abnormality identification during minimally invasive surgery. IEEE Trans. Biomed. Eng. **60**(10), 2735–2744 (2013)
8. Raitt, D.G., Abad, S.A., Homer-Vanniasinkam, S., Wurdemann, H.A.: Soft, stiffness-controllable sensing tip for on-demand force range adjustment with angled force direction identification. IEEE Sens. J. **22** (2022)
9. Jenkinson, G.P., Conn, A.T., Tzemanaki, A.: ESPRESS.0: eustachian tube-inspired tactile sensor exploiting pneumatics for range extension and sensitivity tuning. Sensors **23** (2023)

Real-World Testing of Ultrasonic Beacons for Mobile Robot Radiation Emulation

David Batty[1]($^{\boxtimes}$), Andrew West[2], Ipek Caliskanelli[3], and Paolo Paoletti[1]

[1] School of Engineering, University of Liverpool, Liverpool L69 3GH, UK
d.w.batty@liverpool.ac.uk
[2] Manchester Centre for Robotics and AI, The University of Manchester, Manchester M13 9PL, UK
[3] RACE, UKAEA, Culham Campus, Oxfordshire OX14 3DB, UK

Abstract. The paper describes a sensor system utilising Ultra Wide Band (UWB) ultrasonic beacons to emulate ionising radiation intensity mapping within complex environments. By strategically deploying a UWB beacon onto a robot platform and integrating this with its onboard systems, and by positioning stationary ground beacons around the environment, it is possible to emulate synthetic ionising radiation intensity. This involves setting up a system that can estimate the robot position based on distances to each beacon, and subsequently applying the inverse square law to determine radiation intensity. The system will provide continuous radiation intensity estimates as the robot navigates its environment, providing a suitable simulation alternative to an ionising radiation detector that requires the use of radioactive sources. Through experimental validation and calibration, the accuracy of radiation intensity calculations is verified, showing reliable performance in comparison to currently utilised technologies. Finally, an example deployment and the achieved performance is shown demonstrating the possible application of the technology.

Keywords: Radiation · Robotics · Real-world · Testing · Sensors · Emulation

1 Introduction

The integration of robotics into various real-world applications, including hazardous environment exploration, nuclear facility maintenance, and disaster response, necessitates rigorous testing to ensure system reliability and safety. Among the critical factors affecting robotic performance in such environments is

---

The authors wish to thank Dr. William McGenn and Dr. Linqing Zhang for their support in accessing and use of the anechoic chamber facility. This research was supported by RACE, UKAEA and EPSRC via a GREEN CDT PhD studentship. AW acknowledges support from UK Research and Innovation via the ALACANDRA project EP/V026941/1.

radiation, posing significant challenges for testing and validation. However, the conventional approach of utilising real radiation sources for testing is not only cumbersome and time-consuming but also poses risks to personnel and equipment [6,11]. Furthermore this often limits testing to a single source [2] which may prove to be unrepresentative of a real world environment [14].

Hence, the development of simulation techniques for replicating radiation environments becomes imperative. Simulating radiation enables researchers and engineers to conduct comprehensive testing of robotic systems in a controlled and repeatable manner, without the need for actual radioactive materials [15]. This approach not only mitigates safety concerns but also facilitates faster and more widespread validation of robotic platforms.

In this paper, we investigate an approach to mimic physically realistic ionising radiation source behaviour for real-world robotic testing, and highlight the significance of conducting such tests without relying on real radiation sources. By delving into the challenges associated with traditional testing methods and the benefits offered by hardware analogues, we aim to underscore the critical role of simulated radiation environments in advancing the development and deployment of robotic systems for radiation-prone applications.

2 Current Technologies

Localisation is a crucial aspect of mobile robotics, enabling robots to determine their position within their environment [7]. Various methods are employed to achieve this, each with its own advantages and limitations. This work will borrow from the concepts of beacon based systems, where distances to various points are calculated. The main methods of robot localisation include Ultra Wide Band (UWB) technology, Signal Strength (STS) WiFi detectors, Bluetooth Low Energy (BLE) beacons, vision-based or motion capture systems, and LiDAR-based systems [8]. These methods utilise different technologies and principles to provide accurate positioning information to mobile robots, catering to diverse application requirements and environmental conditions.

Ultra Wide Band (UWB) technology utilises radio waves across a wide frequency spectrum to determine the location of a robot precisely. Its primary advantage lies in its ability to localise other beacons without a specific calibration step or user information. This makes it particularly suitable for applications requiring localisation, such as in industrial settings or complex indoor environments, where infrastructure installation is prohibited. In comparison, these beacons are less expensive than more accurate systems such as motion capture cameras, but at the cost of spatial resolution. A practical example of UWB technology comes from Marvelmind [12]. The Marvelmind SuperBeacons can be used as a robust solution that is able to provide accurate distance measurement between beacons for a relatively low price tag and is being investigated in this work to determine if it can be used as an improvement to current solutions discussed below.

Signal Strength 2.4 GHz WiFi or Zigbee based detectors analyze the strength and quality of 2.4 GHz signals to estimate a robots distance from a point [4,9].

One of the main benefits of this method is its low deployment cost, as WiFi networks are commonly available in indoor environments. However, these radio signals can be susceptible to interference from obstacles, multi-path signal issues, and environmental conditions leading to potential inaccuracies in localisation. Moreover, the precision of WiFi-based methods may not always meet the requirements of certain applications that demand high levels of accuracy. Signal strength detectors feature in work conducted by Groves et al.. [5] investigating radiation-aware navigation. This work utilises the Safety Training Systems (STS) RadEye G-10 system as a simulated gamma radiation detector mounted onto a ground rover; this technology, due to previous deployment, is the main point of comparison for the project.

Bluetooth Low Energy (BLE) beacons transmit signals detected by devices equipped with BLE receivers to determine a robot location. These beacons offer a balance between cost-effectiveness and accuracy, with low power consumption and ease of deployment being notable advantages. A practical example of the use of a BLE system is work conducted by Cannizzaro et al. [3], which shows that even the most advanced algorithms only have a very small accurate functional distance, which for the project application is going to be too small to allow for reliable full environment coverage.

Vision-based systems utilize cameras to track markers or objects, such as those developed by April Tags [13], providing flexibility and adaptability in dynamic environments [1]. While this method can achieve high levels of accuracy, it often requires additional computational resources and optimal lighting conditions to function effectively. Additionally, vision-based systems may encounter challenges such as occlusions or changes in the environment, which can affect the reliability of localisation results.

Relying on the localisation of the robot platform itself, typically through use of LiDAR and odometry based fusion approaches, leads to issues around the reliability of the position estimate over time. Furthermore, an operator must have complete knowledge of the environment to digitally tag a spatial coordinate with pseudo radiation sources, as the localisation approach has no knowledge of installed hardware. This is impractical in unknown or unmapped environments.

In summary, each distance measurement method has its advantages and limitations, influenced by factors such as accuracy requirements, cost constraints, and environmental conditions. The choice of the most suitable method depends on careful consideration of these factors. For the project in question, physical testing of UWB and STS system was deemed most feasible, considering their accuracy, cost-effectiveness, existing use in literature, and suitability for the intended application. The following section conducts an accuracy assessment of the base systems to determine the difference in performance between the proposed system and the current system. The current system, the STS RadEye G-10, is used in a number of works by Groves et al.. [5]. This paper aims to determine which system provides the best performance for the task of radiation simulation.

3 Method

Ionising radiation transport has been implemented for robotic simulations [16], and this approach can be leveraged for hardware simulations also. The radiation intensity at a given point can be computed using the inverse square law, which states that the intensity of radiation decreases as the square of the distance from the source. For the purposes of the project, being a proof of concept, the testing will use a simplified point kernel radiation transport model [10]. To compute the radiation intensity at the current location of the mobile robot, we can use the distances measured to each ground unit $(r_1, r_2, ..., r_n)$ along with their respective radiation source intensities $(K_1, K_2, ..., K_n)$. The total radiation intensity (I_{total}) at the current location can be expressed as the sum of intensities from all ground units:

$$I_{\text{total}} = \sum_{i=1}^{n} \frac{K_i}{r_i^2} \quad (1)$$

Each ground transmitter serves as a radiation source of equal magnitude, emitting radiation uniformly in all directions. Any attenuation due to materials isn't considered in this simplified model, with the distribution being calculated as shown in Eq. 1. Any number of point sources can be used in combination to achieve a range of source types and scenarios when combined with tuning of the source intensity.

Anecoic Chamber Testing
Comparing the proposed Ultra Wide Band solution against the existing STS 2.4 GHz Zigbee distance measurement methods, necessitates thorough testing of accuracy under controlled conditions. Fair comparison is required to determine which system can achieve the best precision. As stated earlier, the STS system can suffer from multi-path effects, but this may also be of concern for the UWB approach. Testing conducted in a radio frequency anecoic chamber offers an optimal environment for eliminating external interference and assessing precisely the performance of each system. The testing setup, illustrated in Fig. 1, depicts the interior of the chamber with the system configuration. The chamber is configured to provide isolation for a large range of frequencies, roughly from 10 Hz up to approximately 40 GHz, meaning both frequency ranges such as ultrasonic and 2.4 GHz WiFi signals will be suitably isolated.

The UWB Marvelmind system is configured to ascertain distances between the mobile and stationary beacons, whereas the G10 STS system outputs estimated radiation intensity values based on received signal strength. In both cases, the base station remains stationary at one end of the chamber, while the mobile receiver is positioned at a predetermined distances. Measurements of estimated distance or received signal strength are recorded to capture any fluctuations in the collected data, with emphasis placed on assessing the stability and usability of the systems.

Uncontrolled Environment Chamber Testing

Whilst testing in an anechoic chamber can provide baseline testing of the technology, testing in a realistic environment will provide an assessment of performance of the systems during a realistic deployment. Figure 2 shows the second testing environment. This features a typical lab setup without any system isolation. Work conducted by Groves et al. [5] found anomalies in the data suspected to be from "multi path propagation or the sensor receiving spurious readings from other equipment" such as wireless networks and local network machinery. Several important elements feature during this testing, such as multiple WiFi networks active, noisy equipment such as running 3d printers and fans and finally reflective irregular surfaces leading to the potential for acoustic reflections. In this environment the same testing procedures will be carried out at the same distance increments as those conducted in the anechoic chamber. The primary goal of this section of testing is to determine if the proposed technology is a more suitable candidate based on reliability in uncontrolled environments similar to those that might be found during real-world testing.

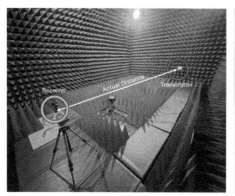 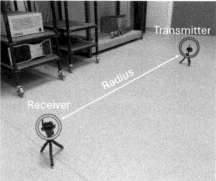

Fig. 1. Experimental testing setup inside an anechoic chamber

Fig. 2. Experimental testing setup in an uncontrolled lab environment

4 Results

Data from both systems were collected using the testing protocols outlined in the previous section. Raw intensity values for the STS G10 unit, and calculated intensity values from the Marvelmind UWB beacons. This includes testing conducted in both the controlled environment (anechoic chamber) and non-controlled environment.

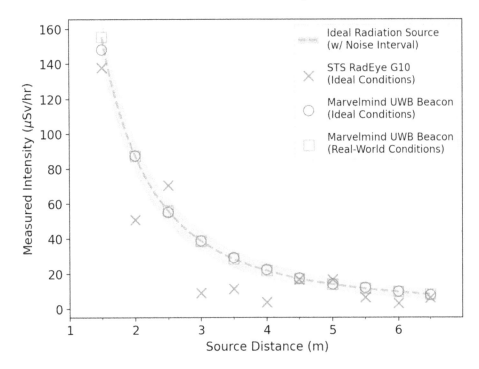

Fig. 3. Radiation intensity as a function of distance for different technologies in comparison to the expected results for an ideal radiation source with an expected noise region.

Figure 3 shows the response from all systems compared to the expected response from a single point radiation source. The noise interval for the ideal case is taken as ± standard deviation of a Poisson distribution with the same mean intensity. Therefore, this represents the likely band of responses from a randomly emitting source during radioactive decay.

The Marvelmind UWB beacon approach tracks within the expected noise interval. During testing in the anechoic chamber, the maximum intensity error across all distances tested was 0.8% occurring at a distance of 1.5 m, this maximum intensity error increased during the uncontrolled environment testing with a value of 4.8% at distance of 6.5 m. The expected noise interval for a true source at these distances are 8% and 35% respectively, therefore, the technique used in this study can accurately replicate the ideal distance-intensity response of a true source.

In contrast, the majority of intensity values for the STS G10 unit was outside the ideal noise interval, with some points falling vastly outside of the ideal noise interval. Though the general qualitative shape of the distance-intensity response is preserved, at 2.5 m it is clear that the value increases, rather an decreases. This occurs at other distances but not as dramatically. Hence, it can be shown that the unreliability of the data can be defined as greater then expected noise.

From a quantitative perspective of accuracy, the cumulative fractional error for the UWB system in the anechoic chamber and real-world conditions are 0.25 µSv/hr and 0.41 µSv/hr respectively. The STS G10 system has a cumulative fractional error of 4.55 µSv/hr. The STS G10 system demonstrates an overall error an order of magnitude worse than the real-world UWB system.

The STS G10 system shows a worse precision of the intensity. With an average standard deviation of 0.3472 µSv/hr it was calculated that the Marvelmind UWB system maintained a considerably lower average standard deviation of 0.0064 µSv/hr throughout each test data set. This pronounced disparity underscores the Marvelmind UWB system's superior reliability, 100 times greater than that of the Radeye STS system.

Overall, it can be seen that even during uncontrolled environment testing the Marvelmind UWB beacons proved to be more reliable than the baseline taken for the Radeye G10 in the anechoic chamber. Based on related work by Groves et al. as mentioned previously, the reliability of the Radeye system decreases significantly when operated in an uncontrolled lab environment similar in nature to that in which the UWB beacons were tested. Hence it can be concluded that both quantitatively and qualitatively the Marvelmind UWB beacons can be deemed as more reliable in the intended testing environment based on testing results.

5 Discussion

Measured intensity of ionising radiation is itself significantly noisy, due to its emission being an inherently stochastic process. For testing purposes, having a simulated system that can replicate almost noise free distributions allows for the possibility of adding noise during testing in a controllable manner, rather than being restricted to already noisy data. This has considerable benefits when evaluating algorithms, which can now be exposed to near identical radiation distributions reliably. The proposed system is therefore of greater benefit to robotics development and testing compared to existing solutions.

The findings reported in the previous section suggest unequivocally that, for the project objectives, the Marvelmind UWB beacons present a vastly superior option in terms of accuracy and reliability for simulating radiation within robotic systems. Additionally, the Marvelmind SuperBeacons' ability to recreate an accurate distribution based on distance allows for more extensive testing through variable noise levels and adds an additional level of capability. The following section will describe deployment of the Marvelmind for emulation of radiation in the real-world testing of mobile ground robots.

6 System Deployment

Following the determination that the Marvelmind beacons' UWB system is significantly more precise than the other suitable systems tested, a deployment method is explored to show how the system can be deployed onto a mobile robot as a simulated ionising radiation detector. For the purposes of testing, a

Unitree Go2 EDU was used as the mobile robot platform, featuring a Jetson Orin single board computer. However, the system is capable of being deployed on any robot with an onboard computer running ROS2.

The Marvelmind SuperBeacons operate in two separate modes, either ground beacon or hedgehog beacons. The ground beacons are assumed to be stationary, they can moved but must be stationary in order for mobile beacon to gain localisation. During testing, these beacons have been mounted on tripods to allow for easier placement around the environment. The hedgehog beacon is Marvelmind's designation for the mobile beacon, to be placed on the mobile robot. This must be done securely to avoid the sensor moving relative to the robot during operation, as due to the accuracy of the sensors this could affect the intensity measurement, particularly when in close proximity to a source. Additionally ensuring the beacons have there relative ground height set will allow for improved accuracy during operation.

The sensors operate using the Marvelmind dashboard. This must be running on a separate computer during the setup of the experiment for the beacons to gain localisation of each other. Once this has occurred this system no longer needs to be operational. This system is not connected in anyway to the robot, it acts only to provide map data to the beacons to gain better distance estimations. The beacon that is mounted onto the robot should be connected to the robot via USB. A ROS2 node has then been created that can read the relative position data from the beacons and determine a linear distance between the source beacons and the hedgehog. These distances can then be utilised calculate the radiation intensity using Eq. 1. This value is then published to a topic as new values are received to emulate the data flow of a real radiation detector. The rate of collection can be set according to the specification of the sensor being simulated (Fig. 5).

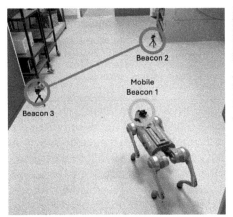
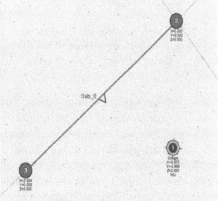

Fig. 4. Unitree Go2 setup with mobile beacon and two Marvelmind Beacons placed as simulated sources

Fig. 5. Marvelmind Dashboard showing the distance data that is available to the ROS2 node via the Marvelmind API

Figure 6 shows an example deployment with a single point source. The receiver was configured as in Fig. 4 and the Go2 was set to walk slowly past the source in the direction of travel indicated. The robot was stopped briefly at each of the three points to distinguish individual distribution curves along its route. Points A and C were 12' away from the beacon and point B was 8" from the source. The Go2 was set to travel at 0.5 ft/s to allow for data capture. However, the system can react reliably at much higher speeds when data capture speed is reduced to a single reading per second to emulate a traditional radiation sensor rate. A ROS2 based node has been developed in conjunction with the Marvelmind driver to provide real-time visualisation of the measured intensity. The advantage of utilising a system with a high level of reliability, is the ability to tune the intensity of the source and apply noise to match specific source characteristics, such as source intensity and attenuation coefficients, in order to change the magnitude and distribution of the intensity measured.

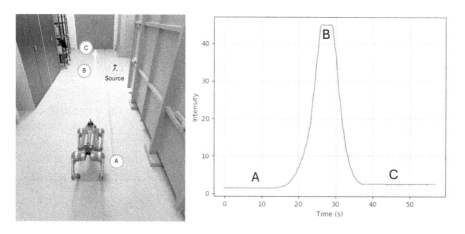

Fig. 6. Example testing data captured during transit past a simulated source beacon. Direction of travel is indicated by the arrow. Robot travelled at a constant speed, briefly stopping at each labelled point.

By deploying Marvelmind SuperBeacons onto a mobile robot and utilising distance measurements to compute radiation intensity using the theory discussed previously, we can enable real-time monitoring and sensing of an environment in the real-world without the use of real radiation sources. This allows for significantly safer testing procedures, thus allowing for more readily available testing scenarios that may have been prohibitively dangerous to be conducted with real radiation sources. This availability is crucial to improve trust in a robot system before deployment into a hazardous environment.

Comparison of the proposed system to real radiation distributions to determine accuracy, was not possible due to safety considerations, with point source simulation being a common technique for simulating possible intensity distributions of environments. However, as mentioned previously, the accuracy of the system can be tuned to represent any radiation source through the adjustment of parameters and combination of multiple point sources to create an extensive range of testing scenarios. In the testing example a simplified conversion of radiation point sources has been applied. The improved reliability of source measurement would allow for more complex custom distributions to be applied easily. Thus greatly increasing the flexibility of the proposed solution in comparison to previously utilised technologies.

7 Conclusion

The integration of Marvelmind SuperBeacons for determining distances to mock radiation sources used to model dose rate intensity in real-world robotic applications, offers a promising solution for enhancing safety in radiation-prone environments. The computational framework based on the inverse square law provides a reliable and physically grounded method for accurately assessing ionising radiation levels at any position. By leveraging distance measurements to nearby SuperBeacons and their radiation constants, real-time monitoring of radiation intensity become feasible, enabling proactive risk mitigation strategies.

Direct comparison of the current state of the art system to a newly proposed technology application found that the new application features significantly higher reliability and adaptability based on performance metrics in both controlled and uncontrolled environments.

The deployment process of Marvelmind SuperBeacons involves strategic positioning, calibration, and integration with the robot's sensing suite. This ensures seamless communication between SuperBeacons and the robot, facilitating accurate radiation intensity calculations during operation. An example deployment method was outlined, illustrating how this technology can be utilised, and preliminary test results showed the level of performance achieved during this example deployment.

The potential applications of this technology span various industries, including nuclear power plants, industrial facilities, and disaster recovery operations, where other physical phenomena could be modelled. The ability to assess radiation intensity enhances situational awareness and enables timely interventions to minimize exposure to hazardous radiation levels.

Moving forward, further research and development efforts are needed to refine the radiation transport model, optimise deployment strategies, and explore synergies with other sensing technologies. Overall, the integration of Marvelmind SuperBeacons for radiation intensity will allow more realistic system testing in nuclear environments without the need to utilise real radiation sources. This implies increased safety during development of robotic systems for these applications, and increased capability for faster, less restricted, and more comprehensive testing of new robotic systems.

References

1. Alkendi, Y., Seneviratne, L., Zweiri, Y.: State of the art in vision-based localization techniques for autonomous navigation systems. IEEE Access **9**, 76847–76874 (2021). https://doi.org/10.1109/ACCESS.2021.3082778
2. Bird, B., et al.: A robot to monitor nuclear facilities: Using autonomous radiation-monitoring assistance to reduce risk and cost. IEEE Rob. Autom. Maga. **26**(1), 35–43 (2019). https://doi.org/10.1109/MRA.2018.2879755
3. Cannizzaro, D., et al.: A comparison analysis of ble-based algorithms for localization in industrial environments. Electronics (Switzerland) **9**(1) (2020). https://doi.org/10.3390/electronics9010044
4. Deasy, T.P., Scanlon, W.G.: Simulation or measurement: the effect of radio map creation on indoor WLAN-based localisation accuracy. Wirel. Pers. Commun. **42**(4), 563–573 (2007). https://doi.org/10.1007/s11277-006-9211-x
5. Groves, K., Hernandez, E., West, A., Wright, T., Lennox, B.: Robotic exploration of an unknown nuclear environment using radiation informed autonomous navigation. Robotics **10**(2), 78 (2021). https://doi.org/10.3390/robotics10020078. https://www.mdpi.com/2218-6581/10/2/78
6. Huntingdon S: The NDA's 'Grand Challenges' for technical innovation - Cleaning up our nuclear past: faster, safer and sooner (2020). https://nda.blog.gov.uk/the-ndas-grand-challenges-for-technical-innovation/
7. Institute of Electrical and Electronics Engineers: 2020 IEEE International Conference on Robotics and Automation (ICRA), Paris, France, 31 May–31 August 2020. IEEE (2020)
8. Jonasson, E.T., Ramos Pinto, L., Vale, A.: Comparison of three key remote sensing technologies for mobile robot localization in nuclear facilities. Fusion Eng. Des. **172** (2021). https://doi.org/10.1016/j.fusengdes.2021.112691
9. Kabiri, M., Cimarelli, C., Bavle, H., Sanchez-Lopez, J.L., Voos, H.: a review of radio frequency based localisation for aerial and ground robots with 5G future perspectives (2023). https://doi.org/10.3390/s23010188
10. Knoll, G.F.: Radiation Detection and Measurement. Wiley, Hoboken (2000)
11. Nuclear Decommissioning Authority: Nuclear Decommissioning Authority NDA Report The NDA's Research and Development Strategy to Underpin Geological Disposal of the United Kingdom's Higher-activity Radioactive Wastes. Nuclear Decommissioning Report (2009)
12. Vagner, M., Palkovics, D., Kovacs, L.: 3D localization and data quality estimation with marvelmind. In: 2022 IEEE 2nd Conference on Information Technology and Data Science, CITDS 2022 - Proceedings, pp. 302–307. Institute of Electrical and Electronics Engineers Inc. (2022). https://doi.org/10.1109/CITDS54976.2022.9914386
13. Wang, J., Olson, E.: AprilTag 2: efficient and robust fiducial detection. Technical report. April Robot Lab (2017). http://april.eecs.umich
14. West, A., et al.: Use of Gaussian process regression for radiation mapping of a nuclear reactor with a mobile robot. Sci. Rep. **11**(1) (2021). https://doi.org/10.1038/s41598-021-93474-4
15. West, A., Wright, T., Tsitsimpelis, I., Groves, K., Joyce, M.J., Lennox, B.: Real-time avoidance of ionising radiation using layered costmaps for mobile robots. Front. Rob. AI **9** (2022). https://doi.org/10.3389/frobt.2022.862067
16. Wright, T., West, A., Licata, M., Hawes, N., Lennox, B.: Simulating ionising radiation in gazebo for robotic nuclear inspection challenges. In: Advances in Robots for Hazardous Environments in the UK (2022). www.mdpi.com/journal/robotics

Robotic Tight Packaging Using a Hybrid Gripper with Variable Stiffness

Michele Moroni[1,2], Ana Elvira Huezo Martin[1], Leonard Klüpfel[1], Ashok M. Sundaram[1], Werner Friedl[1], Francesco Braghin[2], and Máximo A. Roa[1]()

[1] German Aerospace Center (DLR), Münchner Str. 20, 82234 Weßling, Germany
{michele.moroni,ana.martin,leonard.klupfel,ashok.sundaram,
werner.friedl,maximo.roa}@dlr.de
[2] Politecni co di Milano, Via La Masa, 1, 20156 Milano, Italy
michele2.moroni@mail.polimi.it, francesco.braghin@polimi.it

Abstract. The involvement of robots in warehouse automation poses new problems to research in logistic tasks such as tight packaging, in which a container must be completely filled with items, in a regular and ordered manner, leaving minimum clearance between them. This work investigates the effect of a reliable placing strategy using a system with passive compliance to improve robustness and success rate in such a task. The methodology is integrated into a full pipeline to execute the packaging operation and is evaluated in a real robot, using a mechanically compliant hybrid gripper with variable stiffness, exploring the roles of the hand configuration and stiffness level in the task execution. Along different evaluation tasks, the results show an improvement in success rate thanks to a reliable insertion strategy, when compared to a trivial one. They also demonstrate the efficacy of using variable stiffness to reduce error propagation.

Keywords: Robotic Packaging · Variable Stiffness · Hybrid Gripper

1 Introduction

The introduction of robotics in logistic scenarios poses new research challenges, mainly in the development of end-effectors, and methodologies for item manipulation and packaging. For packaging applications, the focus is mainly on checking if an item fits in the bin and finding a suitable placement for it. This aspect is often solved as trivially as lowering vertically the object into its intended position and releasing it [13,19]. Most approaches deal with uncertainty by leaving some margin between objects, which results in loose and irregular packages

---

M. Moroni and A. E. H. Martin—Equal contribution to this work.
This work was partially funded by the European Commission under the Horizon Europe Framework Programme grant number 101070600, project SoftEnable.

© The Author(s), under exclusive license to Springer Nature Switzerland AG 2025
M. N. Huda et al. (Eds.): TAROS 2024, LNAI 15051, pp. 313–326, 2025.
https://doi.org/10.1007/978-3-031-72059-8_27

[2]. A more challenging and realistic application happens when the robot performs tight packaging, meaning that the container is filled with items so that they fit closely together, with minimum clearance between them (Fig. 1). This setting however provides a contact-rich environment, with noise and disturbances coming from either the object localization, the perception of the bin, or the robot and end-effector themselves.

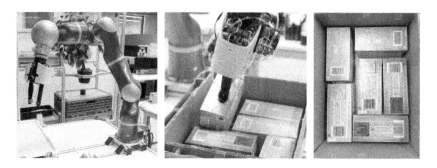

Fig. 1. Robotic tight packaging task and our experimental setup.

The tight packaging task can be subdivided into three smaller operations: object grasping, package planning and item insertion.

Object Grasping. For the robot to get a hold on the object, suction cups have proved a valuable tool, both for grasping [3] and manipulating [4] objects. However, using suction is not always possible, hence we propose to use hybrid fingers to broaden the range of objects the gripper can work with. In our previous work [5], we proposed a planner to select the best grasp modality using hybrid grippers.

Package Planning. Also known as Bin Packing, this problem has been considered since the 80s using different heuristics [1]. More recently, the research has focused toward online packing planning. [17] proposes a methodology to verify if a given item set can fit in the container, regardless of the order of the objects. [7] implements a pipeline to evaluate online the dimension of unknown items arriving with unknown order, and to find a placement for them using a distance-based heuristic; [18] solves the same task with a height-based heuristic.

Item Insertion. Tight packaging is similar to a less constrained Peg-in-Hole problem. Late solutions for peg-in-hole use visual feedback [12], force feedback [16], and some estimation of the contact state [11]. These works highlight the importance of exploiting contacts with the environment and using intrinsic compliance to guide the insertion, an idea that we follow in this paper.

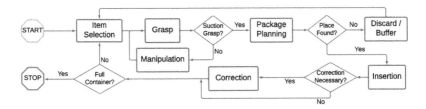

Fig. 2. Scheme of the execution of the task.

A full pipeline was proposed in [15] to solve the tight packaging problem using a suction cup and visual feedback; to place an item, a simple insertion is performed in a free portion of the container, and then the object is moved toward the desired position. Corrective actions are also implemented to fix imprecise placements, pushing and pulling the item. In this way, a tight package is achieved, but it is not possible to fill the whole container, as it is necessary to have space to insert the item and maneuver it. [20] uses instead a clawed gripper to place cubes using force/torque sensing to guide the execution. The approach improves on the previous one by increasing the placement accuracy (reducing the need for corrective actions). However, the chosen gripper also prevents the complete filling of the container.

Our work deals with the tight packaging problem by 1) proposing a package planner integrated with a bio-inspired insertion strategy that only requires the intended place for the item to be free, hence filling the container up to all four walls; 2) using a gripper with mechanical variable stiffness to adapt to uncertainties and deal with error propagation; 3) exploiting the use of a hybrid gripper to execute the task using both pinch and suction grasps with the same end-effector, which benefits both the range of items that can be grasped and their placement in constrained spaces; and 4) We evaluate the methodology in a real robotic system using the DLR Hybrid Compliant Gripper (HCG)[1], integrating planning and insertion strategies into a full tight packaging pipeline. Various experiments are then carried out to test the effectiveness of the approach.

2 Problem Definition and Representation

To perform an efficient package planning, the dimensions of container and items are discretized using a given resolution. The container $C = [P_C^w, G_C]$, is described by the position P_C^w of its top-left corner in world coordinates. It is discretized with a planar grid $G_C = [res_C, g_C]$, characterized by the discretization resolution res_C, and the integer dimensions of the grid $g_C = [g_{C,i}, g_{C,j}]$. The same applies to the buffer $B = [P_B^w, G_B]$, which is represented as a 1D grid $G_B = [res_B, g_B]$.

The item set I is defined as a set of object-types. Each object $o_n = [\widetilde{P_{o_n}^w}, t_n, \{F_m\}_n]$ is an instance of one of the types available in I; it is character-

---

[1] DLR HCG (2021) https://www.dlr.de/rm/en/desktopdefault.aspx/tabid-18061.

ized by its position in world frame $\widetilde{P^w_{o_n}}$, its type t_n, and by a set of faces $\{F_m\}_n$, which define its geometry. In turn, each face $\{F_m\}_n = [P_{F_{n,m}}, b_{n,m}, d_{n,m}]$ is characterized by the relative pose of its centre with respect to the centroid of the object $P_{F_{n,m}}$, a Boolean $b_{n,m}$ which defines if it can be grasped through suction, and the planar discretized dimensions $d_{n,m} = [d_i, d_j]$ that the item has when looked at from the normal along the surface.

The tight packaging task is defined as the collocation of the set of objects O into the container C; each item undergoes a cycle of grasp, planning and placement, as depicted in Fig. 2. In every cycle, a random object from the item set starts in a random pose in the grasp workspace; one object at a time is allowed to be in that region. The item is localized and grasped there. Once the operation is successful, the package planner is invoked to find a position for the item in the bin, using knowledge of the nominal state of the grid. An item can either find a placement, be assigned to the buffer, or be discarded.

In the first case, the planner outputs a target placement for the item in grid coordinates $P^g_{o_n}$, which corresponds to a target pose in world coordinates $P^w_{o_n}$. The object placement is performed along a trajectory that the robot has to follow, defined as a set of poses $T = \{P_k\}$; each pose can be expressed as relative to the target position $P^w_{o_n}$, or in world coordinates, hence called respectively $T^t_{o_n} = \{P^t_k\}$ and $T^w_{o_n} = \{P^w_k\}$. The cycle is repeated until the container is filled, or an unrecoverable error happens. The container is considered filled if an item has been assigned to all the cells, and if the actual pose of each object placed $\widetilde{P^w_{o_n}}$ is within a threshold ϵ_a of the nominal one $P^w_{o_n}$. Similarly, an unrecoverable placement happens when the difference between actual and nominal pose is greater than another threshold ϵ_f.

3 Pipeline for Tight Packaging

Grasp and Manipulation Approach. A prior analysis is performed to define the robot workspace for the tasks, using its capability map [14]. First, the *grasp workspace* is defined as a planar area of the table that the robot can reach and where the vision pipeline can return the poses of the items with sufficient confidence. Then, the best location for the container is found by checking that in the selected position the robot is able to pack items in the whole bin, collision-free and within its reachability. In a similar way, a region of the workspace is assigned to be the buffer.

Due to the constraints of tight packaging, a suction grasp is the preferred modality, as the gripper only needs access to one face of the item. On the contrary, with a pinch grasp a finger will always be in the way of a tight insertion. The grasp algorithm requires knowledge of the object type and its 6D pose, coming from a vision pipeline. With this, the algorithm computes which face of the item is looking upwards, and if it is possible to perform a suction grasp; if not, a pinch grasp is performed and the object is reoriented, in order to expose a different face. The pipeline for this process is described in Algorithm 1.

Algorithm 1. Grasp and Manipulation

Input: Buffer B
1: $items\_number, \{\widetilde{P_{o_n}}, t_n\} \leftarrow call\_vision()$
2: **if** $items\_number \neq 1$ **then**
3:    return exit Error
4: $o_n \leftarrow select\_item(B, [\widetilde{P_{o_n}}, t_n])$
5: Create object: $o_n \leftarrow [\widetilde{P_{o_n}}, t_n, \{F_m\}_n]$
6: $F_{n,m} \leftarrow find\_upwards\_face(o_n)$
7: **if** $b_{n,m}$ **and not** $failure$ **then**
8:    $success \leftarrow plan\_and\_execute\_suction\_grasp(o_n)$
9:    **if** $success$ **then**
10:      $failure \leftarrow$ False
11:      return exit $\leftarrow$ Package Planner
12:    **else**
13:      $failure \leftarrow$ True
14:      return exit $\leftarrow$ Grasp and Manipulation
15: **else**
16:    $success \leftarrow plan\_and\_execute\_pinch\_grasp(o_n)$
17:    **if** $success$ **then**
18:      $execute\_manipulation()$
19:      return exit $\leftarrow$ Grasp and Manipulation
20:    **else**
21:      return exit $\leftarrow$ Grasp and Manipulation

Package Planner and Item Selection. After a successful grasp, the planner (shown in Algorithm 2) selects where to place the item using as input the nominal state of the container, and the planar dimensions of the object $d_{n,m}$. First, the planner checks if the item should be buffered; this applies to small items, which can be useful in the latest stages of execution, where their size allows to fill in the irregular and small gaps remaining. This strategy is used until the buffer is full, or until it contains enough items to fill the remaining empty gaps. The Item Selection works in sync with this strategy: if the buffer needs to be emptied, the first item selected is the largest one in it; otherwise, the robot grasps the item laying in the grasp workspace. If the item is not buffered, the planner uses a Bottom-Left heuristic to find a placement for the item in the bin; the cost function implemented defines a *packing direction*, from the top-left corner towards the opposite one. If no placement is possible, the item is discarded.

Contact Analysis. Our packaging methodology relies on the detection of corners, defined by the edges of the items and the walls of the bin. A correctly oriented pushing action can move an item towards one corner, where it can fit guided by the constraints that it finds (Fig. 3-A). Moreover, as the bin fills up, the items constrain each other, reducing their freedom of movement.

Algorithm 2. Package Planner

Input: o_n, C

1: **if** $o_n \in small\_items$ **and not** $full\_buffer$ **then**
2:     $return\ exit \leftarrow$ Buffer
3: $found, P^g_{o_n} \leftarrow find\_placement(d_{n,m}, G_C)$
4: **if not** found **then**
5:     $d_{n,m} \leftarrow [d_j, d_i]$
6:     $found, P^g_{o_n} \leftarrow find\_placement(d_{n,m}, G_C)$
7: **if** $found$ **then**
8:     $return\ exit \leftarrow$ Insertion
9: **else**
10:     $return\ exit \leftarrow$ Discard

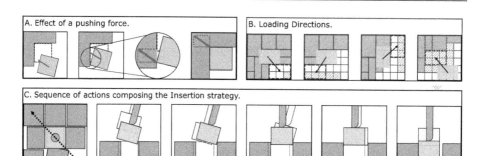

Fig. 3. A. Strategical selection of the pushing force drives the item toward a constraint and then complies into its target position. B. The four possible *loading directions*, depending on the target position (dashed). C. Sequence of actions composing the *insertion strategy* (loading from right to left in the side view).

Insertion Strategy. By properly choosing a *loading direction* it is possible to keep piling objects in the container, correcting the arising uncertainties through a placing-by-pushing action. The insertion starts by selecting a loading direction depending on the current situation in the container (represented by the container grid), and the intended position of the item. As shown in Fig. 3-B, four directions are possible, defined as an angle in the container plane; each of them points from the center of the container towards one of the corners. The item is loaded preferentially towards the walls of the container; if not possible, against the packaging direction enforced by the planner. To insert the object in its slot, a set of actions is performed along the found direction, as depicted in Fig. 3-C. The item is first inclined backwards, to expose a corner, which is used to pierce between the items underneath; the held object is then pulled backwards, to clear some space, and straightened, pushing eventual items out of the way. Then, the item is moved forwards, onto its target position, and then lowered. This sequence of pushing and pulling does not disrupt the state of the bin achieved up to this moment. These actions are performed by moving the item along a trajectory defined by poses in space; these points are computed by a distance from the target position of the item in the container along the loading direction, a height from the container floor, and an orientation. Algorithm 3 details the execution.

Algorithm 3. Insertion

Input: Bin C, item dimensions $d_{n,m}$, target pose $P^g_{o_n}$

1: $P^w_{o_n} = from\_relative\_to\_world\_coordinates(P^g_{o_n}, P^w_C)$
2: $\theta = compute\_loading\_direction(P^g_{o_n}, d_{m,n}, G_C)$
3: $T^w_{o_n} = compute\_trajectory(T^t_{o_n}, P^w_{o_n}, \theta)$
4: $execute\_trajectory(T^w_{o_n})$
5: $return\ exit = $ Correction

Algorithm 4. Correction

Input: Container grid G_C, target pose $P^w_{o_n}$, acceptance threshold ϵ_a

1: $\widetilde{P^w_{o_n}} = query\_item\_position()$
2: **if** $\widetilde{P^w_{o_n}} - P^w_{o_n} < \epsilon_a$ **then**
3: $\quad return\ exit = $ Grasp and Manipulation
4: $\theta = compute\_correction\_direction(\widetilde{P^w_{o_n}}, P^w_{o_n})$
5: **if** $verify\_feasibility(G_C, P^w_{o_n}, \theta)$ **then**
6: $\quad plan\_and\_execute\_correction(P^w_{o_n}, \theta)$
7: $return\ exit = $ Grasp and Manipulation

Hybrid Grasp and Stiffness Level. To select a suitable hand configuration and stiffness level, the requirements of the task are analyzed, following our previous work in [10]. To accomplish the *insertion* it is necessary, on the one hand, to retain compliance in the horizontal plane, so to allow the item to look for contacts and to adapt to the surrounding despite the uncertainties. On the other hand, the robot needs to exert vertical forces to pierce and to insert the item, requiring a higher stiffness along said direction. The benefits of the hybrid gripper are here exploited: the item is grasped using suction, and the other fingertip is simply sitting on top of the item, to resist the vertical forces while allowing it to slide under its tip, thus keeping the required compliance (Fig. 4-A).

Corrective Actions. To minimize the chances of having objects occluding potential slots, it is possible to perform *corrective actions* after a placement. Given the actual and intended position of the item, a correction is triggered if their distance exceeds a given threshold. The correction is a pushing action along a direction, computed as a vector from $\widetilde{P^w_{o_n}}$ to $P^w_{o_n}$. It is performed by inserting a straight finger and moving it until a force exceeding a given threshold is sensed, signaling that a constraint has been found. Therefore, before executing the action, it is necessary to check if there is enough space to safely insert a finger. This is done considering the nominal situation in the container, verifying that a sufficient number of cells are empty in front of the item along the pushing direction. Algorithm 4 details the execution.

4 Experimental Setup

For the experimental task we use the redundant DLR LWR III with the mounted DLR Hybrid Compliant Gripper (HCG) and a F/T sensor placed between arm and gripper (Fig. 1). The HCG is a tendon-driven, underactuated, and hybrid gripper. Its fingers are the thumb modules of the CLASH hand [6], with 3 DoFs, slightly modified to accommodate a suction cup at the tip; there is an additional DoF at the base of each finger for enhancing the width of the grasp. The stiffness of each finger can be independently set, on a scale from 0 to 100% (the higher the value, the stiffer the finger), and independently of the finger configuration.

We capture RGB and depth images using an Azure Kinect DK camera. For computing the 3D position of an item, we first run GroundingDINO [9] to detect the item and pass the corresponding bounding box to a Segment Anything model [8] to extract the respective segmentation mask. Second, we derive the depth value of the segmentation mask's centroid and unproject the position with known camera intrinsics. Eventually, we return the position of the item with respect to the robot through a prior robot-to-camera calibration.

The packaging pipeline is implemented as a state machine, and relies on the DLR motion planning library *RMPL*. The DLR middleware *Links and Nodes* is used to enable communication among all the components. The user is involved in the execution, as the corrective actions are manually triggered, since this feedback is unavailable in the current implementation. The same setup is replicated in a simulation environment using Python, with PyBullet as physics engine.

5 Experimental Evaluation: Results and Discussion

Three experiments are designed to test the proposed pipeline:

1. Blocks Packaging: Mega Blocks have to be packed upright in a half-opened container, both in simulation and reality, with nominal clearance between items of 1 mm (Fig. 5-A). The blocks are graspable through suction only on the sides, with a grasp error of up to 15 mm. Hence, this task emphasizes the grasp and manipulation aspect of the pipeline. A simplified insertion strategy is tested, since there is no need to incline the item and pierce. The hybrid grasp uses stiffness values of 10% and 40% of the maximum stiffness.
2. Blocks Stacking: follows the previous task, but now the blocks are stacked onto a base layer, rather than laying them on a flat surface (Fig. 5-A). This is performed in reality only, with a nominal clearance of 1 mm, and fingers with stiffness of 10% and 60%.

3. Boxes Packaging: a closed container is to be filled with boxes, with a nominal clearance up to 15 mm. The full insertion strategy is tested here (Fig. 5-B). In simulation, 5 shapes of boxes are used, and a buffer is present; in reality, the task is performed with tea boxes, with only one shape available. The grasp strategy uses a stiff finger (value equal to 30%) in the rear of the object, to pierce and push, and a soft one (value equal to 10%) in the front, to comply to the surrounding; the object is held with suction, due to its weight, with a grasp error that can be up to 25 m.

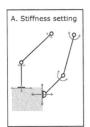 

Fig. 4. A. HCG uses different stiffness levels in each finger: a suction cup with soft setting (blue) grasps the object while providing adaptability, and a fingertip with stiff setting (red) allows application of forces for insertion. B. A block being grasped and released in a pose that allows suction. (Color figure online)

 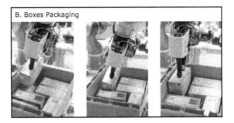

Fig. 5. A. Blocks packaging (left) and stacking (right). Note the hybrid grasp used for supporting the object. B. Frames of the insertion of a box.

Grasp and Manipulation. The success rate for the grasp and manipulation portion of the pipeline is shown in Fig. 6-A. Finger grasping proves to be reliable and robust to errors. On the other hand, the main reason behind a failed Suction grasp is an imprecise object localization, and a lack of local dexterity in the robot; this is usually due to the robot working close to its joint limits. An unsuccessful manipulation is caused by either slippage, that lead to no toppling, or to the object falling in a pose that prevents its identification with the vision pipeline.

Package Planner. To measure the performance of the planner, reported in Table 1, a batch of simulations is run with different container dimensions, size of the objects in the item set, and presence of a buffer. In each one, the number of items entering the system and their outcome is counted and averaged over the batch; similarly, the total area of the items for each outcome is summed and normalized with respect to the container area. The results show how the planner is able to run the task until completion, but having a large amount of items that end up *discarded*. The presence of a buffer improves the results, but performance drops drastically when increasing the variety in the item set.

Table 1. Planner and buffer performance. Highlight: Settings for the experiments.

Grid Size	Itemset	Buffer	Spawned	Placed	Discarded	Leftovers
6x4	1x1 to 3x2	no	14.367 (179.62%)	9.217 (100.00%)	5.150 (79.62%)	–
6x4	1x1 to 3x2	yes	11.833 (147.90%)	8.437 (100.00%)	2.672 (40.96%)	0.724 (6.94%)
12x8	1x1 to 5x4	yes	39.345 (306.94%)	14.751 (100.00%)	22.307 (192.78%)	2.287 (14.17%)

Insertion Strategy: Blocks Packaging. Figure 6-B plots the overall success rate of the tasks; this is evaluated by measuring the maximum distance in which the obtained package exceeds the intended area of the container. Similarly, it is also possible to evaluate in which threshold does the error of each placement fall, and correlate the number of unsuccessful ones to the position in the grid. Such analysis shows that the error is less than 1 cm (30% of the cell size) in 63.93% of the individual placements. Both simulation and reality show that the error is skewed along one direction. This is due to a lack of local dexterity, and at times, to the gripper getting in the way of a correct placement.

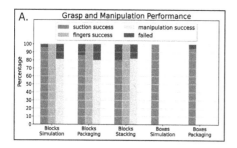
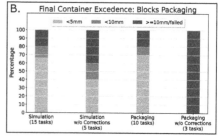

Fig. 6. A. Success ratio of the Grasp and Manipulation experiments. B. Success rate for blocks packaging.

The presence of a grasp error is also correlated with the final placement; no significant difference is noticed, meaning that the pipeline is partially able to absorb such error, but also that its magnitude can be shadowed by the other error sources mentioned. Corrective actions are effective to reduce the misplacement inside the 5 mm threshold in 73.80% of the times. Repeating the experiment without them shows that the error increases at the end of the task, but not after each placement; this proves their necessity only for the containment of the outer layer, while the packaging strategy is able to deal with misplacements up to 1cm. The experiment is then repeated, first setting both fingers stiff, and then both soft. The results do not particularly worsen, but some behaviours can be observed. First, the softer the hand, the more the fingers bend under the weight of the objects, approaching the target with residual inclination; moreover, the items tend to get more stuck against the container walls due to friction. With stiffer joints, higher forces are transmitted due to the higher rigidity of the system; in this way, an imprecise positioning of the item can squeeze another item out of its position, rather than comply to it.

Insertion Strategy: Blocks Stacking. Along the 10 executions, the blocks are successfully placed in the right cell 92.3% of times; in the remaining, the item gets stuck on the wrong one, and leads to a failed task. In 47.9% of times, the block is pressed in its spot; in the remaining it is only released on it.

One more source of error is observed, which leads to these partial failures: the hand needs to release the block and reconfigure itself before pressing it in its position, to avoid having a finger colliding with the base-heads; in half of the executions, the hybrid grasp is not able to partially insert the block, which then tilts and is missed by the pressing finger (as shown in Fig. 7-A).

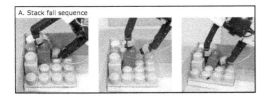

Fig. 7. A. Even after the block is released in the correct position, a tilted item produces failure at pressing down with the finger. B. Exemplary results in simulation and in experiments.

Insertion Strategy: Boxes Packaging. The full insertion strategy is performed in these tasks; some examples of the outcomes are shown in Fig. 7-B.

Out of the 11 execution, 7 are successful, meaning that the container is filled completely. All the failures happen because the piercing action fails, meaning that the corner of the grasped box gets stuck onto the top of the ones underneath and is unable to enter between them, leading to the loss of the suction constraint. This happens because of the uncertainty in the relative position between object and robot, due to *grasp inaccuracy*. In presence of this error, even if the arm reaches the nominal position, the box does not, but is rather shifted back; hence the piercing action is not performed as intended and fails.

The task is then repeated using a simple insertion strategy, consisting in a top-down insertion of the item, directly in its intended position. This strategy fails 5 times out of 5, often before reaching the tight regions of the container; this happens because of the inability to deal with the large error coming from the grasping portion of the pipeline (up to 3 times the nominal clearance). The repetitions with low and high stiffness present a reduced success rate. The suction cups are quite compliant, hence their softness shadows the stiffness commanded to the fingers. In case of two soft fingers, the excessive compliance makes the execution even more vulnerable to getting stuck on the top of the boxes due to friction; making both fingers stiffer does not improve the behaviour, since only the rear finger actually utilizes its stiffness to pierce and push (Table 2).

Table 2. Summary of experimental results.

	Achieved clearance [mm] (% of item size)	Extra space required	Success Rate
Blocks Packaging	5 (16.1%)	yes	73.8%
Blocks Stacking	1 (3.2%)	yes	92.3%
Boxes Packaging	15 (10.3%)	no	63.3%

6 Conclusions

This paper presented a complete pipeline to do tight packaging. We focused on developing a strategy for inserting an item requiring only its intended area to be free; hence, our approach can fill a container up to all four walls. The use of a hybrid and variable-stiffness gripper ensures versatility while grasping, and allows suitable system behaviours to properly perform the item placement. There is scarce work so far on the tight packaging problem. The closest works in [15] and [20] use a gripper that prevents them from filling up the full container. However, our HCG and planning approach allow achieving this goal.

The results show an improvement in the ability to deal with uncertainty, with respect to a trivial strategy. Corrective actions are effective in reducing the errors, but are not necessary for smaller objects, as the insertion method can deal with them; moreover they cannot recover errors such as toppling and failed placements. The hybrid grasp and the stiffness settings prove useful in achieving the desired compliant behaviour necessary to perform each operation.

Still, improvements can be done in future work to further improve the efficacy of the proposed method. A fully trained visual pipeline could be used to localize the items, reducing the grasping uncertainty. The planner can be improved to search in multiple layers and in 3D; its efficacy would greatly improve with knowledge of the set of items. So that the planner can efficiently avoid discarding the items that do not fit in a given step. As a next step, we will complement the placement strategy with sensor feedback, so to find an actual contact by sensing it, in spite of possible shifts in the grasp. The corrections could be expanded to a set of recovery actions, so to tackle the different failures that can be encountered; the detection of such failures could be automatized with visual feedback on the container, making the pipeline completely autonomous.

References

1. Berkey, J.O., Wang, P.Y.: Two-dimensional finite bin-packing algorithms. J. Oper. Res. Soc. **38** (1987)
2. Chen, V.K., Chin, L., Choi, J., Zhang, A., Rus, D.: Real- time grocery packing by integrating vision, tactile sensing, and soft fingers. IEEE RoboSoft (2024)
3. Cheng, X., Hou, Y., Mason, M.T.: Manipulation with suction cups using external contacts. In: Robotics Research. Springer, Cham (2022). https://doi.org/10.1007/978-3-030-95459-8_42
4. Correa, C., Mahler, J., Danielczuk, M., Goldberg, K.: Robust toppling for vacuum suction grasping. In: IEEE CASE (2019)
5. D'Avella, S., Sundaram, A.M., Friedl, W., Tripicchio, P., Roa, M.A.: Multimodal grasp planner for hybrid grippers in cluttered scenes. IEEE RA-L **8** (2023)
6. Friedl, W., Roa, M.A.: CLASH—a compliant sensorized hand for handling delicate objects. Front. Rob. AI **6**, 138 (2020)
7. Hong, Y.D., Kim, Y.J., Lee, K.B.: Smart pack: online autonomous object-packing system using RGB-D sensor data. Sensors **20** (2020)
8. Kirillov, A., et al.: Segment anything. arXiv:2304.02643 (2023)
9. Liu, S., et al.: Grounding dino: marrying dino with grounded pre-training for open-set object detection. arXiv:2303.05499 (2023)
10. Martin, A.E.H., Sundaram, A.M., Friedl, W., Ruiz, V., Roa, M.A.: Task-oriented stiffness setting for a variable stiffness hand. In: IEEE ICRA (2023)
11. Morgan, A.S., Bateux, Q., Hao, M., Dollar, A.M.: Towards generalized robot assembly through compliance-enabled contact formations. IEEE ICRA (2023)
12. Morgan, A.S., Wen, B., Liang, J., Boularias, A., Dollar, A.M., Bekris, K.E.: Vision-driven compliant manipulation for reliable, high-precision assembly tasks. Rob. Sci. Syst. (RSS) (2021)
13. Pan, J.Y., et al.: Sdf-pack: towards compact bin packing with signed-distance-field minimization. In: IEEE/RSJ IROS, pp. 10612–10619 (2023)
14. Porges, O., Stouraitis, T., Borst, C., Roa, M.A.: Reachability and capability analysis for manipulation tasks. In: ROBOT2013. Springer, Heidelberg (2014)
15. Shome, R., et al.: Tight robot packing in the real world: a complete manipulation pipeline with robust primitives (2019)
16. Su, J., Liu, C., Li, R.: Robot precision assembly combining with passive and active compliant motions. IEEE Trans. Ind. Electron. (2021)
17. Wang, F., Hauser, K.: Robot packing with known items and nondeterministic arrival order. IEEE Trans. Autom. Sci. Eng. **18** (2019)

18. Wang, F., Hauser, K.: Dense robotic packing of irregular and novel 3D objects. IEEE Trans. Rob. **38** (2022)
19. Yang, S., et al.: Heuristics integrated deep reinforcement learning for online 3D bin packing. IEEE Trans. Autom. Sci. Eng. **21** (2024)
20. Zhou, Z., Zhang, Z., Xie, K., Zhu, X., Huang, H., Cao, Q.: A method of tight placement for robotic dense packing. In: ICARM (2022)

Design, Fabrication and Calibration of an Embroidery Textile Tactile Sensor Array

Ningzhe Hou[1]([✉]), Marco Pontin[1], Leone Costi[1], and Perla Maiolino[1,2]

[1] Oxford Robotics Institute, University of Oxford, Oxford, UK
exet5825@ox.ac.uk
[2] University of Genoa, Liguria, Genoa, Italy

Abstract. This paper presents the development and evaluation of an embroidery textile tactile sensor, designed for advanced pressure mapping applications. Utilizing a unique embroidery technique with conductive threads and incorporating a Velostat piezoresistive layer, the sensor array is configured in an 8×8 grid, offering multiple contact detection over large areas. The sensor's performance is validated through several experiments, including quasi-static loading and pressure mapping with various weights and hand gestures. This research proposes a flexible, scalable solution for tactile sensing that is both cost-effective and high-performing. The findings indicate significant potential for the sensor in real-world applications, promising advancements in how tactile information is captured and utilized.

Keywords: Tactile Sensing · Textile Sensors · Embroidering

1 Introduction

Soft tactile sensing is a fundamental sensory modality in biological systems such as mammals, with the skin being the largest organ in humans [24]. Because of the importance of tactile sensing in everyday tasks, the development of effective, cheap, and scalable tactile sensing technology has been a primary objective for the scientific literature [13,15]. Moreover, with recent advancements in sensing technology, soft tactile sensors have seen application in a number of tasks such as medical monitoring [22], wearables, and human-machine interface [8]. The main role of tactile sensors is to detect contact, in the form of pressure or force, and transduce it into an electronic output [7]. In general, such sensors can be categorized in base of the physical transduction principle, the sensor's materials, or the sensor design.

In terms of transduction principles, state-of-the-art literature showcases piezoelectric, piezoresistive, capacitive, and triboelectric sensors. More recently,

---

This work was supported by the Engineering and Physical Sciences Research Council (EPSRC) Grant EP/V000748/1.

© The Author(s), under exclusive license to Springer Nature Switzerland AG 2025
M. N. Huda et al. (Eds.): TAROS 2024, LNAI 15051, pp. 327–338, 2025.
https://doi.org/10.1007/978-3-031-72059-8_28

studies have also reported soft flexible tactile sensors that rely on the optical properties of the sensor's material [15].

Another possible classification can be made on the nature of the material used: in order to achieve a soft flexible device, the three most used materials are flexible PCBs [5], elastomers [14], or textiles [27]. Concerning PCBs, they offer high customization of the sensor and a well-established and low-cost manufacturing process. However, their limited flexibility, on top of the complex design needed for large areas over non-gaussian surfaces, limits their usability in fields like wearable technologies and medical devices [25]. When it comes to elastomers, they provide the ability to integrate several fillers to achieve smart functional hybrid materials [4], but they often result in bulkier devices, with relatively high sensor thickness and limited bending radius [19]. Tactile sensors can also be created in textile form. Mainly thanks to the increase in quality and availability of conductive threads, recent state-of-the-art works have shown how textile-based tactile sensors can achieve good sensing performance, whilst maintaining extreme flexibility [11].

Lastly, current scientific literature showcases two main design strategies to achieve large area sensing: either in sensing element for taxel [17] or grid-architecture and multiplexing [2].

The most commonly adopted sensing principles in textile sensing are piezoresistive [26] and capacitive [23]. However, capacitive sensing relies on the contact area, meaning that higher spatial resolution results in smaller changes in signal, posing challenges in signal processing and readout electronics. In this work, we focus on piezoresistive textile sensing, as we prioritize the flexibility and spatial resolution of the sensor. Textile sensors are produced through the insertion of a conductive element, usually a conductive thread, within a non-conductive textile substrate in one of three ways: stitching [6], crocheting [2], and embroidering [1]. Of those three techniques, embroidering results in higher transduction, hence a greater change in resistance, due to the higher tension in the thread and lower distance among anchor points between the substrate and conductive pattern [3].

As a result, state-of-the-art embroidered sensors are characterized by a high piezo-resistive response, and a wide range of possible designs, that can be roughly divided between single-layer [26] and sandwich design [18], depending on the number of textile layers in the final device.

In particular, *Aigner et al.* [3] designed a single-layered piezoresistive tactile sensor array using an off-the-shelf embroidery machine. Utilizing the surface resistance of EeonTex fabric instead of traditional top and bottom contact fabrics significantly simplified the sensor design. The experiment also demonstrated a promising response, with up to approximately 80% reduction in resistance under pressure. *Zhang et al.* [26] also developed a single-layer tactile sensor array. The sensing mechanism relies on the deformation of conductive thread fibers, which results in an increase in the conductivity of the yarn and, consequently, a decrease in resistance. The work by *Rocha et al.* [20] presents a single pressure sensor with a sandwich structure, involving two top and bottom electrodes, a middle EeonTex material as piezo-resistive layer, and embroidery encapsulation.

However, aside from the few aforementioned papers, most textile-based sensors require complex material processing [21], which hinders their commercialization and scalability.

In this paper, we propose an cost-effective embroidered textile tactile sensor that can be fabricated using readily available materials and traditional embroidery techniques, The sensor is composed of 64 sensing units arranged in an 8 × 8 grid architecture and is manufacturing through embroidered patterns in the two external layers of the device, with Velostat as an internal piezoresistive layer. The sensitivity of the cells is characterized by performing a force calibration on five taxels, showcasing high repeatability both intra-taxels and inter-taxels. Then, the ability to reproduce spatial resolution tactile maps is showcased using calibrated weights and through hand interaction both on a flat and a curved surface.

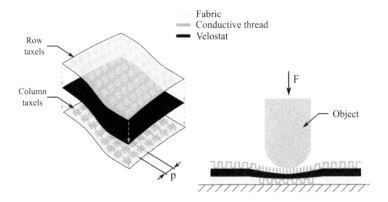

Fig. 1. Design of the tactile sensor array. The sensor consists of a layered structure with a Velostat sheet sandwiched in between embroidered textile. Each textile sheet square has 64 taxels embroidered on it, in a raw-column arrangement. When a force is applied to any of the taxels, the resistance between the top and bottom embroidered patterns that constitute the taxel decreses. This change, which is proportional to the force being applied, is due to the piezoresistive velostat layer, which changes its resistivity when deformed.

2 Material and Methods

2.1 The Fabrication of the Sensor

As introduced in Sect. 1, the tactile sensor array adopts a sandwich structure, where the top and bottom fabric layers embed the column and row electrodes. The embroidery was achieved using a domestic embroidery machine (Brother Innov-Is 880E) and conductive thread (Shieldex HC12) on woven 240 g cotton calico. The zigzag pattern is used in the sensing area to ensure maximum crossing contact between the conductive threads on the top and bottom electrodes through Velostat, as this pattern results in the top and bottom threads at 90-degree angles to each other, enhancing the sensor's sensitivity. The total sensing area measures 80×80 mm^2, with each taxel sized at 5 mm by 5 mm. For each electrode, the density of the zigzag pattern is 0.5 mm, which means the conductive thread traverses back and forth 10 times to form the electrode area. Additionally, the distance between the center of each taxel (p in Fig. 1) is 10 mm.

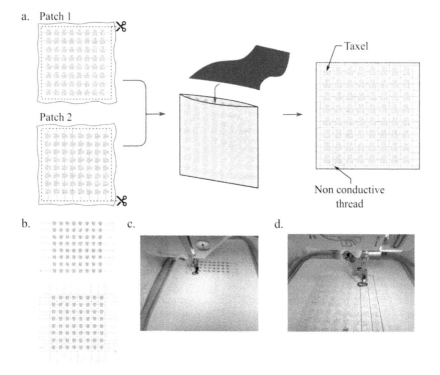

Fig. 2. Manufacturing of the sensor. a) The manufacturing consists of three phases. First, two square textile sheets are embroidered using conductive thread. They are then layered one on top of the other, with the conductive patterns facing each other, and a Velostat sheet is inserted in between them. The final step consists in partially electrically isolating each taxel by embroidering straight non-conductive traces. b) Visualization of the two embroidery cycles used to create the sensor array. c) Photo taken during the first step of the manufacturing process. d) Photo taken during the fourth and final step of the manufacturing. The machine is embroidering the separation patterns using non-conductive thread.

After embroidering, the top and bottom fabric, featuring the same patterns, were cut into square shapes to facilitate alignment (Fig. 2a). The two fabric patches were arranged to form the row and column pattern of the array, with the conductive patterns facing each other. After careful alignment of the 64 electrode-pairs, the layers were temporarily stapled together and a thin Velostat sheet was inserted in between the two. The assembled sensor was then placed back in the embroidery frame and a laser pointer in the embroidery machine was used to reposition the pattern. The non-conductive thread was then embroidered to separate the columns and rows (Fig. 2d). This procedure allows us to reduce taxels' cross-talk and consistent responses to pressure, in addition to permanently fixing the layers together. Once unmounted from the frame, the staples are removed, completing the manufacturing of the sensor array.

2.2 Acquisition Circuit

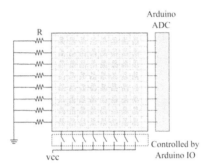

Fig. 3. Circuit used for acquiring the readout of the sensor array. A multiplexed voltage divider topology is used to read each taxel in succession. Digital outputs on an Arduino Mega board are used to sequentially connect each row with each column and a voltage reading is taken for the corresponding taxel. The voltage is then converted into a resistance value by applying the voltage divider equation with the known resistor value $R = 1.6$ kΩ.

Figure 3 shows the multiplexed voltage divider circuit used to perform the sensor readout taxel by taxel. Each row of the array is connected to a 1.6 kΩ resistor R, and each column is connected to one digital output of an Arduino Mega 2560 Microcontroller. The Arduino analog inputs are connected between the array sensor and the 1.6 kΩ resistor, one input for each row in the array. As a result, the Arduino ADC measures the variable voltage across the 1.6 kΩ resistor. During the measurement, one of the digital outputs is set to High, activating one column of the sensor, while the others are set to open circuit. In sequence, each row is read by the ADC which samples the corresponding voltage across all of the 1.6 kΩ resistors. The column digital output is then set to Low, and the process is repeated for the remaining seven columns.

Through this setup, for each measured taxel, the circuit configuration simplifies to a variable resistor and a 1.6 kΩ resistor connected in series, and the voltage is 5 V. The voltage across the 1.6 kΩ resistor is measured, allowing the resistance of the sensor to be expressed by the following equation:

$$R_v = 1600\left(\frac{5 - V_{\text{ADC}}}{V_{\text{ADC}}}\right) \tag{1}$$

2.3 Experimental Setup

A quasi-static loading experiment was conducted to determine the resistance-force response of the sensor's taxels. The experiment utilizes a servo lead screw (HIWIN Technology), a 3 kg load cell (DAYBENBOR, Inc.), and a 3D-printed indenter, which is designed to match the dimensions of a single taxel. The data from both the Arduino and the load cell were transmitted to a laptop via the serial protocol and processed in MATLAB. Since the data were sampled at different rates and from different devices, the absolute time of receipt was added to each data point. The Arduino used a 10-bit ADC to convert the analog signals to digital format. These digital values were then converted back to voltages, and Eq. 1 was used to convert these voltages to resistance values. Linear Time Series Imputation was used to merge the force and resistance data into a single dataset.

At the beginning of the experiment, the tactile sensor array was manually positioned to align the taxel to be measured directly underneath the indenter. The array was then left to rest for 10 minutes to allow the sensor response to stabilize. After that, the servo lead screw was manually controlled to press the indenter at a constant velocity of 0.01 mm/s, while the load cell continuously recorded the force applied to the sample. This process was maintained until the sensor saturation. At the same time, the Arduino kept recording the resistance of measured taxel. A total of five taxels are selected: four at each corner and one in the middle, to cover the range of the entire tactile sensor. Three repetitions of loading is conducted for each taxel.

The pressure mapping experiment was conducted in two parts. First, calibrated weights of 20, 50, and 100 g were placed in exactly the same position on the tactile array, both with and without the 3D-printed soft mat. The pressure maps from these scenarios were recorded and visualized through MATLAB. Second, as mentioned in the introduction, one of the key use-cases for these sensors is in wearable devices, we used a human hand to press on the pressure-sensing mat with two different gestures on the table, and recorded the visualized pressure maps. Then, the tactile array was placed on the surface of a cylindrical water bottle with a 12 cm diameter. We adopted the same pressing gesture and recorded the pressure map.

3 Results

3.1 Tactile Sensor Calibration

The image of the experimental setup is shown in Fig. 4a. As shown in Fig. 4b, when no force is applied, the baseline resistance varies, ranging from 5.7 kΩ to 6.8 kΩ. Upon application of pressure, the resistance of all the taxels decreases hyperbolically. The most sensitive taxel of the five tested plateaued at around

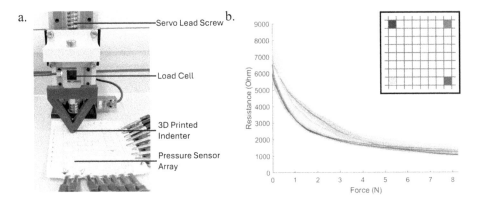

Fig. 4. The quasi-static loading experiment. a) Experimental Setup. The setup includes a servo lead screw, a load cell, a 3D-printed indenter, and the pressure sensor array. b) The graph illustrates the relationship between the applied force and the resistance at selected taxels on the array, including standard deviation and an indication of their positions.

6 N, showing only minimal changes in resistance beyond this point. In contrast, the least sensitive taxel continued to change until reaching 9 N, before plateauing. The final resistance values of the tested taxels were similar, with only a slight difference of 296.1 Ω between the maximum (top left corner) and the minimum (top right corner). It is also worth noting that the standard deviation of the resistance values among different taxels varies considerably. For instance, the middle taxel, shown in light green, has a significantly higher standard deviation (up to 515 Ω) compared to the top left corner (up to 70 Ω).

The relationship between force and resistance across the five taxels (shown in Fig. 5) is analyzed through curve fitting to derive a function that converts resistance to pressure in the pressure mapping experiment. The result of the power function curve fitting can be expressed as follows:

$$f(x) = 1.048 \times 10^4 \cdot x^{-0.1187} - 7011 \tag{2}$$

This curve fitting relationship is in accordance with the resistance/force characteristic of commercially available piezoresistive Force Sensitive Resistor (FSR) [12]. The curve fitting equation models resistance as a function of force changes and must be inverted when converting resistance back into force measurements.

3.2 Pressure Mapping Experiment

Although crosstalk between neighboring taxels has been reduced through the final fabrication step, where individual taxels are separated by a running stitch with non-conductive thread, crosstalk can still be observed at low pressures. This is due to the relationship between resistance and pressure, which shows that the sensor is highly sensitive at low pressures. An example of this crosstalk

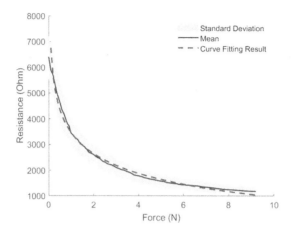

Fig. 5. The relationship between applied force and mean resistance over five taxels including standard deviation, and the result of curve fitting.

can be seen in Fig. fig:6a, where taxels with no weight applied, but close to the weighted area, have a non-zero readout. This could affect the accuracy of the pressure mapping.

To mitigate crosstalk without compromising the sensor's flexibility, an additional layer of a 3D-printed soft mat was strategically placed on top of the sensor. The mat, depicted in Fig. 6b, features a round dot positioned directly above each taxel, with all dots interconnected, enhancing the geometric design to channel force more effectively onto the sensing areas. This design also helps to prevent load dispersal caused by the raised, non-conductive embroidery patterns surrounding each taxel. The mat was fabricated using Agilus30 material on a J735 3D printer (Stratasys, Inc.), chosen for its flexibility.

It can be seen from the comparison between the pressure maps in Fig. 6a and 6b that when the mat is placed on top of the sensor, only the taxels with the weight directly applied show activation. In contrast, the pressure map of the sensor without the mat displays significant crosstalk, particularly among the neighboring taxels around the 50 and 100-gram weights. Despite the improvement provided by the mat, the large variability in taxel sensitivity is evident, with taxels on the periphery of the 20 g weight appearing deactivated. Another problem with the setup is linked to the simple readout structure used. In the presence of multi-touch events, only the relative intensity of the contacts can be reliably assessed, while information about the absolute magnitude is lost. To correct for these problems, a taxel-by-taxel calibration procedure should be used in the future together with more advanced readout strategies [9,10,16].

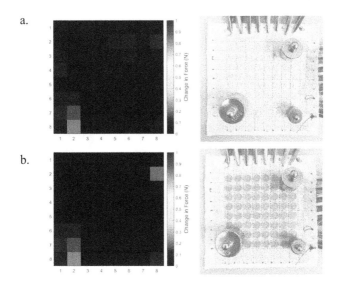

Fig. 6. The pressure mapping results of placing weights of 100 g, 50 g, and 20 g on the same position on the sensor, a) without and b) with the 3D-printed filter mat.

The results depicted in Fig. 7 demonstrate the sensor's capability to track different hand gestures effectively for wearable devices. It can be observed in Fig. 7 that after placing the mat on top, the tactile sensor array can precisely reconstruct the pressure map of different hand gestures. The pressure map clearly indicates the positions and pressures of two fingers in gesture 1 and the curved little finger in gesture 2. It is also worth mentioning that as the joints of the fingers are stiffer, the joint areas exhibit higher pressure than other parts of the fingers, further demonstrating the precision of the pressure mapping.

Figure 7c shows the pressure map of gesture 1, but with the sensor adhering to the curved surface of a water bottle. The pressure image shows no significant difference compared to the pressure image in Fig. 7a. The only noticeable difference is that, on the curved surface, the pressure at the joint position is not higher. This can be explained by the fact that when the finger is curved, the effect of the joint is not as significant as when the finger is straight, because the joint is no longer prominent.

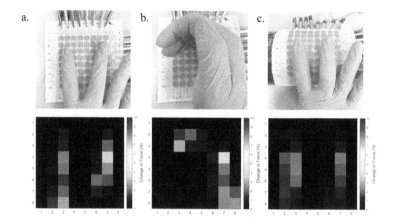

Fig. 7. The pressure mapping results of pressing the pressure sensor array with a hand. a) Gesture 1. b) Gesture 2. c) Gesture 1 on a curved surface.

4 Conclusion

This research has developed a low-cost embroidered textile tactile sensor and demonstrated a method to increase the precision of pressure mapping capabilities by means of a 3D printed textured interfacing layer. Our experimental results confirmed the sensor's capability to accurately detect and map pressure distributions, effectively differentiating between various hand gestures and weights. The addition of a 3D-printed soft mat atop the sensor array further minimized crosstalk, proving especially effective in scenarios involving low mechanical loads. This feature was critical in maintaining the precision of pressure mapping across different experimental setups, including curved surfaces like water bottles, which simulated more complex, real-world applications.

The sensor's low fabrication cost and flexible, textile-based design make it particularly suitable for applications requiring nuanced tactile feedback, such as robotic sensing and wearable technologies. Future work will focus on further optimizing the sensor's design and exploring its integration into more complex systems.

References

1. Abdullah al Rumon, M., Shahariar, H.: Fabrication of interdigitated capacitor on fabric as tactile sensor. Sens. Int. **2**, 100086 (2021). https://doi.org/10.1016/j.sintl.2021.100086. https://www.sciencedirect.com/science/article/pii/S2666351121000073
2. Ahn, Y., Song, S., Yun, K.S.: Woven flexible textile structure for wearable power-generating tactile sensor array. Smart Mater. Struct. **24**(7), 075002 (2015). https://doi.org/10.1088/0964-1726/24/7/075002

3. Aigner, R., Pointner, A., Preindl, T., Parzer, P., Haller, M.: Embroidered resistive pressure sensors: a novel approach for textile interfaces. In: Proceedings of the 2020 CHI Conference on Human Factors in Computing Systems, CHI '20, pp. 1–13. Association for Computing Machinery, New York (2020). https://doi.org/10.1145/3313831.3376305
4. Charalambides, A., Bergbreiter, S.: A novel all-elastomer mems tactile sensor for high dynamic range shear and normal force sensing. J. Micromech. Microeng. **25**(9), 095009 (2015). https://doi.org/10.1088/0960-1317/25/9/095009
5. Cheng, M.Y., Huang, X.H., Ma, C.W., Yang, Y.J.: A flexible capacitive tactile sensing array with floating electrodes. J. Micromech. Microeng. **19**(11), 115001 (2009). https://doi.org/10.1088/0960-1317/19/11/115001
6. Choudhry, N.A., Shekhar, R., Rasheed, A., Arnold, L., Wang, L.: Effect of conductive thread and stitching parameters on the sensing performance of stitch-based pressure sensors for smart textile applications. IEEE Sens. J. **22**(7), 6353–6363 (2022). https://doi.org/10.1109/JSEN.2022.3149988
7. Cutkosky, M.R., Provancher, W.: Force and Tactile Sensing, pp. 717–736. Springer, Cham (2016). https://doi.org/10.1007/978-3-319-32552-1_28
8. Dempsey, S., Szablewski, M., Atkinson, D.: Tactile sensing in human–computer interfaces: the inclusion of pressure sensitivity as a third dimension of user input. Sens. Actu. A **232**, 229–250 (2015). https://doi.org/10.1016/j.sna.2015.05.025. https://www.sciencedirect.com/science/article/pii/S0924424715300212
9. Domínguez-Gimeno, S., Medrano-Sánchez, C., Igual-Catalán, R., Martínez-Cesteros, J., Plaza-García, I.: An optimization approach to eliminate crosstalk in zero-potential circuits for reading resistive sensor arrays. IEEE Sens. J. **23**(13), 14215–14225 (2023)
10. Hidalgo-Lopez, J.A., Romero-Sánchez, J., Fernández-Ramos, R.: New approaches for increasing accuracy in readout of resistive sensor arrays. IEEE Sens. J. **17**(7), 2154–2164 (2017)
11. Honda, S., Zhu, Q., Satoh, S., Arie, T., Akita, S., Takei, K.: Textile-based flexible tactile force sensor sheet. Adv. Func. Mater. **29**(9), 1807957 (2019). https://doi.org/10.1002/adfm.201807957. https://onlinelibrary.wiley.com/doi/abs/10.1002/adfm.201807957
12. Industries, A.: Force sensitive resistor (fsr) guide (2022). https://cdn-learn.adafruit.com/assets/assets/000/010/126/original/fsrguide.pdf. Accessed 18 May 2024
13. Lee, M., Nicholls, H.: Review article tactile sensing for mechatronics-a state of the art survey. Mechatronics **9**(1), 1–31 (1999). https://doi.org/10.1016/S0957-4158(98)00045-2. https://www.sciencedirect.com/science/article/pii/S0957415898000452
14. Li, W., Konstantinova, J., Noh, Y., Ma, Z., Alomainy, A., Althoefer, K.: An elastomer-based flexible optical force and tactile sensor. In: 2019 2nd IEEE International Conference on Soft Robotics (RoboSoft), pp. 361–366 (2019). https://doi.org/10.1109/ROBOSOFT.2019.8722793
15. Mandil, W., Rajendran, V., Nazari, K., Ghalamzan-Esfahani, A.: Tactile-sensing technologies: trends, challenges and outlook in agri-food manipulation. Sensors **23**(17) (2023). https://doi.org/10.3390/s23177362. https://www.mdpi.com/1424-8220/23/17/7362
16. Oballe-Peinado, Ó., Vidal-Verdú, F., Sánchez-Durán, J.A., Castellanos-Ramos, J., Hidalgo-López, J.A.: Improved circuits with capacitive feedback for readout resistive sensor arrays. Sensors **16**(2), 149 (2016)

17. Pang, Y., et al.: Skin-inspired textile-based tactile sensors enable multi-functional sensing of wearables and soft robots. Nano Energy **96**, 107137 (2022). https://doi.org/10.1016/j.nanoen.2022.107137. https://www.sciencedirect.com/science/article/pii/S221128552200218X
18. Pizarro, F., Villavicencio, P., Yunge, D., Rodríguez, M., Hermosilla, G., Leiva, A.: Easy-to-build textile pressure sensor. Sensors **18**(4) (2018). https://doi.org/10.3390/s18041190. https://www.mdpi.com/1424-8220/18/4/1190
19. Robinson, S.S., et al.: Integrated soft sensors and elastomeric actuators for tactile machines with kinesthetic sense. Extreme Mech. Lett. **5**, 47–53 (2015). https://doi.org/10.1016/j.eml.2015.09.005. https://www.sciencedirect.com/science/article/pii/S235243161500108X
20. Goveia da Rocha, B., Tomico, O., Markopoulos, P., Tetteroo, D.: Crafting research products through digital machine embroidery. In: Proceedings of the 2020 ACM Designing Interactive Systems Conference, DIS 2020, pp. 341–350. Association for Computing Machinery, New York (2020). https://doi.org/10.1145/3357236.3395443
21. Tan, Y., Yang, K., Wang, B., Li, H., Wang, L., Wang, C.: High-performance textile piezoelectric pressure sensor with novel structural hierarchy based on zno nanorods array for wearable application. Nano Res. **14**(11), 3969–3976 (2021). https://doi.org/10.1007/s12274-021-3322-2
22. Tiwana, M.I., Redmond, S.J., Lovell, N.H.: A review of tactile sensing technologies with applications in biomedical engineering. Sens. Actu. A **179**, 17–31 (2012). https://doi.org/10.1016/j.sna.2012.02.051. https://www.sciencedirect.com/science/article/pii/S0924424712001641
23. Truong, T., Kim, J.S., Yeun, E., Kim, J.: Wearable capacitive pressure sensor using interdigitated capacitor printed on fabric. Fash. Text. **9**(1), 46 (2022). https://doi.org/10.1186/s40691-022-00320-w
24. Yousef, H., Alhajj, M., Sharma, S.: Anatomy, skin (integument), epidermis. In: StatPearls [Internet] (2022). https://www.ncbi.nlm.nih.gov/books/NBK470464/. Accessed 14 Nov 2022
25. Yuan, Y., et al.: Flexible wearable sensors in medical monitoring. Biosensors **12**(12) (2022). https://doi.org/10.3390/bios12121069. https://www.mdpi.com/2079-6374/12/12/1069
26. Zhang, H., Tao, X.M.: A single-layer stitched electrotextile as flexible pressure mapping sensor. J. Text. Inst. **103**(11), 1151–1159 (2012). https://doi.org/10.1080/00405000.2012.664868
27. Zhang, Y., Lin, Z., Huang, X., You, X., Ye, J., Wu, H.: A large-area, stretchable, textile-based tactile sensor. Adv. Mater. Technol. **5**(4), 1901060 (2020) https://doi.org/10.1002/admt.201901060. https://onlinelibrary.wiley.com/doi/abs/10.1002/admt.201901060

SmartAntenna: Enhancing Wireless Range with Autonomous Orientation

Michael Swann, Pedro Machado(✉)[ID], Isibor Kennedy Ihianle[ID], Salisu Yahaya[ID], Farbod Zorriassatine, and Andreas Oikonomou[ID]

Computational Intelligence and Applications Research Group, Clifton Campus, Nottingham Trent University, Nottingham NG11 8NS, UK
n1020368@my.ntu.ac.uk, {pedro.machado,isibor.ihianle,salisu.yahaya, farbod.zorriassatine,andreas.oikonomou}@ntu.ac.uk
https://www.ntu.ac.uk/research/groups-and-centres/groups/computational-intelligence-applications-research-group

Abstract. The SmartAntenna proposes a novel approach to extend wireless communication, focussing on autonomous orientation to extend range and optimise performance. Through meticulous evaluation, various aspects of its functionality were evaluated, revealing both strengths and areas for improvement. In particular, the antenna tracking mechanism exhibited remarkable efficacy. The SmartAntenna demonstrated robust functionality throughout extensive testing, underscoring its reliability even amidst complex operational scenarios. However, challenges emerged during target tracking, particularly evident in 360° sweeps, necessitating further refinement to enhance accuracy. Despite reliance on the HC-12 module, Long Range Wide Area (LoRa), performance limitations surfaced, raising concerns about its suitability for production systems, especially within noisy frequency bands. Nevertheless, the SmartAntenna's adaptability across various wireless technologies holds promise, opening avenues for extended communication ranges and diverse applications. SmartAntenna research contributes valuable insights into optimizing wireless communication systems, paving the way for enhanced performance and expanded capabilities in diverse operational environments.

Keywords: Yagi antenna · omnidirection antenna · autonomous orientation

1 Introduction

In 1896, Guglielmo Marconi, inspired by Heinrich Hertz's pioneering work on radio waves, secured a patent for the world's inaugural wireless telegraph [4]. His groundbreaking experiment in 1901, spanning Cornwall, UK, to Newfoundland, Canada, marked a historic achievement in communication [7]. The success of the transmission sent ripples of concern through established cable telegraph companies of the era, which had invested heavily in underwater cable that connected

Europe and United States (U.S.) [4]. Over a century later, the electromagnetic waves central to Marconi's innovation continue to underpin some of today's most sophisticated wireless technologies. As recognised by the cable telegraph companies of Marconi's time, these wireless systems offer numerous advantages over their wired counterparts, including mobility, cost-effective coverage of expansive areas, and, in contemporary contexts, data rates that rival those of wired connections [2]. Wireless communication still presents various challenges such as signal weakening over distances, interference from obstacles, and regulatory constraints on radio frequencies [1]. Despite these hurdles, multiple wireless solutions exist, each prioritising different features like data rates, energy efficiency, or range [2]. For example, while Wireless Fidelity (WiFi) offers high data rates but consumes significant power, Bluetooth excels in security and energy efficiency but has limited range, and the HC-12[1], a LoRa device provides extensive range at the expense of transmission rates. LoRa transceivers have gained traction in wireless systems, particularly in applications that require lower transmission rates, such has Internet-of-Things (IoT). However, discrepancies arise between claimed and actual performance, with reported ranges varying significantly [9,13]. Moreover, factors such as latency are crucial for applications such as robotic control [2]. The SmartAntenna project pioneers an innovative autonomous antenna orientation system, prioritising critical facets of wireless communication: latency, throughput, and range. The innovative technology holds promise for deployment across diverse and demanding environments, including underwater, mining, and offshore operations. Furthermore, it could be instrumental for emergency response services, facilitating the extension or restoration of compromised wireless communication infrastructures in the aftermath of natural disasters, conflicts, or other natural crises such as wildfires, floods, and storms. The structure of the article is as follows: the relevant research is covered in Sect. 2, the design and implementation of the SmartAntenna are detailed in Sect. 3, the analysis of results is conducted in Sect. 4, and the final conclusions and prospects for future work are addressed in Sect. 5.

2 Background Research

Wireless communication technologies play a pivotal role in IoT systems, facilitating data transmission wirelessly between devices across varying distances. The selection of a wireless technology often depends on the specific application domain of the IoT system, ranging from short-range connections to long-distance transmissions. The market's heterogeneity and the variety of available technologies necessitate categorisation based on factors such as electromagnetic wave length and operating frequencies. These technologies encompass Radio Frequency Transmission, Infrared Transmission, Microwave Transmission, and Light wave Transmission [11]. Bluetooth, developed by the Bluetooth Special Interest Group (SIG), primarily facilitates short-range communication between devices,

---

[1] Available online, https://statics3.seeedstudio.com/assets/file/bazaar/product/HC-12_english_datasheets.pdf, last accessed: 15/05/2024.

particularly prevalent in mobile and computer peripheral connections. Bluetooth Low Energy (BLE) emerged as a variant, offering reduced power consumption and setup time, catering to devices with low bandwidth requirements [11]. Near-Field Communication (NFC), engineered by Sony and Philips, enables two-way data transfer between devices placed in close proximity, typically within a 10 cm range. Widely utilised in mobile devices for contactless payment and data access, NFC operates on magnetic coupling principles and Radio-Frequency Identification (RFID) technology [3]. WiFi, standardised by the Institute of Electrical and Electronics Engineers (IEEE)) as 802.11, represents a ubiquitous communication standard renowned for its broad coverage, high data transfer rates, and extensive market penetration. Utilising frequencies in the 2.4 to 5 GHz range, WiFi requires an access point and wireless network adapter for network establishment [12]. Cellular networks, encompassing generations from 1G to 5G Long Term Evolution (LTE), form large-scale communication infrastructures divided into cells, each serviced by a transceiver. Characterised by low energy consumption and expansive coverage, cellular networks cater to high-bandwidth requirements and are prevalent in mobile phones and IoT devices. The latest iterations, 5G and LTE, offer enhanced speed, capacity, and device connectivity [8]. LoRa and Long Range Wide Area Network (LoRaWAN) represent distinct components within the realm of wireless communication [10]. LoRa serves as the physical layer technology developed by Semtech, enabling long-distance wireless communication. It operates by modulating data onto radio waves, offering robust performance even in challenging environments. On the other hand, LoRaWAN encompasses the higher layers of the communication stack, serving as a protocol for managing communication between LoRa-equipped devices and network gateways. LoRaWAN defines the network architecture, addressing, and data rates, facilitating efficient and scalable communication over LoRa's physical layer. LoRa provides the underlying technology for long-range communication, LoRaWAN adds the necessary network infrastructure and protocols for creating wide-area IoT networks, enabling devices to securely transmit data over long distances to centralised network servers [10]. LoRaWAN protocol, compatible with European and North American regulatory standards, operates on specific frequency bands facilitating data transmission [10]. Many commercial and industrial solutions leverage LoRaWAN, communicating with backend systems through network and application servers, formerly named the LoRa Network and LoRa Application Server [10]. In contrast, NB-IoT, belonging to the LTE-IoT network category, operates on 4G mobile networks, offering extensive coverage and reduced power consumption. ZigBee, standardised by IEEE in 2003, remains distinguished for its longevity, often exceeding 10 years, and is commonly used in sensor networks owing to its reliability and cost-effectiveness, albeit with a limited transmission distance of up to 100 m [11]. For the SmartAntenna project, it was decided to focus on LoRa. Marpaung et al. [9] developed an early warning system for peat land fires that relied on the HC-12 to transmit temperature readings from remote devices to a base device. They stated in their description of the HC-12 that it has a maximum range of 1800 m, while in their testing achieved a range of 870 m in the long-range setting (FU4) and 460 m in the default mode (FU3). A prototype wearable device that utilises the HC-12 was developed by Hassaballah et al. [6] for the medical industry, to enable doctors to wirelessly monitor patients. The accompanying report stating that the devices small size and ease of use have made it popular and the frequency band it uses is safe for use in the medical industry. It is, however, important for the SmartAntenna project and other future projects that wish to utilise the HC-12, to know accurately what the performance is of the device. As a clear answer has not been obtained from the literature review, it would be necessary to perform tests on the device.

3 Methodology

To facilitate testing, two devices were developed that utilised the HC-12 module. A base device consisting of a Raspberry Pi Model 4b (RPi4)[2] which could be powered by mains electricity or a battery pack and a portable device using an Arduino Nano that was battery powered. Tests for throughput, latency, and range were conducted. As the HC-12 can be configured in various modes, a number of these would be tested to see how they affected performance. For the latency and throughput tests, a wired connection was also tested for comparison. Furthermore, the HC-12 connects to the Trasmitter (Tx) and Receiver (Rx) pins of the Universal Asynchronous Receiver-Transmitter (UART) port.

Figure 1 showcases a system designed for real-time communication and data collection. At its core lies the Base Unit, a powerful RPi4 that acts as the central hub for communication and data processing. ZMQ, a high-performance messaging library, facilitates seamless data exchange between the Base Unit and other system components. ZMQ ensures smooth communication throughout the system. Web Sockets enable real-time, two-way communication between the Base Unit and a web server. Web sockets enable constant data exchange and updates, crucial for remote monitoring and control via a web browser. The Web UI, a user interface developed using Django, provides users with a web-based interface for interacting with the system. Users can view real-time sensor data (potentially from the portable device), track the current antenna position, and send control commands to both the antenna and the portable device. The system boasts an Offline Mode, ensuring functionality even without an internet connection. Enabling the reliable operation in diverse environments where internet access might be limited. The Barometric Dongle, an external device, gathers atmospheric pressure data. The data can be used for various purposes, such as providing additional environmental context for sensor readings or enhancing the accuracy of other sensors. Another external device, the GPS Dongle, supplies real-time location information. Global Positioning System (GPS) is essential for accurate tracking and positioning of the antenna relative to the portable device. The system operates through a two-way data flow. Sensor data from the portable

---

[2] Available online, https://www.raspberrypi.com/products/raspberry-pi-4-model-b/, last accessed: 15/05/2024.

SmartAntenna: Enhancing Wireless Range with Autonomous Orientation 343

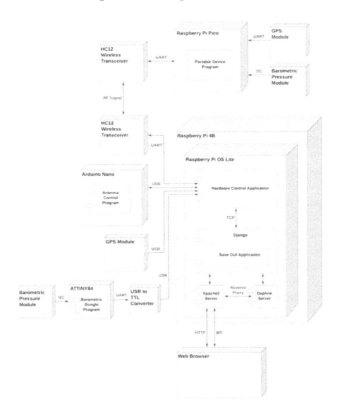

Fig. 1. Deployment diagram illustrating the various hardware and software components of the proposed system.

device likely travels through web sockets to the Base Unit. The Base Unit then processes the data and presents it for real-time monitoring through the web UI. Control commands issued from the web UI are likely transmitted via web sockets to the Base Unit, which then relays these commands to the antenna and/or portable device.

To measure the amount of throughput the device is capable of, the Arduino device would print 10 Kb of data to the serial output. A series of characters at the start and end of the data would signal the receiver to record the time. The same data would be used every time and had been saved to a text file on the RPi4, so it could be compared with the data received.

The SmartAntenna main Algorithm 1 controls a servo motor for a compass rotator system. It utilises two key functionalities: calibration and validation. During calibration, the system iteratively moves the servo through a full rotation while recording compass readings. The calibration establishes a baseline for the compass's response across different orientations. Validation involves moving the servo to specific angles and comparing the measured compass readings with the expected values based on the calibration data. Any significant discrepancies indicate potential errors in the system. By performing these procedures, the program ensures the accuracy and reliability of the compass readings as the servo motor rotates. Another important performance metric is latency. To measure latency, a device would send a single character (1 byte) to another device, which would immediately return the character to the sender. The sending device would time how long it took from sending the character to receiving it. Again, the tests were first conducted with a wired connection to help isolating the additional time penalties caused by the processor on the HC-12 module. To evaluate range performance a test was devised that required the portable device to transmit its current GPS location to the base device. The coordinates of the furthest transmission received could then be used to measure the distance. The portable device consisted of an Secure Digital (SD) card reader, GPS module, and HC-12 module connected to an Arduino Nano. The GPS module would receive new coordinates every second. This would be increased to a transmission at 10 s intervals (later 5 s), due to the time needed to process the GPS data. At the end of each interval, if the National Marine Electronics Association (NMEA) sentence was valid then it was written to the SD card. If the write command successful, then it is transmitted via the HC-12 module.

The reason for validating the coordinates before transmitting them was because the NMEA sentence checksum would be used later to confirm successful transmission. If the portable device were to transmit invalid messages, which would be picked up as an error in the transmission from the HC-12, when it is in fact an error from the GPS module. It was also important to log transmissions into the SD card on the sending device to record exactly what was sent.

If an absence of transmissions was tested at the receiver only, then the performance of the HC-12 could again be affected by other components. For example, the Arduino freezing, or power loss, would look to the receiver as if a transmission had been lost. A python program was developed to process the NMEA sentences from both the SD card and retrieve the file from the base device using Secure Copy Protocol (SCP). Coordinates that were present on both devices were plotted on a map as successful transmissions, and coordinates only present on the portable device plotted as lost transmissions.

The antenna development involved creating a controllable unit based on a gimbal design, actuated by a positional servo and a motor. The Arduino Nano was chosen as the microcontroller due to its small form factor and ease of communication via USB. PlatformIO was used for programming, leveraging its cross-platform capabilities and integration with VSCode for enhanced development features. The servo's speed was controlled via software to avoid mechanical wobble, enhancing stability during operation.

4 Results Analysis

The implemented SmartAntenna system was composed of the base unit (see Fig. 2) and a portable unit (see Fig. 3).

The results in Table 2 show that the wired results are only slightly slower than the calculated minimum for each baud rate, which proves that the test

Algorithm 1. SmartAntenna main pseudocode

Require: Libraries: Arduino, compass.h, rotator.h
Ensure: Functional compass and rotator
    Define Constants: (SERVO_PIN, etc.)
    Initialise Objects: Compass, Rotator
    Procedure Calibrate()
Reset compass, set angle/stability (omit details)
while angle $\neq$ 360 **do**
    Move servo, update stability (omit details)
    Update angle based on direction (omit details)
end while
Store calibration, move servo back to 0, print result
    Procedure Validate()
Initialize stability, prev_reading, max_error
for angle increments from 0 to MAX_ROTATION **do**
    Move servo, get stable heading, update stability (omit details)
    if angle $\neq$ 0 **then**
        Update error based on difference from expected increment (omit details)
    end if
    Update prev_reading
end for
Move servo back to 0, print result
    Procedure DecipherCommand(buffer, bufLen)
Extract command value (omit details)
Switch first_char
Case 'V': Validate
Case 'C': Calibrate
Case 'P': Move servo if valid command value
    Procedure GetCommandValue(buffer, bufLen)
if bufLen < 2 OR second character is '0' **then**
    Return -1 (invalid)
end if
Extract and convert buffer to integer, return (handle 0)
    Procedure Setup()
Initialise serial, compass, rotator, move servo to initial position
    Procedure Loop()
Read compass heading with stability, print values
Call Listen to receive commands
    Procedure Listen()
if Data available **then**
    Read and append character to buffer until newline, call DecipherCommand
end if

programme was acting correctly. The test was then performed on the wireless configurations chosen using the HC-12. The differences of FU modes are listed in Table 1.

Some configurations that were tested use the same output baud rate but with differing modes, which can affect performance.

Fig. 2. Portable device powered by Raspberry Pi Pico

Fig. 3. Antenna system comprising of tracking antenna, base control unit, and positional USB devices.

Table 1. HC-12 operation modes

Mode	Transmission Range	Data Rate	Power Consumption	Latency	Use Case
FU1	Long	Low	High	High	Long-range communication
FU2	Medium	Medium	Medium	Medium	Balanced range and data rate
FU3	Short	High	Low	Low	Short-range, high-speed data
FU4	Configurable	Configurable	Configurable	Variable	Custom configurations, varied use cases

Most have a very similar result to the wired equivalent, meaning the HC-12 is mostly not impeding the transfer of data. The bottleneck for data transfer in most cases seems to be the UART connection with the device. Tests 1 and 3 are the only results with significant time differences to the wired test. The modes used in these tests prioritise low power and long range respectively. They differ from the other modes in that the data transfer over the air is slower than the UART connection speed. If data is continuously sent to the device in these modes, the input buffers will overflow, and data will be lost. The HC-12 user manual suggests sending packets no bigger than 20 bytes for FU2 and 60 bytes for FU4 mode, with ¿2 s between each packet. In some of the tests the wired connection was very slightly slower, which seems like it should be impossible. It is possible that the length of the signal wires in the wired connection may have influenced the result due to their higher capacitance. The difference is less than 1% in any case, so it is not of significant concern.

The latency test results in Table 3 indicate that there are additional overheads created by the device.

Again, tests using the FU2 and FU4 modes are showing the most significant difference from the wired speeds. Tests 4 and 7 using mode FU1 have a very similar time despite very different baud rates. In the user manual it states in mode FU1, the air baud rate is a uniform 250,000 bps regardless of the UART baud rate. However, in mode FU3 the air baud rate changes with the UART baud rate, explaining why these speeds vary in comparison.

Table 2. Results from throughput test measuring the time taken to transfer 10Kb of data.

ID	Baud rate	Mode	Time taken [s]	Wired equivalent time [s]	Min. Calculated
1	1200	FU4	194.436825	83.649479	83.33333333
2	2400	FU3	41.884853	41.877711	41.66666667
3	4800	FU2	50.904906	21.011591	20.83333333
4	9600	FU1	10.547779	10.559831	10.41666667
5	9600	FU3	10.560724	10.559831	10.41666667
6	38400	FU3	2.701640	2.700745	2.604166667
7	115200	FU1	0.924378	0.931097	0.868055556
8	115200	FU3	0.938599	0.931097	0.868055556

Table 3. Averaged results of latency test on different modes and baud rates.

ID	Baud rate	Mode	Avg response (ms)	Wired equivalent (ms)
1	1200	FU4	2592.411621	14.996183
2	2400	FU3	189.5862802	7.5204258
3	4800	FU2	524.0615822	3.8166278
4	9600	FU1	32.6835997	1.9512919
5	9600	FU3	86.7270886	1.9512919
6	38400	FU3	35.6899399	0.5407718
7	115200	FU1	30.6233627	0.244038
8	115200	FU3	15.8306192	0.244038

The farthest distance from a successful transmission of each mode tested is shown in Table 4.

Table 4. HC-12 Range test results of various modes.

Baud rate	Mode	Furthest distance (metres)	Test ID
1200	FU4	60.84	184
4800	FU2	9.15	187
9600	FU1	12.14	185
9600	FU3	37.55	255
38400	FU3	17.6	180
115200	FU3	13.85	182

The range tests on test site were conducted up to 100 m, and it was expected that most configurations would manage the distance. The results of the tests revealed that none of the modes achieved 100 m and the maximum experimental range was 60 m in the long-range mode (FU4). Unfortunately, not all configurations could be tested due to issues in the field that could not be resolved. Given that the long-range mode and the default mode were successfully tested, it was not necessary to conduct these tests again.

The most surprising result of these tests is the range in all modes. A maximum of 60.84 m is significantly less than what has been claimed. The results also show that modes FU2 and FU4 have too high a latency to be really considered for the application. Mode FU1 was even poorer range and as power consumption is not a priority for the project it is not worth considering. Mode FU3 at 9600 baud rate has the most balanced performance and most likely why it is the default mode. Lowering the baud rate in the mode also lowers the air baud rate, leading to an increased range. Lower baud rates in FU3 mode could be considered to extend the range, if it is found that data transfer speeds are not as important.

Two more antennas were purchased, which could potentially improve range performance. The Unity Gain omnidirectional antenna and a Yagi Uda directional antenna (see Fig. 4). The Unity Gain antenna works the same way as the previous antenna, just with a much longer antenna length (see Fig. 5). A Yagi-Uda antenna has the potential to drastically increase the range of a radio signal by utilising additional components. In addition to the driven element or dipole, there also exists a reflector element that sits behind the dipole and a director element in front of it. The director and reflector are not electrically connected to the dipole and are known as parasitic elements [7]. The way these parasitic elements are excited by incoming radio waves causes them to emit electromagnetic radiation, and when arranged correctly, the radiation from these parasitic elements can strengthen the signal as it reaches the dipole [5]. Table 5 shows the results from each antenna. The Yagi Uda antenna was also tested while pointing perpendicular to the receiver, to show the directional effect. It should also be noted that for the Yagi Uda tests, the Unity Gain antenna was used as the antenna for the receiver. The results show that the range is indeed increased by a substantial amount by utilising these antennas. The farthest distance for the Unity Gain antenna is slightly misleading, which can be seen by studying the GPS results shown in Appendix A. Although the unity gain antenna reached 131.1 m, it only had consistently successful transmissions up to about 90 m. What should also be noted, the Yagi Uda antenna reached 137.95 m, but that is the maximum

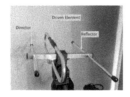

Fig. 4. 433 MHz Yagi Uda antenna

Fig. 5. Dolla Tek Unity Gain omnidirectional antenna.

Table 5. Range tests conducted on two antennas designed for increased range performance.

Baud rate	Mode	Furthest distance (metres)	Test ID	Antenna
9600	FU3	131.1	344	Unity Gain omni-directional
9600	FU3	137.95	345	Yagi Uda
9600	FU3	46.99	346	Yagi Uda (aimed perpendicular)

distance tested and it may have reached further because no signals were dropped in the entire test. To demonstrate the directional properties of the Yagi Uda antenna another test was performed with the antenna aiming perpendicular to the receiver, which yielded much worse results. The range improvements achieved by the Yagi-Uda antenna were deemed good enough for the project to continue with the HC-12 module as the wireless technology. There is an obvious drawback with using a directional antenna to communicate with a moving receiver, however. Therefore, the moving target will move out of the optimal signal strength. The solution involves complex algorithms to calculate the direction of arrival of a signal and then electronically adapt the radiation pattern of the antenna array to maximise the signal (Bazan, Kazi and Jaseemuddin, 2021). However, in addition to being very difficult to implement, the HC-12 does not provide any measurement of signal strength that could be used in algorithms. Another solution is to physically move a directional antenna at the target, known as a tracking antenna. There is another similar design by Airborne innovations[3], who do not explicitly say how the antenna tracks but state that it has integrated GPS and compass. Their design also includes an additional omni-directional antenna for close range operations. A tracking antenna seems like a more viable option and as GPS had already been implemented successfully for range testing.

5 Conclusions and Future Work

The antenna tracking mechanism demonstrated unexpected efficacy, notably outperforming other motors despite being 3D printed. Despite comprising numerous components, the system exhibited robust functionality without notable issues during extensive testing sessions. However, challenges arose during target tracking, particularly with 360° sweeps, resulting in some inaccuracies. Yet, the single-director antenna utilised displayed adequate accuracy for its purpose. Latency, averaging 253.05 ms and peaking at 505ms, while acceptable for many applications, proved slower for robotic control, warranting optimisation for more direct command signal transmission. However, the SmartAntenna's adaptability for various wireless technologies offers promising prospects, particularly if paired with a more reliable wireless module. Additionally, hardware selections such as the Raspberry Pi Pico and GPS receivers showcased commendable performance

---

[3] Available online, https://www.airborneinnovations.com/ai/datalinks/tracking-antennas/, last accessed: 18/05/2024.

and affordability, underscoring their suitability for projects requiring positioning accuracy and microcontroller capabilities. Moreover, deploying a web application on the RPi4 proved an effective solution for scalable interface access, albeit requiring consideration for remote deployment in areas lacking Wi-Fi access, suggesting potential integration of touch screen functionality for improved usability in future iterations.

Future work for the SmartAntenna involves addressing performance limitations associated with the HC-12 module, which raise concerns about its suitability for production systems due to issues and potential defects, especially within the noisy 433 MHz band. Suggestions for enhancement include integrating an omnidirectional antenna alongside the directional one to mitigate positional errors at close ranges, as proposed by Airborne Innovations. Moreover, the SmartAntenna's potential application in extending aerial, terrestrial, or underwater communications range at a reasonable cost presents an intriguing use case.

Disclosure of Interests. The authors have no competing interests to declare that are relevant to the content of this article.

References

1. Beard, C., Stallings, W.: Wireless Communication Networks and Systems. Pearson, Boston (2015)
2. Celebi, H.B., Pitarokoilis, A., Skoglund, M.: Wireless communication for the industrial iot. In: Industrial IoT: Challenges, Design Principles, Applications, and Security, pp. 57–94 (2020)
3. Curran, K., Millar, A., Mc Garvey, C.: Near field communication. Int. J. Electr. Comput. Eng. **2**(3), 371 (2012)
4. Granatstein, V.L.: Physical Principles of Wireless Communications. Auerbach publications, Boca Raton (2007)
5. Hartnagel, H.L., Quay, R., Rohde, U.L., Rudolph, M.: Fundamentals of RF and Microwave Techniques and Technologies. Springer, Heidelberg (2023)
6. Hassaballah, H.J., Fayadh, R.A.: Implementation of wireless sensor network for medical applications. In: IOP Conference Series: Materials Science and Engineering, vol. 745, p. 012089. IOP Publishing (2020)
7. Huang, J., Encinar, J.A.: Reflectarray Antennas. A John Wiley & Sons. Publication, Inc., Hoboken (2008)
8. Manam, S., Yashwanth, P., Telluri, P.: Comparative analysis of digital wireless mobile technology: a survey. Int. J. Innov. Technol. Explor. Eng. **8**, 268–273 (2019)
9. Marpaung, N., Amri, R., Ervianto, E.: Analysis of wireless fire detector application to detect peat land fire based on temperature characteristic. In: IOP Conference Series: Materials Science and Engineering, vol. 846, p. 012051. IOP Publishing (2020)
10. Shah, S.H., Yaqoob, I.: A survey: internet of things (iot) technologies, applications and challenges. In: 2016 IEEE Smart Energy Grid Engineering (SEGE), pp. 381–385 (2016)

11. Sikimić, M., Amović, M., Vujović, V., Suknović, B., Manjak, D.: An overview of wireless technologies for iot network. In: 2020 19th International Symposium INFOTEH-JAHORINA (INFOTEH), pp. 1–6 (2020). https://doi.org/10.1109/INFOTEH48170.2020.9066337
12. Song, S., Issac, B.: Analysis of wifi and wimax and wireless network coexistence. arXiv preprint arXiv:1412.0721 (2014)
13. Winasis, M.M.: Prayer guide tool for deaf using gyroscope sensor and HC-12. Ph.D. thesis, Universitas Muhammadiyah Yogyakarta (2022)

Machine Vision

Unsupervised Clustering with Geometric Shape Priors for Improved Occlusion Handling in Plant Stem Phenotyping

Katherine Margaret Frances James[1,2]($\boxtimes$)[iD] and Grzegorz Cielniak[1][iD]

[1] University of Lincoln, Brayford Pool, Lincoln LN6 7TS, UK
kajames@lincoln.ac.uk
[2] NIAB East Malling, New Road, East Malling, Kent ME19 6BJ, UK

Abstract. Automated robotic phenotyping has the potential to measure traits at a hitherto unprecedented level of detail, assisting in both crop monitoring and breeding programs for the production of new varieties. The use of 3D point cloud data allows for the measurement of geometric traits, however, occlusion results in partial point clouds. The measurement of stem length requires the computation of a skeleton through the medial axis of the point cloud, however, existing skeletonization methods are highly influenced by missing points, resulting in off-centre, non-biologically relevant skeletons. In this study, we propose a method which exploits a cylindrical shape prior during the generation of skeleton points which results in a more centred skeleton, accounting for missing points. We evaluate the k-cylinder clustering skeletonization method on real-world point clouds of strawberry petioles (stems) and demonstrate greater robustness at increased occlusion levels than the state-of-the-art.

Keywords: Robotic phenotyping · Point cloud · Skeletonization · Occlusion

1 Introduction

Automated in-field phenotyping has the potential to address the bottleneck in the selection process for new crop varieties, providing detailed data on numerous traits [10] which would otherwise be too time-consuming to measure manually [8]. Furthermore, automated phenotyping tools coupled with mobile robots would facilitate in-field crop monitoring [11]. A promising technological solution to the labour problems in the soft-fruit industry is the introduction of fruit-picking robots [4]. Such robots require detailed information about different organs within the plant, for steps such as path planning and grasping [21]. As such, automated phenotyping will also form an important component of robotic harvesters.

Recent developments in phenotyping have shown that phenotyping using 3D data provides more precise measurements and gives greater insights into the

---

Supported by the Collaborative Training Partnership (CTP) for Fruit Crop Research.

geometry and structure of the plant [16]. A challenge within this domain, however, pertains to the 3D representation of fine structures such as stems, the point clouds of which are often partial due to occlusion during data capture by other organs. Such structures form a major part of the plant architecture, impacting the overall structure or 'habit' of the plant as well as the 'display' - a term relating to the ease or difficulty of accessing the fruit for harvest. It is important to address the issue of occlusion and partial point cloud data to allow for accurate phenotyping trait extraction, in particular, the measurement of length. The length of a stem can be computed by computing the line which represents the medial axis of the stem - a representation called a skeleton. Skeletonization in the general 3D domain has become a fairly well-studied area [17], and is relatively successful through methods such as L1-median [7], but the application of these methods to small herbaceous plants and overcoming the challenges they pose is insufficiently investigated.

In this study, we propose a skeletonization technique that exploits a geometric prior combined with unsupervised clustering to produce skeleton points that are biologically correct, remaining centred in the presence of occlusion. The use of a cylindrical prior ensures that skeleton points remain centred despite partial point clouds. Furthermore, the connection of skeleton points to form a graph usually uses simple methods such as nearest neighbour or minimum spanning tree to connect these points, irrespective of the architecture. Our proposed method generates a series of graph fragments along the cylinder axes, connecting pairs of skeleton points at either end of the cylinder. This approach by its nature thus supplies more information for the process of connecting the skeleton points to form a graph, which is crucial for interpreting the skeleton in terms of length.

2 Related Work

Skeletonization algorithms require that the point cloud is broken into a set of clusters, from which a skeleton point can be computed. One way in which this can be achieved is through the approach of iterative contraction of sample set means [7]. This approach makes no assumptions about geometry and is applicable to general 3D skeletonization, but has also been applied to tomato [2]. Clusters can also be computed using level-set approaches exploiting geodesic distances and a root node [3,6]. In plant-specific skeletonization, the idea of a root node is common. A widely used method, initially developed for application to tree skeletonization, combines a root node weighted graph of local neighbouring points, from which Dikstra's method is then applied to find the shortest path [19]. Points are clustered into bins and the centroids of these form the skeleton points. This approach has been used as a comparison algorithm for the skeletonization of trees and *Arabidopsis* [1].

The notion of extracting a biologically relevant skeleton is important for making measurements for biological or phenological applications. To this end, refinement of pre-computed skeleton points is sometimes performed, in which steps tend to be focused on using either existing points in a local region to modify the position of the point [1] or exploit a geometric prior such as a cylinder for adjustment [6]. The fitting of a single cylinder to a point cloud, or portion thereof, is an accepted approach for approximating the skeleton of objects which are roughly cylindrical in geometry. Applications extend beyond plants and include pole fitting for signposts and other pole-like objects in street settings [15]. Fitting tree trunks for forestry (estimation of diameter at breast height) is a main use of cylinder fitting for plant skeletonization, for example [13,18]. Curving or branching stems cannot, however, be approximated well using this approach given the lack of a dominant direction for clustering points and such cases must be addressed either through subsequent refinement or by grouping points to form appropriate clusters before fitting a cylinder [20].

In contrast to the approaches described above, the method introduced in this paper combines the use of a cylindrical shape prior into the process of computing the initial skeleton point, involving the prior in the clustering process. This means that a subsequent refinement step is not needed to recenter the skeleton point. Furthermore, the combination of the geometric prior and unsupervised clustering eliminates the need for a manually selected root node or level-set approach, as there is no need for a predetermined dominant direction.

3 K-Cylinder Clustering Skeletonization

Skeletonization through k-cylinder cluster clustering computes skeleton points by modifying the k-means unsupervised clustering algorithm to make use of a cylindrical shape prior, and updating centroids using cylinder centres. K-cylinders are fit using least-squares, minimising the error across all cylinders, and the endpoints of each cylinder, connected along the medial axis of each, form graph fragments which are connected using a minimal spanning tree (MST) and smoothing applied.

3.1 Least-Squares Fitting

Least-squares fitting is a standard statistical procedure for finding the best fit through a set of data points. In this study, we applied least-square fitting to fit a cylinder to points in 3D space, using the procedure outlined in [5]. Spherical coordinates are used to compute each new direction vector $\mathbf{w}$, with increments of both ϕ and θ (in this study, $\Delta\phi=25$ and $\Delta\theta=50$). As output, we obtain the fit error for each cylinder, its radius squared and the cylinder centre $\mathbf{C}$.

3.2 Adapting K-Means

K-means is an established unsupervised clustering algorithm, which intrinsically uses a spherical prior by calculating the mean of the points as the centre. We break this assumption and instead prescribe that points should fall within a cylinder, and substitute the calculation of the mean point with that of the calculation of the cylinder centroid through least-squares fitting.

The approach is summarised as follows:

- Select k seed points - these are cylinder centroids $\mathbf{C}_i$
- Assign each point to the nearest $\mathbf{C}_i$
- Fit a cylinder to each cluster of points
- Update $\mathbf{C}_i$ for each cluster

Fitting is performed across n iterations, with early stopping when the change in error is less than δ ($n = 10$, $\delta = 0.01$ in our experiments). Total error ϵ is calculated as cumulative across all k cylinders. This yields a set of k cylinder centres $\mathbf{C}$ and associated parameters (r and $\mathbf{w}$). The clustering result is demonstrated in Fig. 1a. For this study, we set k to be the square root of the length of the diagonal across the point cloud between its maximum and minimum bounds, a rough indicator of the complexity of the stem versus its length. This value was found to give generally good results for samples in the tuning dataset, although tuning k per sample can give more optimal results.

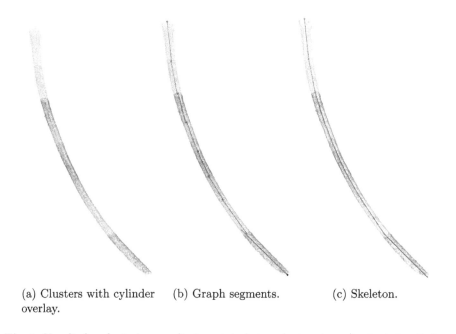

(a) Clusters with cylinder overlay. (b) Graph segments. (c) Skeleton.

Fig. 1. K-cylinder clustering results in points being clustered as a) oriented cylinders and yields b) graph segments through the cylinders which are then connected using a Kruskal's algorithm [12] and finally merged to produce c) a skeleton.

3.3 Graph Construction

Graph Fragment Formation. For each cluster point, we extract the top and bottom of the cylinder along the medial axis through the centroid, finding the height of the cylinder $|y_2|$ (see Eq. 1), by iterating through all the points and finding the point which lies most in the direction of **w**. Subsequently, we adjust the coordinates along the cylindrical axis away from the midpoint in both directions by half the height to compute the endpoints. These endpoints can then be connected to form a small, two-point graph. Thus, we obtain k graph fragments (see Fig. 1 b).

$$|y_2| = |\mathbf{w} \cdot (\mathbf{x} - \mathbf{C})| \leq h/2 \tag{1}$$

Graph Construction. We construct a cost matrix for the nodes in the graph fragments, assigning a zero cost to the connections between nodes in the individual fragments to ensure connection. For all other nodes, the Euclidean distance is computed as the cost. A minimal spanning tree is then computed from this matrix using Kruskal's algorithm [12], which produces the remaining edge connections between the nodes. The nodes in the graph forming the beginning and end of adjacent cylinders are close together. For this reason and to increase smoothness, we then merge these adjacent points to produce the final skeleton (see Fig. 1c).

4 Evaluation

To quantitatively evaluate the k-cylinder clustering skeletonization algorithm against state-of-the-art skeletonization algorithms, we apply a graph matching procedure. Point clouds of real-world data are used, with each point cloud representing a single stem instance and synthetic occlusion applied to examine the performance of the method as the level of occlusion increases.

4.1 Graph Matching

We make use of graph matching for the quantitative comparison of ground truth and estimated skeletons, following the same procedure as in [9]. We set the matching threshold t to 0.792 mm, which will consider any point within a radius equal to a quarter width of a standard stem to be matched. A dense graph of both the ground truth and estimated skeletons is computed, with a maximum spacing of 0.3 mm. We then report the matched precision and recall of skeleton nodes, the error in the total matched length (ϵ_L), the absolute percentage error between the measured lengths (irrespective of matching with the ground truth), the number of segments and the number of endpoints.

4.2 Datasets

We have performed experiments on two real-world datasets of strawberry petioles. The first dataset, used for tuning and validation, contains 26 point clouds with minimal noise. The second dataset comprises of 267 individual petiole instances from the LastStraw dataset [9]. This dataset is more challenging than the first, as it contains more imperfections, such as noise and scanning artefacts. Petiole instances in this dataset in some cases also widen at the top as they begin to branch into the support structure for the leaf. The tuning dataset was used during the development of the method and for the tuning of parameters. The validation dataset was used for occlusion analysis in the ablation study, while the LAST-Straw dataset was used to determine the generalisation of the methodology.

4.3 Generation of Synthetic Occlusion

We performed synthetic occlusion on the validation dataset using the following process, whereby we set a viewpoint V and utilised the ground truth skeleton S_g to include points within an allowed deviation of θ across the vector from the cylinder axis to V. For each edge pair we then identify the set of points to retain, falling within a particular range of theta values to obtain a) an entire point cloud, b) a three-quarter point cloud, c) half point cloud and d) a quarter-point cloud, respectively: $\theta = [2\pi, \frac{3}{2}\pi, \pi, \frac{\pi}{2}]$. To achieve this we utilise the endpoints A and B of each segment of the line set to determine whether a point P in the point cloud falls within the allowed θ region.

A sample demonstrating the effects of synthetic occlusion is shown in Fig. 2. Each point cloud and line set was pre-processed by translating it to be centred on the origin. The viewpoint V was set to (-3,0,0), which was manually determined to fall outside of the point clouds.

Fig. 2. Example of a synthetically occluded point cloud, starting with the original point cloud at the top and increasing occlusion by a quarter in each of the subsequent point clouds. The ground truth skeleton (shown in green) is held constant for all levels of occlusion. (Color figure online)

5 Results

We demonstrate skeletonization results for two studies. Firstly, in an ablation study to illustrate the impact of increasing occlusion and secondly a generalisation study. We compare our results to three state-of-the-art skeletonization algorithms, namely the method of [19], the skeletonization application of self-organising maps (SOMs) used by [14] and the L1-medial skeletonization approach of [7], all of which have publicly available open-source implementations.

5.1 Ablation Study

A study was performed to examine the performance of the methods in the presence of increasing occlusion, using synthetically occluded point clouds. The ground truth skeleton of the original point cloud was used for each iteration of synthetic occlusion for comparison against each of the evaluated methods. The synthetically occluded dataset consists of 26 point clouds of individual stems, resulting in a total of 104 point clouds following synthetic occlusion. We assess the performance of the methods across this dataset, with a histogram of results for each metric shown in Fig. 3.

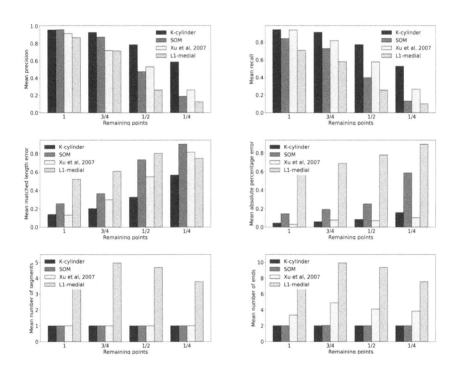

Fig. 3. Histograms of metric distribution for the ablation study.

Examining the precision and recall in this figure, we observe that although other methods exhibit similar performance when there is no to low levels of synthetic occlusion, as this increases the precision of our method does not drop off to the same extent as other methods. This means that the nodes within the skeleton fall within the matching threshold of the ground truth with greater reliability. Considering the mean matched length error ϵ_L, k-cylinder clustering provides the best length estimate at all levels of occlusion. The mean percentage error in length (not considering which portions are matched) indicates that the method of [19] can produce a marginally lower error in length. However, this comes at the cost of a high number of endpoints, indicating the presence of erroneous branching. Finally, we observe that all methods except L1-medial produce fully connected skeletons, while the L1-medial method produces a fragmented skeleton with a high number of segments. These observations indicate that overall the k-cylinder method provides more robustness at higher levels of occlusion, allowing for better skeletons to be created in such cases. At high levels of occlusion, points in some samples became extremely sparse ($<$ 100 points), resulting in the L1-medial method failing to produce a skeleton for four samples, and the SOM method for two samples. Figure 4 visually demonstrates how as occlusion increases, the results of the other methods are skewed away from the ground truth skeleton by the mass of the points, whereas the skeleton produced using the k-cylinder method holds its position more robustly.

Fig. 4. Point clouds with skeletonization results for a stem with increasing occlusion from right to left.

5.2 Generalisation Study

The occlusion study demonstrated that when holding the ground truth skeleton constant, but increasing occlusion, the k-cylinder clustering algorithm was better able to maintain centredness than other methods. To determine the generalisation of this method to other datasets and the limitations thereof, we make use of the LAST-Straw dataset. First, we report results on a subset of this dataset, in which 51 samples meet the criteria: 1) no branching at the end, 2) low levels of noise and 3) occlusion is present. These samples are selected as they all demonstrate occlusion, yet have sufficient points such that the annotation confidence for the centredness of the ground truth skeleton is sufficiently high. This ensures that the metrics reported give an accurate reflection of the geometric matching.

Results for this comparison can be seen in Table 1, with metrics computed cumulatively across the dataset. From this table, it is clear that the performance of the method generalises well to these samples, achieving the best score in all metrics. Particularly importantly for phenotyping and in terms of the measurement of length, the matched length error ϵ_L and absolute percentage error are particularly low. This is very important, as these metrics indicate the measurement error, which is the purpose of the exercise of skeletonization for phenotyping.

Table 1. Metrics calculated cumulatively across the LAST-Straw dataset.

Method	Precision	Recall	F1-score	ϵ_L	Absolute % error
K-cylinder	**0.83**	**0.85**	**0.84**	**0.29**	**0.03**
SOM [14]	0.59	0.47	0.52	0.61	0.22
Xu [19]	0.43	0.48	0.45	0.61	0.07
L1-medial [7]	0.22	0.32	0.26	0.77	0.75

To examine the behaviour of the k-cylinder clustering method across less ideal samples, we report metrics across the entire dataset in Table 2. This includes samples with relatively few points, noise and scanning artefacts, branching near the end of the petiole and extremely occluded samples for which the annotation confidence is low for centredness. Due to this last factor, we only report the F1-score and absolute percentage error to limit reliance on the samples which have low annotation confidence for centredness. Skeletonization results for some challenging examples can be seen in Fig. 5.

Table 2. Metrics calculated cumulatively across the entire LAST-Straw dataset.

Method	F1-score	Absolute % error
K-cylinder	0.60	0.11
SOM [14]	0.71	0.22
Xu [19]	**0.74**	**0.08**
L1-medial [7]	0.43	0.79

From the results, we see that in a more noisy dataset, the advantages of the k-cylinder method diminish. The method is susceptible to noise, as cylinders are skewed by this and scanning artefacts. However, it is noted that the absolute percentage error is not substantially higher than that of the method of [19], which indicates that this method can still deliver acceptable length measurements even in more challenging cases. As it stands, the k-cylinder clustering skeletonization approach is only suited to the skeletonization of individual stems, that do not branch. Another limitation of the method is the necessity to specify k, although this limitation is shared by the SOM skeletonization method [14] and both the shortest path method of [19] and L1-median sampling method of [7] require the specification of particular parameters as well. Selecting a k which is too low will result in a jagged skeleton while setting it too high results in mis-orientation of cylinders. However, in the case where occlusion is the main challenge, noise is relatively low and a suitable k is selected, k-cylinder clustering yields results which produce a biologically relevant stem skeleton, from which accurate length measurements can be extracted.

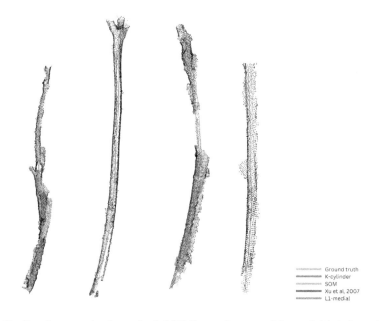

Fig. 5. Challenging samples from the LAST-Straw dataset with overlaid skeletonization results.

We additionally compare the average computational runtime for the methods, run on a PC with an Intel core i7 12th gen CPU and 32 GB of RAM, without GPU acceleration. In order of efficiency, these were the Xu method (0.02 s), SOM method (0.34 s), k-cylinder method (2.90 s) and L1-medial method (11.39 s). However, the Xu method does additionally require manual placement of the root node, which is a necessary inefficiency to ensure an accurate start point, upon which the algorithm is dependent. This indicates that computational time may be another factor to consider in addition to the trade-off between robustness to occlusion and noise.

6 Conclusion and Future Work

In this work, we proposed a method which exploited a cylindrical prior for the skeletonization of stems to allow for more robust skeletonization in the presence of occlusion. The combination of this geometric prior with unsupervised clustering results in oriented cylinders along the length of the stem, overcoming the requirement of a manually specified root node or dominant direction for use in level-set approaches. The result is skeletons that better retain their centredness, particularly in cases where points are missing along a particular side of a stem due to occlusion, which is often the case due to limited access to viewpoints for cameras and sensors when blocked by other plant organs during data collection. The effectiveness of this approach was demonstrated in an ablation study where we synthetically occluded samples in the validation dataset - increasing the existing level of occlusion from none to three-quarters. The results of this study demonstrate the robustness of this method, with the k-cylinder method producing a skeleton that was the most invariant to increasing occlusion. Application of the method to a larger dataset, to which synthetic occlusion was not applied, indicates that this holds true for samples where occlusion is present and there is minimal noise. The generalisability of the method is, however, limited by its susceptibility to noise and scanning artefacts. Data quality remains a critical factor affecting the success of automated phenotyping and further research into developing methods which are robust to both occlusion and noise is highlighted as a direction for future research.

References

1. Chaudhury, A., Godin, C.: Skeletonization of plant point cloud data using stochastic optimization framework. Front. Plant Sci. **11** (2020). https://doi.org/10.3389/fpls.2020.00773
2. Chebrolu, N., Läbe, T., Stachniss, C.: Spatio-temporal non-rigid registration of 3D point clouds of plants. In: 2020 IEEE International Conference on Robotics and Automation (ICRA), pp. 3112–3118 (2020). https://doi.org/10.1109/ICRA40945.2020.9197569
3. Delagrange, S., Jauvin, C., Rochon, P.: PypeTree: a tool for reconstructing tree perennial tissues from point clouds. Sensors **14**(3), 4271–4289 (2014). https://doi.org/10.3390/s140304271. https://www.mdpi.com/1424-8220/14/3/4271
4. Duckett, T., Pearson, S., Blackmore, S., Grieve, B.: Agricultural robotics: the future of robotic agriculture. Technical report (2018). https://doi.org/10.31256/WP2018.2
5. Eberly, D.: Least squares fitting of data by linear or quadratic structures (1999). https://www.geometrictools.com/Documentation/LeastSquaresFitting.pdf. Accessed 2021
6. Fu, L., Liu, J., Zhou, J., Zhang, M., Lin, Y.: Tree skeletonization for raw point cloud exploiting cylindrical shape prior. IEEE Access **8**, 27327–27341 (2020). https://doi.org/10.1109/ACCESS.2020.2971549
7. Huang, H., et al.: L1-medial skeleton of point cloud. ACM Trans. Graph. **32**(4), 65 (2013). https://doi.org/10.1145/2461912.2461913

8. James, K., Sargent, D., Whitehouse, A., Cielniak, G.: High-throughput phenotyping for breeding targets - current status and future directions of strawberry trait automation. In: Plants, People, Planet, pp. 1–12 (2022). https://doi.org/10.1002/ppp3.10275
9. James, K., Heiwolt, K., Sargent, D., Cielniak, G.: Lincoln's annotated spatio-temporal strawberry dataset (LAST-Straw). ArXiv (2024). arXiv:2403.00566
10. Jiang, Y., Li, C.: Convolutional neural networks for image-based high-throughput plant phenotyping: a review. Plant Phenomics **2020**, 4152816 (2020). https://doi.org/10.34133/2020/4152816
11. Jin, Y., Liu, J., Xu, Z., Yuan, S., Li, P., Wang, J.: Development status and trend of agricultural robot technology. Int. J. Agric. Biol. Eng. **14**(4), 1–19 (2021). https://doi.org/10.25165/j.ijabe.202n404.6821
12. Kruskal, J.B.: On the shortest spanning subtree of a graph and the traveling salesman problem. Proc. Am. Math. Soc. **7**(1), 48–50 (1956)
13. Liang, X., Kankare, V., Yu, X., Hyyppä, J., Holopainen, M.: Automated stem curve measurement using terrestrial laser scanning. IEEE Trans. Geosci. Remote Sens. **52**(3), 1739–1748 (2014). https://doi.org/10.1109/TGRS.2013.2253783
14. Magistri, F., Chebrolu, N., Stachniss, C.: Segmentation-based 4D registration of plants point clouds for phenotyping. In: 2020 IEEE/RSJ International Conference on Intelligent Robots and Systems (IROS), pp. 2433–2439 (2020). https://doi.org/10.1109/IROS45743.2020.9340918
15. Nurunnabi, A., Sadahiro, Y., Lindenbergh, R.: Robust cylinder fitting in three-dimensional point cloud data. Int. Arch. Photogram. Remote Sens. Spat. Inf. Sci. **XLII-1/W1**, 63–70 (2017). https://doi.org/10.5194/isprs-archives-XLII-1-W1-63-2017. https://isprs-archives.copernicus.org/articles/XLII-1-W1/63/2017/
16. Paulus, S.: Measuring crops in 3D: using geometry for plant phenotyping. Plant Methods **15**, 103 (2019). https://doi.org/10.1186/s13007-019-0490-0
17. Tagliasacchi, A., Delame, T., Spagnuolo, M., Amenta, N., Telea, A.: 3D skeletons: a state-of-the-art report. Comput. Graph. Forum **35**(2) (2016). https://doi.org/10.1111/cgf.12865
18. Xie, Y., Zhang, J., Chen, X., Pang, S., Zeng, H., Shen, Z.: Accuracy assessment and error analysis for diameter at breast height measurement of trees obtained using a novel backpack lidar system. Forest Ecosyst. **7**, 1–11 (2020). https://doi.org/10.1186/s40663-020-00237-0
19. Xu, H., Gossett, N., Chen, B.: Knowledge and heuristic-based modeling of laser-scanned trees. ACM Trans. Graph. **26**(4), 19–es (2007). https://doi.org/10.1145/1289603.1289610
20. Yan, D.M., Wintz, J., Mourrain, B., Wang, W., Boudon, F., Godin, C.: Efficient and robust reconstruction of botanical branching structure from laser scanned points. In: 2009 11th IEEE International Conference on Computer-Aided Design and Computer Graphics, pp. 572–575 (2009). https://doi.org/10.1109/CADCG.2009.5246837
21. Zhou, H., Wang, X., Au, W., Kang, H., Chen, C.: Intelligent robots for fruit harvesting: recent developments and future challenges. Precis. Agric. **23**, 1856–1907 (2022). https://doi.org/10.1007/s11119-022-09913-3

Depth Priors in Removal Neural Radiance Fields

Zhihao Guo[✉] and Peng Wang

Department of Computing and Mathematics, Manchester Metropolitan University,
Manchester, UK
`zhihao.guo@stu.mmu.ac.uk, p.wang@mmu.ac.uk`

Abstract. Neural Radiance Fields (NeRF) have achieved impressive results in 3D reconstruction and novel view generation. A significant challenge within NeRF involves editing reconstructed 3D scenes, such as object removal, which demands consistency across multiple views and the synthesis of high-quality perspectives. Previous studies have integrated depth priors, typically sourced from LiDAR or sparse depth estimates from COLMAP, to enhance NeRF's performance in object removal. However, these methods are either expensive or time-consuming. This paper proposes a new pipeline that leverages SpinNeRF and monocular depth estimation models like ZoeDepth to enhance NeRF's performance in complex object removal with improved efficiency. A thorough evaluation of COLMAP's dense depth reconstruction on the KITTI dataset is conducted to demonstrate that COLMAP can be viewed as a cost-effective and scalable alternative for acquiring depth ground truth compared to traditional methods like LiDAR. This serves as the basis for evaluating the performance of monocular depth estimation models to determine the best one for generating depth priors for SpinNeRF. The new pipeline is tested in various scenarios involving 3D reconstruction and object removal, and the results indicate that our pipeline significantly reduces the time required for the acquisition of depth priors for object removal and enhances the fidelity of the synthesized views, suggesting substantial potential for building high-fidelity digital twin systems with increased efficiency in the future.

Keywords: Neural Radiance Fields · Monocular Depth Estimation · 3D Editing · 3D Reconstruction

1 Introduction

Neural Radiance Fields (NeRF) have remarkably transformed the domain of 3D reconstruction and novel view synthesis [1]. The advancements are promising in revolutionizing sectors from augmented reality to robotic environment perception and modeling. Particularly, it would help to develop high-fidelity digital twins of human-robot collaboration systems in manufacturing settings [2], facilitating high-consequence decision-making and safe and flexible human-robot collaboration. Despite the advancements in NeRF techniques, a persistent challenge

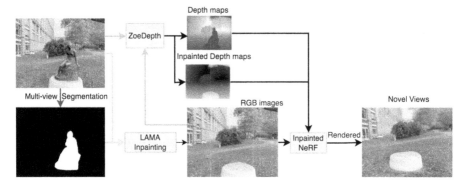

Fig. 1. Overview of the proposed object removal pipeline. Starting with sparse views and corresponding masks as inputs, multi-view segmentation (indicated by the red arrow) is employed to generate consistent masks. Depth maps and inpainted depth maps as priors are produced by ZoeDepth (depicted by the blue arrow). The green arrow highlights the multi-view consistent inpainting process, which integrates inpainted depth maps and RGB images into the updated NeRF model to render novel views. (Color figure online)

within the framework is editing the scenes rendered by NeRF seamlessly, such as removing objects of interest and inpainting the removed area in coherence with the background. This capability is particularly vital in contexts such as human-robot collaborative environments, where the strategic removal of certain obstacles is essential for the planning of safe navigation through environments for robots.

Recent works have addressed this issue by introducing depth priors into NeRF, to supervise the rendering process and improve the rendering quality [3–5]. Typically, depth is provided by LiDAR point clouds or depth completion techniques trained with sparse COLMAP depth, which are either expensive economically or time-consuming to get [6]. Nevertheless, depth priors or depth maps help preserve the fidelity of edited scenes, such as when objects are removed. They play a vital role in maintaining consistency in lighting and shadows, which are key to realistic rendering and accurate perception. Therefore, it is crucial to thoroughly evaluate how different depth priors affect the performance of Removal NeRF.

This paper proposes a novel removal NeRF pipeline which employs SpinNeRF as the foundational reconstruction and removal model. Depth priors generated by various monocular depth estimation models, such as ZoeDepth [7], are next integrated into SpinNeRF to enhance its performance and accelerate the speed in object removal tasks, which are evaluated across a range of metrics such as PSNR (Peak Signal-to-Noise Ratio) and SSIM (Structural Similarity Index Measure). It is worth noting that monocular depth estimation models are exploited due to their efficiency and the potential to bring SpinNeRF into real-time systems.

One challenge facing performance evaluation of monocular depth estimation models is the absence of ground truth depth data. This lack of verifiable depth information necessitates finding substitutes for true depth values where they are unavailable. In response to this issue, an extensive evaluation of COLMAP's capabilities for dense depth reconstruction was performed using various KITTI datasets [8]. The evaluation focused on two key metrics: the precision of the depth reconstruction and the completeness of the depth maps produced. These metrics help assess how accurately and fully COLMAP can infer depth information from monocular images.

Our research findings demonstrate that COLMAP exhibits superior performance in depth estimation, positioning it as a viable substitute for conventional ground truth depth acquisition methods, particularly in scenarios where acquiring true depth information can be prohibitively expensive. Taking depth generated by COLMAP as ground truth, we then evaluated several state-of-the-art monocular depth estimation models on the SpinNeRF dataset [4] to find the optimal model for providing depth priors. ZoeDepth emerges as the leading monocular depth estimation method, delivering high-quality depth priors while concurrently reducing computational overhead. This makes the integration of ZoeDepth with SpinNeRF intuitive to create our proposed pipeline, which substantially improves both the robustness and the quality of the rendered NeRF scenes and the object removal processes. These findings underscore the transformative potential of monocular depth estimation in enhancing NeRF applications. Figure 1 shows an overview of the proposed pipeline.

The contributions of this work include 1). A new pipeline that integrates depth priors from monocular depth estimation models such as ZoeDepth with SpinNeRF enables SpinNeRF to handle complex object removal scenarios and reduce computational overhead. 2). A comprehensive evaluation of COLMAP's dense depth reconstruction capabilities on KITTI dataset helps establish COLMAP as a viable, cost-effective alternative to conventional ground truth depth data acquisition methods. 3). A systematic evaluation of various monocular depth estimation models, guiding the selection of the optimal depth estimation model to improve reconstruction and object removal performance of NeRF-based methods.

The remainder of the paper is organised as follows. Some related works are introduced in Sect. 2. Section 3 elaborates on the impacts of depth priors on the removal performance of NeRF, and the methodology integrating SpinNeRF with ZoeDepth is presented. Experiments, discussions, and an ablation study are provided in Sect. 4, and the paper is concluded in Sect. 5.

2 Related Works

This section explores NeRF-based 3D reconstruction and depth-enhanced NeRF for object removal. We will also delve into various depth estimation methods to discuss their potential to provide depth priors for the removal NeRF.

2.1 Neural Radiance Fields

NeRF. NeRF [1] utilise implicit representation for the first time to achieve photo-realistic perspective synthesis, subsequently inferring the 3D structure of the scene. It utilises a limited number of input views to train a neural network to represent a continuous volumetric scene, enabling the generation of new perspectives of the scene after neural network training. To be specific, given the 3D location $\mathbf{p} = (x, y, z)$ of a spatial sampling point (to be rendered) and 2D view direction $\mathbf{d} = (\theta, \phi)$ of the camera, NeRF predicts the color $\mathbf{c} = (r, g, b)$ of the sampling point and the volume density σ through a multi-layer perception (MLP) neural network [9] which can be represented as

$$F_\Theta : (\mathbf{p}, \mathbf{d}) \rightarrow (\mathbf{c}, \sigma), \tag{1}$$

where F_Θ is the MLP parameterised by Θ. The color is estimated by volumetric rendering via quadrature, which can be formulated as

$$C(\mathbf{r}) = \int_{t_n}^{t_f} T(t)\sigma(\mathbf{r}(t))\mathbf{c}(\mathbf{r}(t), \mathbf{d})dt, \tag{2}$$

where $T(t) = \exp\left(-\int_{t_n}^{t} \sigma(\mathbf{r}(s))\,ds\right)$, $C(\mathbf{r})$ is the sampling pixel value and is calculated by integrating the radiance value $\mathbf{c}$ along the ray $\mathbf{r}(t) = \mathbf{o} + t\mathbf{d}$, in which $\mathbf{o}$ is the camera position, $\mathbf{d}$ is the direction from the camera to the sampled pixel, within near and far bounds t_n and t_f, and the function $T(t)$ denotes the accumulated transmittance along each ray from t_n to t. While NeRF is renowned for its ability to produce highly detailed and coherent 3D renderings from a relatively sparse set of images, it can be computationally intensive and slow to train and render new perspectives, particularly for complex scenes. In addition, in cases where insufficient input images are used for training, the rendering quality of NeRF can be fairly poor.

Depth Supervised NeRF. Depth priors provide explicit information about the distance between scene surfaces and the camera. Recent advancements have effectively integrated depth priors into NeRF reconstructions to enhance their photorealism and details from sparse inputs. NerfingMVS [10] leverages depth from Multi-View Stereo (MVS) to train a depth predictor, significantly enhancing the scene's detail and guiding the NeRF sampling process. This approach effectively uses depth priors to refine the sampling strategy within the NeRF pipeline, thereby improving the model's efficiency and output quality. Similar to NerfingMVS, DONeRF [11] introduces a depth oracle trained to pinpoint significant surfaces in 3D space, optimising rendering speeds and computational resource allocation.

Furthermore, Depth-Supervised NeRF (DSNeRF) [6] employs depth priors obtained from 3D point clouds estimated via Structure from Motion (SfM). DSNeRF introduces a novel loss function that aligns the termination of a ray with specific 3D key points, incorporating reprojection errors to gauge uncertainty. This approach ensures that novel view renderings are not only more accurate but also incorporate a measure of confidence in the depth estimation, which is crucial for realistic outputs. Similarly, the study presented in Dense Depth Priors for NeRF [12] develops a network that enriches sparse depth data from SfM, resulting in dense depth maps that guide both the optimisation process and the scene sampling methodology. By providing depth information where it is typically missing, this approach provides a more robust pipeline for NeRF to produce high-quality reconstructions from limited data. However, one significant challenge for DSNeRF is the substantial time consumption required to generate dense depth information, which can substantially decelerate the process, impeding the practical deployment of this technology in real-time applications where rapid processing is essential.

Removal NeRF. Removal NeRF is a specialized variant of NeRF designed for editing and removing objects synthesized by NeRF models. This capability is crucial for applications requiring dynamic scene manipulation such as in virtual reality, augmented reality, and potentially digital twins. The techniques developed in this area can be broadly classified into two main categories, each addressing the challenges of object removal with distinct approaches:

1) Object Compositional NeRF: This category focuses on learning decomposable components of a scene, enabling selective editing of individual elements. Object-NeRF [3] approaches scene decomposition by learning separate NeRF for different objects and allows for the manipulation of specific scene components without altering the rest of the environment. Despite its effectiveness in isolated edits, maintaining consistency across different viewpoints remains a challenge, particularly when dealing with complex interactions between light and geometry. ObjectSDF [13] uses signed distance functions to represent each object within the NeRF pipeline. While ObjectSDF enhances the geometric definition of edited areas, ensuring photorealistic blending and view consistency in diverse lighting conditions continues to pose significant difficulties.
2) Inpainting-Based Removal: This approach integrates 2D inpainting techniques to reconstruct the scenes following object removal, using depth information to ensure spatial consistency. SPinNeRF [4] leverages the LAMA inpainting method [14] combined with depth priors to enhance the quality of the regions from which objects are removed. This method focuses on generating depth-consistent inpainting to seamlessly blend the edited areas with their surroundings. NeRF-Object Removal [5] builds upon this concept by adapting the inpainting process based on depth cues from the surrounding environment, improving the quality and realism of the final rendered images.

2.2 Depth Estimation

COLMAP Dense Depth. COLMAP is an end-to-end multi-view image-based 3D reconstruction technique that is built on top of SfM [15] and MVS reconstruction [16]. In addition, it can also achieve depth estimation, which is required by 3D synthesis models such as removal NeRF. We will briefly go through the pipeline of COLMAP as it is used for depth ground truth generation in this paper for datasets where true depth values are unavailable.

SfM mainly entails the following procedures: feature extraction F_i, feature matching M_{ij}, robust model estimation, and bundle adjustment. Feature extraction F_i aims at extracting features such as SIFT from a given image I_i. The features from image I_i will be next matched with features from another image(s) I_j, and we denote the matching process as M_{ij}. The matched features will be used to create a sparse 3D point cloud using triangulation [17], which is denoted as

$$P_{ij} = T\Big(M_{ij}, C_i, C_j\Big), \qquad (3)$$

where P_{ij} are the 3D points triangulated from matches M_{ij}, and C_i, C_j are the camera parameters (intrinsic and extrinsic) associated with images I_i and I_j, respectively. The bundle adjustment procedure refines camera parameters and 3D points simultaneously to minimise the re-projection error, and each minimization term individually considers camera parameter C_i and C_j, as well as the 3D points P_{ij} to simplify the optimization problem and expedites the computational process, which is denoted as

$$\min_{C_i, C_j, P_{ij}} \sum_{i,j} d(Q(M_{ij}, C_i, C_j), P_{ij}), \qquad (4)$$

where $d(\cdot, \cdot)$ is the Euclidean distance, and $Q(\cdot, \cdot, \cdot)$ is the projection function that projects 3D points P_{ij} onto the image plane using camera parameters C_i and C_j.

Monocular Depth Estimation. Depth estimation from monocular images has led to a variety of approaches. Early monocular depth estimation works [18,19] relied on handcrafted features, which lack generality to unpredictable and variable conditions in natural scenes. The limitations of these methods have prompted the exploration of deep learning methods [7,20–23] in depth estimation.

Some recent works focus on encoder-decoder models [7,21], for instance, ZoeDepth [7] introduces a novel architecture combines relative and metric depth, following MiDaS depth estimation framework [24] to predict relative depth, and add heads to the encoder-decoder model for metric depth estimation, which is achieved by attaching a metric bins module to the decoder. The final depth of a pixel i is obtained by a linear combination of the bin centers weighted by the corresponding predicted probabilities,

$$d(i) = \sum_{k=1}^{N_{\text{total}}} p_i(k) c_i(k), \qquad (5)$$

where $d(i)$ is the predicted depth at pixel i, N_{total} is the total number of depth bins, $p_i(k)$ is the probability of the $k-th$ bin for pixel i, and $c_i(k)$ is the center of the $k-th$ depth bin.

DepthAnything [21] is another deep learning-based model depth estimation model. It learns from both unlabeled image datasets $\mathcal{D}^u = \{u_i\}_{i=1}^{N}$ and labeled datasets $\mathcal{D}^l = \{(x_i, d_i)\}_{i=1}^{M}$. A teacher model T is first learned from $\mathcal{D}^l$, which is next used to assign pseudo depth labels for $\mathcal{D}^u$. The depth estimation model will then be trained on the combination of labeled datasets and pseudo-unlabeled datasets.

Recently, EcoDepth [20] introduced a new architecture module called Comprehensive Image Detail Embedding (CIDE) and formulated monocular depth estimation as a dense regression problem:

$$\mathbb{P}(\mathbf{y} \mid \mathbf{x}, \mathcal{E}) = \mathbb{P}(\mathbf{y} \mid \mathbf{z}_0) \, \mathbb{P}(\mathbf{z}_0 \mid \mathbf{z}_t, \mathcal{C}) \, \mathbb{P}(\mathbf{z}_t \mid \mathbf{x}) \, \mathbb{P}(\mathcal{C} \mid \mathbf{x}), \qquad (6)$$

where $\mathbf{x}$ and $\mathbf{y}$ denote the input image and the output depth, respectively. The terms in the equation are defined as follows: $\mathcal{C}$ represents the semantic embedding derived from the CIDE module.

$$\mathbb{P}(\mathcal{C} \mid \mathbf{x}) = \mathbb{P}(\mathcal{C} \mid \mathcal{E})\mathbb{P}(\mathcal{E} \mid \mathbf{x}), \qquad (7)$$

where $\mathcal{E}$ is the embedding vector from a ViT module [25], and $\mathbb{P}(\mathcal{C} \mid \mathcal{E})$ is implemented using downstream modules in CIDE consisting of learnable embeddings. $\mathbb{P}(\mathbf{z}_t \mid \mathbf{x})$ is implemented using the encoder of a variational autoencoder [26]. $\mathbb{P}(\mathbf{z}_0 \mid \mathbf{z}_t, \mathcal{C})$ is implemented using conditional diffusion. $\mathbb{P}(\mathbf{y} \mid \mathbf{z}_0)$ is implemented through the depth regressor module, which is a two-layer convolutional neural network. The final depth map $\mathbf{y}$ is obtained through an upsampling decoder. This decoder includes deconvolution layers and performs a resolution transition, specifically transitioning from a lower resolution of the concatenated feature map back to the original dimensions of the input image.

3 Removal NeRF with Monocular Depth Priors

Depth priors are instrumental in delivering essential spatial metrics that substantially influence the performance of NeRF models. These priors provide precise distance measurements between scene objects and the camera, crucial for the accurate reconstruction of 3D scenes. By streamlining the rendering process and facilitating model training, depth priors not only expedite convergence but also minimize computational demands, thereby optimizing the overall efficiency of the NeRF framework. This paper investigates how depth priors affect the object removal performance of NeRF models. To this end, the architecture of SpinNeRF [4] is employed as the base NeRF model. Depth priors obtained by different models such as ZoeDepth are next introduced to SpinNeRF, and the new model's performance on object removal is then evaluated against a series of metrics.

SpinNeRF is a cutting-edge NeRF model designed to enhance the rendering of 3D scenes from sparse viewpoints. Unlike traditional NeRF models, SpinNeRF introduces a unique capability to interpolate and extrapolate views dynamically by learning the underlying structure of a scene through sparse image inputs. To be specific, given a set of RGB images $\mathcal{I} = \{I_i\}_{i=1}^{n}$, with the corresponding camera poses $\mathcal{G} = \{G_i\}_{i=1}^{n}$ and camera intrinsic matrix K estimated by SfM, and the corresponding removal object mask $\mathcal{M} = \{M_i\}_{i=1}^{n}$ provided by users or learned by a different model, SpinNeRF leverages a deep learning framework to encode the volumetric properties of a scene into a neural network, allowing for highly realistic and computationally efficient 3D rendering. During the rendering process, SpinNeRF utilises LAMA [14] to achieve image inpainting, and the inpainted images are denoted as $\left\{\widetilde{I}_i\right\}_{i=1}^{n}$, with $\widetilde{I}_i = \text{INP}(\mathcal{I}, \mathcal{M})$ denoting the inpainting process.

To improve the overall removal and rendering performance, SpinNeRF uses the perceptual loss, LPIPS [27] to optimise the masked regions during rendering, while MSE is still used for unmasked regions,

$$\mathcal{L}_{\text{LPIPS}} = \frac{1}{|\mathcal{B}|} \sum_{i=1}^{|\mathcal{B}|} \text{LPIPS}\left(\widehat{I}_i, \widetilde{I}_i\right), \tag{8}$$

where $\mathcal{B}$ is a subset of the entire dataset of indices from 1 to $|\mathcal{B}|$, $\widehat{I}_i$ is the $i-th$ view rendered using NeRF, $\widetilde{I}_i$ is the $i-th$ view that has been inpainted.

To achieve high-quality image inpainting, depth information on objects of interest is required. In the SpinNeRF pipeline, for each pixel along ray r, depth priors $D(r)$ are calculated using

$$D(r) = \sum_{i=1}^{N} T_i \Big(1 - \exp\left(-\sigma_i \delta_i\right)\Big) t_i. \tag{9}$$

where N is the number of sample points along each ray r from the camera, t_i is the depth of $i-th$ sample point, σ_i is the density at the i-th sample, indicating how likely the light is to scatter at this point, δ_i is the segment length between consecutive samples along the ray, and T_i is the cumulative transmittance to the $i-th$ sample, reflecting the probability that light reaches this point without being obstructed.

Depth priors are then passed to INP to generate inpainted depth maps $\left\{\widetilde{D}_i\right\}_{i=1}^{n}$, where $\widetilde{D}_i = \text{INP}(D_i, M_i)$. The depth maps will be used to supervise the inpainted NeRF, and the inpainting process is optimised using the ℓ_2 distance between the rendered depths D_i and the inpainted depths $\widetilde{D}_i$

$$\mathcal{L}_{\text{depth}} = \frac{1}{R} \sum_{r=1}^{R} |D(r) - \widetilde{D}(r)|^2, \tag{10}$$

where $\widetilde{D}(r)$ and $D(r)$ are the inpainted depths and rendered depths, R is the total number of rays.

4 Experiments and Analysis

4.1 Dataset

We have conducted a comprehensive evaluation of dense depth maps reconstructed utilizing COLMAP on a select subset of the KITTI dataset, which includes ground truth depth data. This data is pivotal in verifying the accuracy and practicality of employing COLMAP-generated depth in real-world scenarios, particularly for 3D view synthesis where depth serves as a prior. The collection includes 1,048 frames, which are strategically selected to represent a diverse range of conditions and settings.

Our evaluations on the KITTI dataset verifies the hypothesis that dense depth maps produced by COLMAP attain the accuracy required to function as ground truth for applications such as 3D view synthesis. Concurrently, we employed a secondary dataset used by SpinNeRF [4], comprising 10 distinct scenes accompanied by camera parameters and sparse reconstructed points from SfM. The ground truth depth maps for this dataset were generated through COLMAP's dense reconstruction process.

Building on the affirmative conclusion that dense depth maps generated by COLMAP can be treated as ground truth, we subsequently assess how depth priors derived from alternative sources, such as monocular depth estimation methods, impact object removal in NeRF models. To validate our assumptions, a comprehensive suite of experiments has been conducted.

4.2 Evaluation Metrics

Depth Metrics. The disparities between depth generated by COLMAP (or real depth from KITTI when available) and monocular depth estimation models are evaluated using different metrics such as Root Mean Squared Error (RMSE) and the logarithmic scale error [28]. Additionally, the widely adopted threshold-based accuracy metric δ_1 is used. This metric specifically measures the proportion of predicted depth values that fall within predefined error thresholds relative to the ground truth. The three metrics are defined as follows

$$\text{RMSE} = \sqrt{1/N \sum_{i=1}^{N} (D_{\text{gt}_i} - D_{\text{pred}_i})^2}, \tag{11}$$

$$\delta_1 = \max\left(D_{\text{gt}_i}/D_{\text{pred}_i}, D_{\text{pred}_i}/D_{\text{gt}_i}\right) < 1.25, \tag{12}$$

$$\log_{10} \text{Error} = 1/N \sum_{i=1}^{N} |\log_{10}(D_{\text{gt}_i}) - \log_{10}(D_{\text{pred}_i})|, \tag{13}$$

where N is the number of images, and D_{gt_i} and D_{pred_i} are the $i-th$ depth ground truth and the corresponding predicted depth, respectively.

Considering the challenges in calculating the metrics introduced by very sparse ground truth depth values from KITTI, we have designed Algorithm 1 to calculate these values.

NeRF Metrics. The widely used metrics for synthesized image quality evaluation such as PSNR and SSIM [29,30] are used to evaluate the performance of

Algorithm 1. Dense Depth Evaluation

1: **Inputs:** Predicted depth maps D_{pred_i} and ground truth depth maps D_{gt_i}
2: **Outputs:** Mean values of each metric across all frames
3: Initialize metrics lists for RMSE, δ_1, δ_2, δ_3, and $\log_{10}$
4: **for** each frame in the dataset **do**
5:     Load ground truth depth map D_{gt}
6:     Load predicted depth map D_{pred}
7:     Align dimensions of D_{gt} and D_{pred} by padding
8:     Compute scaling factor s and adjust D_{pred} by s
9:     Crop to valid region in both D_{gt} and D_{pred}
10:    Compute performance metrics for the frame
11: **end for**
---

object removal using NeRF with different depth priors. The definitions of PSNR and SSIM are as follows.

$$\text{PSNR} = 20 \cdot \log_{10}\left(\frac{\text{max}_I}{\sqrt{\text{MSE}}}\right), \quad (14)$$

where max_I is the maximum possible pixel value of the image, and MSE stands for the mean squared error between the original and reconstructed images.

$$\text{SSIM}(x, y) = \frac{(2\mu_x\mu_y + c_1)(2\sigma_{xy} + c_2)}{(\mu_x^2 + \mu_y^2 + c_1)(\sigma_x^2 + \sigma_y^2 + c_2)}, \quad (15)$$

where x is the original depth map pixel, y is the corresponding predicted depth map pixel. μ_x is the average of x, μ_y is the average of y, σ_x^2 and σ_y^2 are the variances of x and y, σ_{xy} is the covariance of x and y, c_1 and c_2 are variables to stabilize the division with a weak denominator.

4.3 Results and Discussions

Depth Evaluation on KITTI Datasets. This set of experiments is meant to support our first assumption that the depth generated by COLMAP can be taken as depth ground truth in scenarios lacking ground truth information. This will further support our second assumption that we can take depth generated by COLMAP as ground truth to evaluate depth obtained by monocular depth estimation models. All experiments were conducted using an NVIDIA A100 GPU.

We conducted rigorous evaluations of depth maps from COLMAP's dense depth reconstruction and various monocular depth estimation methods using the selected KITTI subsets with ground truth depth maps. The results, presented in Table 1, reveal that COLMAP consistently outperforms (or achieves on-the-par results) monocular methods across all metrics, with a δ_1 value of 91%, highlighting its closer approximation to ground truth. In scenarios where acquiring ground truth is cost-prohibitive, COLMAP's dense depth reconstruction proves to be a viable alternative. Additional details and visual comparisons of depth maps generated by different methods are illustrated in Fig. 2. For instance, the

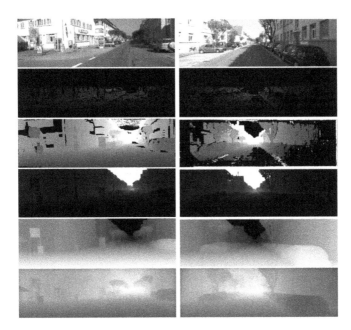

Fig. 2. Depth Map Estimation Comparison. From top to bottom: the raw image, ground truth depth map, COLMAP dense depth map, EcoDepth, Depth Anything, and ZoeDepth.

Table 1. Different Methods on Selected KITTI for Depth Estimation

Methods	$\delta_1 \uparrow$	RMSE $\downarrow$	$\log_{10} \downarrow$
COLMAP	**0.910**	**0.017**	**0.042**
ZoeDepth	0.807	0.024	0.064
EcoDepth	0.670	0.017	0.080
Depth Anything	0.184	0.044	0.357

COLMAP dense depth map, as shown in the second row of Fig. 2, reveals a more precise delineation and representation of object edges compared to maps produced by monocular depth estimation methods.

Depth Evaluation on Removal NeRF Dataset. Leveraging findings from COLMAP depth studies, we generated ground truth depth maps for the Removal NeRF Dataset from SpinNeRF. We conducted a comprehensive evaluation using several monocular depth estimation methods, with detailed metrics presented in Table 2. Density in Table 2 is defined as the proportion of pixels that contain non-zero values, which serves as a quantitative measure of the coverage and completeness of depth across the scene. Due to the significant variation in depth distances within our dataset, we mainly use the δ_1 metric to evaluate the adaptability and robustness of methods across different depth scales. Our results highlight ZoeDepth as the best method for depth map acquisition in the Removal NeRF dataset, achieving over 94% overlap with true depth values, significantly outperforming EcoDepth and DepthAnything. The depth maps results in Fig. 3 further illustrate ZoeDepth's enhanced detail capture, notably in object shape and edge delineation.

Table 2. Monocular Depth Estimation. 'D' stands for 'Dataset'.

Models	Metric	D-1	D-2	D-3	D-4	D-5	D-6	D-7	D-8	D-9	D-10
ZoeDepth	$\delta_1 \uparrow$	**0.53**	**0.99**	**1.00**	**0.95**	**0.98**	**1.00**	**0.99**	**0.99**	**0.99**	**0.99**
	RMSE $\downarrow$	0.10	0.03	0.03	0.07	0.04	0.05	0.03	0.06	0.04	0.06
	$\log_{10} \downarrow$	0.15	0.03	0.02	0.29	0.02	0.02	0.02	0.03	0.02	0.02
	Density	0.97	0.99	0.99	0.89	0.98	0.99	0.98	0.85	0.91	0.81
Depth Anything	$\delta_1 \uparrow$	0.06	0.26	0.23	0.33	0.36	0.27	0.25	0.15	0.22	0.26
	RMSE $\downarrow$	0.17	0.31	0.29	0.26	0.45	0.33	0.33	0.64	0.36	0.50
	$\log_{10} \downarrow$	0.64	0.21	0.21	0.20	0.19	0.16	0.25	0.34	0.22	0.22
	Density	0.97	0.99	0.99	0.89	0.98	0.99	0.98	0.85	0.91	0.81
ECoDepth	$\delta_1 \uparrow$	0.50	0.95	0.99	0.82	0.96	0.92	0.62	0.89	0.94	0.63
	RMSE $\downarrow$	0.08	0.05	0.03	0.09	0.07	0.07	0.11	0.12	0.06	0.17
	$\log_{10} \downarrow$	0.13	0.05	0.02	0.06	0.04	0.03	0.08	0.05	0.04	0.08
	Density	0.97	0.99	0.99	0.89	0.98	0.99	0.98	0.85	0.91	0.81

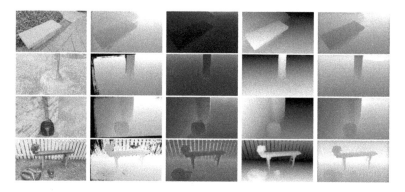

Fig. 3. Estimated Depth Map Comparison. From left to right columns: input images; depth maps from COLMAP; depth maps from EcoDepth; depth maps from Depth Anything; depth maps from ZoeDepth.

Different Depth Priors Comparison in Removal NeRF. To further assess the impact of various depth estimation models on object removal within NeRF, we continued our experiments using SpinNeRF as a benchmark, rigorously evaluating its performance with both its original pipeline where DSNeRF was used for complete depth acquisition and when augmented with ZoeDepth monocular depth estimation. Detailed comparison results are presented in Table 3. We found that using ZoeDepth instead of the complete depth provided by DSNeRF not only improved the PSNR by 6.87% but also significantly reduced the depth estimation time consumption from 44.5 s per frame to 0.58 s per frame, making our pipeline appealing to real-time applications such as human-robot collaboration. Additionally, comparisons of inpainting results using different depth priors are shown in Fig. 4. Our findings indicate that ZoeDepth excels in capturing the shape, position, and depth of objects, both in estimating the depth of the original image and after removing the object.

Table 3. The Effect of Different Depth Priors in Removal NeRF

Model	Depth Priors	PSNR ↑	SSIM ↑	Depth Estimation Time ↓
SpinNeRF	Complete Depth	21.943	0.192	44.5 s/per image
	ZoeDepth	**23.451**	**0.192**	**0.58 s/per image**

Fig. 4. Depth Map Comparison on Input Image and Inpainted Image. *Top Row*, from left to right: the input image; the depth map obtained by DSNeRF; the inpainted image; the inpainted depth map. *Bottom Row*, from left to right: the input image; the depth map obtained by ZoeDepth; the inpainted image; the inpainted depth map

We analyzed the training dynamics of the proposed pipeline that integrates different depth priors into SpinNeRF for object removal, with details illustrated in Fig. 6. This analysis shows a decrease in the model's loss values and an increase in the PSNR values as the number of iterations grows. Specifically, Fig. 6(b) demonstrates that incorporating ZoeDepth into SpinNeRF results in a lower and more consistent loss profile and higher PSNR at the same iteration counts compared to using DSNeRF depth priors. A detailed comparison of rendering quality is presented in Fig. 5, where each column represents the same sequence of rendered images. The top row displays the views rendered using DSNeRF depth priors, while the bottom row presents results using ZoeDepth depth priors. Notably, the rendering quality improves significantly with ZoeDepth and depth priors generator, since it helps to retain details such as the walls in the images.

Fig. 5. Rendered views. *Top Row*: depth priors from DSNeRF; *Bottom Row*: depth priors form ZoeDepth.

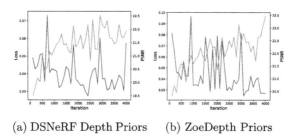

(a) DSNeRF Depth Priors   (b) ZoeDepth Priors

Fig. 6. The new pipeline training process with different depth priors.

5 Conclusion

This paper introduces a new removal NeRF pipeline that innovatively incorporates depth priors, specifically introducing monocular depth estimation models such as ZoeDepth into the SpinNeRF architecture to enhance object removal performance. This approach not only improves SpinNeRF's capability in handling complex removal scenarios but also significantly reduces computational overhead. Additionally, our extensive evaluation of COLMAP's dense depth reconstruction on the KITTI dataset positions COLMAP as a cost-effective and scalable alternative to traditional ground truth depth data acquisition methods, a crucial advantage in scenarios constrained by budget limitations. Furthermore, our detailed comparative analysis highlights ZoeDepth as the leading monocular depth estimation method, offering high-quality depth priors with reduced computational demands. The contributions of this paper enhance understanding and further development in NeRF technologies, paving the way for future advancements in digital twin systems and other applications that require robust, detailed 3D reconstructions.

References

1. Mildenhall, B., Srinivasan, P.P., Tancik, M., Barron, J.T., Ramamoorthi, R., Ng, R.: NeRF: representing scenes as neural radiance fields for view synthesis. Commun. ACM **65**(1), 99–106 (2021)
2. Wang, S., Zhang, J., Wang, P., Law, J., Calinescu, R., Mihaylova, L.: A deep learning-enhanced digital twin framework for improving safety and reliability in human-robot collaborative manufacturing. Robot. Comput.-Integr. Manuf. **85**, 102608 (2024)
3. Yang, B., et al.: Learning object-compositional neural radiance field for editable scene rendering. In: Proceedings of the IEEE/CVF International Conference on Computer Vision, pp. 13779–13788 (2021)
4. Mirzaei, A., et al.: Spin-NeRF: multiview segmentation and perceptual inpainting with neural radiance fields. In: Proceedings of the IEEE/CVF Conference on Computer Vision and Pattern Recognition, pp. 20669–20679 (2023)
5. Weder, S., et al.: Removing objects from neural radiance fields. In: Proceedings of the IEEE/CVF Conference on Computer Vision and Pattern Recognition, pp. 16528–16538 (2023)
6. Deng, K., Liu, A., Zhu, J.-Y., Ramanan, D.: Depth-supervised NeRF: fewer views and faster training for free. In: Proceedings of the IEEE/CVF Conference on Computer Vision and Pattern Recognition, pp. 12882–12891 (2022)
7. Bhat, S.F., Birkl, R., Wofk, D., Wonka, P., Müller, M.: ZoeDepth: zero-shot transfer by combining relative and metric depth (2023)
8. Geiger, A., Lenz, p., Urtasun, R.: Are we ready for autonomous driving? The KITTI vision benchmark suite. In: 2012 IEEE Conference on Computer Vision and Pattern Recognition, pp. 3354–3361. IEEE (2012)
9. Riedmiller, M., Lernen, A.: Multi layer perceptron. Machine Learning Lab Special Lecture, University of Freiburg, p. 24 (2014)
10. Wei, Y., Liu, S., Rao, Y., Zhao, W., Lu, J., Zhou, J.: NerfingMVS: guided optimization of neural radiance fields for indoor multi-view stereo. In: Proceedings of the IEEE/CVF International Conference on Computer Vision (ICCV), pp. 5610–5619 (2021)
11. Neff, T., et al.: DoNeRF: towards real-time rendering of compact neural radiance fields using depth oracle networks. In: Computer Graphics Forum, vol. 40, pp. 45–59. Wiley Online Library (2021)
12. Roessle, B., Barron, J.T., Mildenhall, B., Srinivasan, P.P., Nießner, M.: Dense depth priors for neural radiance fields from sparse input views. In: Proceedings of the IEEE/CVF Conference on Computer Vision and Pattern Recognition, pp. 12892–12901 (2022)
13. Wu, Q., et al.: Object-compositional neural implicit surfaces. In: European Conference on Computer Vision, pp. 197–213. Springer (2022)
14. Suvorov, R., et al.: Resolution-robust large mask inpainting with Fourier convolutions. In: Proceedings of the IEEE/CVF Winter Conference on Applications of Computer Vision, pp. 2149–2159 (2022)
15. Schönberger, J.L., Frahm, J.-M.: Structure-from-motion revisited. In: Conference on Computer Vision and Pattern Recognition (CVPR) (2016)
16. Schönberger, J.L., Zheng, E., Frahm, J.-M., Pollefeys, M.: Pixelwise view selection for unstructured multi-view stereo. In: Leibe, B., Matas, J., Sebe, N., Welling, M. (eds.) Computer Vision – ECCV 2016, pp. 501–518. Springer, Cham (2016)

17. Triggs, B., McLauchlan, P.F., Hartley, R.I., Fitzgibbon, A.W.: Bundle adjustment—a modern synthesis. In: Vision Algorithms: Theory and Practice: International Workshop on Vision Algorithms Corfu, Greece, 21–22 September 1999, pp. 298–372. Springer (2000)
18. Hoiem, D., Efros, A.A., Hebert, M.: Recovering surface layout from an image. Int. J. Comput. Vision **75**, 151–172 (2007)
19. Liu, C., Yuen, J., Torralba, A., Sivic, J., Freeman, W.T.: Sift flow: dense correspondence across different scenes. In: Computer Vision–ECCV 2008: 10th European Conference on Computer Vision, Marseille, France, 12–18 October 2008, Part III, pp. 28–42. Springer (2008)
20. Patni, S., Agarwal, A., Arora, C.: EcoDepth: effective conditioning of diffusion models for monocular depth estimation (2024)
21. Yang, L., Kang, B., Huang, Z., Xu, X., Feng, J., Zhao, H.: Depth anything: unleashing the power of large-scale unlabeled data (2024)
22. Bhat, S.F., Alhashim, I., Wonka, P.: AdaBins: depth estimation using adaptive bins. In: Proceedings of the IEEE/CVF Conference on Computer Vision and Pattern Recognition (CVPR), pp. 4009–4018 (2021)
23. Gasperini, S., Morbitzer, N., Jung, H., Navab, N., Tombari, F.: Robust monocular depth estimation under challenging conditions. In: Proceedings of the IEEE/CVF International Conference on Computer Vision (ICCV), pp. 8177–8186 (2023)
24. Ranftl, R., Lasinger, K., Hafner, D., Schindler, K., Koltun, V.: Towards robust monocular depth estimation: mixing datasets for zero-shot cross-dataset transfer. IEEE Trans. Pattern Anal. Mach. Intell. **44**(3), 1623–1637 (2020)
25. Dosovitskiy, A., et al.: An image is worth 16x16 words: Transformers for image recognition at scale. arXiv preprint arXiv:2010.11929 (2020)
26. Takagi, Y., Nishimoto, S.: High-resolution image reconstruction with latent diffusion models from human brain activity. In: Proceedings of the IEEE/CVF Conference on Computer Vision and Pattern Recognition, pp. 14453–14463 (2023)
27. Zhang, R., Isola, P., Efros, A.A., Shechtman, E., Wang, O.: The unreasonable effectiveness of deep features as a perceptual metric. In: Proceedings of the IEEE Conference on Computer Vision and Pattern Recognition, pp. 586–595 (2018)
28. Godard, C., Aodha, O.M., Firman, M., Brostow, G.J.: Digging into self-supervised monocular depth estimation. In: Proceedings of the IEEE/CVF International Conference on Computer Vision (ICCV) (2019)
29. Wang, Z., Bovik, A.C., Sheikh, H.R., Simoncelli, E.P.: Image quality assessment: from error visibility to structural similarity. IEEE Trans. Image Process. **13**(4), 600–612 (2004)
30. Lin, Y., Wang, P., Wang, Z., Ali, S., Mihaylova, L.: Towards automated remote sizing and hot steel manufacturing with image registration and fusion. J. Intell. Manuf. 1–18 (2023)

YOLOv8-LiDAR Fusion: Increasing Range Resolution Based on Image Guided-Sparse Depth Fusion in Self-Driving Vehicles

Ahmet Serhat Yildiz[(✉)] [iD], Hongying Meng [iD], and Mohammad Rafiq Swash [iD]

Department of Electronic and Electrical Engineering, Brunel University London,
London UB8 3PH, UK
ahmetserhat.yildiz@brunel.ac.uk

Abstract. Self-driving vehicles are significant in industrial and commercial applications, primarily driven by the development of environmental awareness systems. The need for real-time object recognition, segmentation, perception, projection, and position has significantly increased in object and line tracking, obstacle avoidance, and route planning. The primary sensors used are high-resolution cameras, Light Detection and Ranging (LiDAR), and high-precision GPS/IMU inertial navigation systems. However, out of all these sensors, LiDARs and cameras have a vital function in perception and comprehensive situations. Although LiDAR is capable of providing precise depth information, its resolution is constrained. On the other hand, cameras provide abundant semantic information but do not offer precise assessments of the distance to objects. This work presents the incorporation of YOLOv8, an advanced object identification method, into the fusion process. We specifically investigate the notion of Camera-LiDAR Projection and provide a thorough explanation of the process of projecting LiDAR point clouds onto an image coordinate frame. This is achieved by utilizing transformation matrices that establish the relationship between the LiDAR and the camera. This project aims to improve the range resolution and perception capabilities of autonomous driving systems by combining YOLOv8-based object recognition with LiDAR point cloud data by using the KITTI object detection benchmark.

Keywords: LiDAR-camera fusion · YOLOv8 · perception enhancement · sparse point cloud · object detection · Self-Driving Vehicles

1 Introduction

Over the last decade, the proliferation of the processing of visual and depth information has become indispensable for perception systems in automobiles worldwide [12]. With the improvement of autonomous vehicles, RGB-camera and LiDAR are two of the most important sensors that provide sufficient information about the surrounding environment needed in autonomous driving [10]. LiDAR and cameras are highly valuable sources of data, as they provide distinct information that is essential for enabling diverse algorithms and deep learning models to comprehend and infer significant observations regarding their surroundings [2]. LiDAR is an active remote sensing technology that employs electromagnetic waves within the optical spectrum to not only detect objects and accurately measure the distance between these targets (objects) and the range (sensor). Additionally, it effectively records crucial physical characteristics of the surface being observed, including scattering and reflection phenomena. The distance between the target objects and the sensor is determined by calculating the time it takes for the emitted laser beam to travel to the objects and return to the sensor, and this calculation takes into account the speed at which the laser beam propagates, which is equivalent to the speed of light. However, the sparse depth map produced by the LiDAR point cloud has low resolution. In contrast, cameras have exceptional proficiency in capturing intricate visual details, encompassing elements such as colour and texture. Nevertheless, their capacity to accurately assess distances is restricted [1].

Plenty of research exists in producing a dense depth map by combining the sparse depth which is produced by LiDAR and the image captured by the camera [3] [5]. However, current research is based on focusing on complementing the depth information rather than improving the performance of the sparse depth image. Instead of improving the resolution of the full sparse-depth image which takes a lot of time, this work aimed to focus on improving the specific objects such as persons, bicycles, cars, buses, and trucks. To do this, objects were detected by using YOLOv8s. The coordinates of bounding boxes and semantic masks on images were produced by YOLOv8s and the resolution of the sparse depth map points in the coordinates of the detected objects is increased by the linear interpolation method.

You Only Look Once, generally known as YOLO has emerged as an essential real-time object detection technology for robots and autonomous vehicles applications. The YOLOv8 version is a computer vision model that has the capability to do many tasks like object detection, segmentation, pose estimation, tracking, and classification [15]. The YOLOv8 utilizes a novel design that has integrated Feature Pyramid Network (FPN) and Path Aggregation Network (PAN) modules, to showcase an exceptional equilibrium between speed and accuracy, and this allows for swift and dependable object detection in images. FPN is employed to produce feature maps at various dimensions and resolutions, whilst PAN is utilized to consolidate features from different network levels in order to enhance accuracy [13].

In this research, YOLOv8 was chosen for its ability to recognize and segment objects simultaneously in a cohesive framework. Unlike Mask Region-based Convolutional Neural Network (Mask R-CNN), YOLOv8 can effectively identify and segment without a complicated two-stage approach. This effective approach simplifies implementation and enhances processing efficiency, making it ideal for real-time applications [14]. The Single Shot MultiBox Detector (SSD) is known for its quick detection, however YOLOv8 has better accuracy and segmentation. SSDs' speed often reduces accuracy, especially with little objects or complex backgrounds [7]. Overall, YOLOv8's extensive object detection and segmentation approach, precision, speed, and user-friendliness make it the best option for turning sparse depth maps into bounding boxes and segmentation masks.

YOLOv8 which introduces five different scaled versions, was trained using a combination of the COCO dataset and multiple other datasets [6]. KITTI's object detection benchmark provides precise 3D and 2D bounding boxes for several object classes, such as cars, vans, trucks, pedestrians, and bicycles [4]. Due to the existence of these object classes in the COCO dataset, the previously trained version of YOLOv8s was used. YOLOv8s (small) is optimized for faster inference times and uses fewer computing resources (e.g., memory, GPU) than larger models, making it better for real-time applications. YOLOv8n (nano) has lower accuracy than YOLOv8s due to its smaller size and streamlined architecture. Moreover, YOLOv8s quickly provides coordinates to indicate the precise position of the identified items inside the image and converts them to [left, top, right, bottom].

2 Related Work

Integrating LiDAR and video sensors is crucial in the field of autonomous vehicle perception as it helps to understand the surrounding environment. This section examines relevant literature and approaches that use comparable technology and processes.

Researchers are concentrating on integrating LiDAR and video data in realtime to accurately determine the distances to objects in autonomous driving. The technology employs geometric modifications and projections to accomplish sensor fusion at a low level. The method enables autonomous cars to improve their perceptual capabilities, hence assuring a dependable and consistent understanding of the surrounding environment [8]. Integrating sparse depth maps with object recognition and identification in autonomous cars via the fusion of 3D LiDAR and vision camera data has shown encouraging outcomes. This approach provides exact object region recommendations using 3D LIDAR. These recommendations are transferred to image space and processed by a CNN. Combining sensor data to identify objects increases autonomous vehicle safety and performance [16]. Makarov et al. investigated methods to enhance the accuracy of estimating dense depth maps from sparse depth data using RGB-D pictures. RGB-D cameras are restricted to indoor use and have restrictions on the maximum distance they can capture [11].

This section synthesizes various relevant efforts to explain the suggested strategy within autonomous vehicle sensor fusion and perception technology. Understanding current methods emphasizes the uniqueness and importance of merging LiDAR and video sensors to improve autonomous driving. The suggested approach for autonomous vehicle perception using YOLOv8-LiDAR Fusion will be highlighted in this organized approach, which will cover previous research and emphasize its unique contributions.

3 Methods

This section provides a comprehensive explanation of the proposed approach for autonomous cars. The approach is able to perceive and comprehend its immediate surroundings by utilizing LiDAR and camera sensors to capture and analyze the different physical characteristics of the environment. Sparse depth maps are generated using various depth-sensing technologies, and this depth map comprises depth data pertaining to certain points or areas inside the picture, and the remaining pixels may either be left unspecified or filled with a default value, such as zero. RGB cameras are exceptionally proficient at recording images with a remarkable level of intricacy, which is crucial for precisely ascertaining the precise positioning of things. The initial stage of the technique involves retrieving the camera's intrinsic parameters and the camera-to-LiDAR transformation matrix from the calibration data. Subsequently, the intrinsic parameters of the camera and the transformation matrix are multiplied together to determine the LIDAR-to-camera transformation matrix. Afterward, this depth map point is mapped onto the bounding box and segmentation mask locations of items identified by YOLOv8, including car, trucks, pedestrians, and bicycles.

3.1 The Process of Extracting Features from 3D Point Cloud for Sparse Depth Map

LiDAR sensors, also known as Light Detection and Ranging sensors, use laser beams to determine the distances between themselves and other objects present in their immediate environment. The LiDAR data is acquired in the format of a three-dimensional point cloud, whereby each point represents a three-dimensional coordinate (x, y, z) and a reflecting value (r) within the spatial domain.

The coordinate frame in the 3D LiDAR point cloud is defined by an origin and a position vector $\boldsymbol{P}$, which has three elements representing the mutually orthogonal axes x, y, and z.

$$\boldsymbol{P} = \begin{bmatrix} \boldsymbol{Px} & \boldsymbol{Py} & \boldsymbol{Pz} \end{bmatrix}^T \qquad (1)$$

The coordinates of the transformed point, relative to the original axes, are determined by the rotation of the coordinate frame using matrix R around the origin. $R_X(\theta)$, $R_Y(\alpha)$ and $R_Z(\gamma)$ rotate the coordinate system around the x-axis, y-axis, and z-axis by angle $(\theta), (\alpha)$ and (γ) respectively. A 3D rotation matrix is constructed by using the primary, secondary, and tertiary rotation matrices to define the relation between the P_{XYZ} and $P_{X_{\theta\alpha\gamma}Y_{\theta\alpha\gamma}Z_{\theta\alpha\gamma}}$ coordinate systems. Also, the reference camera's rectifying rotation matrix is represented by $R_{\text{rect}}^{(0)}$ while the translation vector from the velodyne to the reference camera is denoted by T_{velo}^{cam}.

$$\mathbf{P}_{\text{xy}} = \boldsymbol{P}(\theta\alpha\gamma) \; \mathbf{R}_{\text{rect}}^{(0)} \; \mathbf{T}_{\text{velo}}^{cam} \; \boldsymbol{P} \qquad (2)$$

Equation (2) indicates the translation between the position of the LiDAR origin coordinate point and the position of the camera origin coordinate, and the location of the LiDAR sensor is described with respect to the camera [9] (Fig. 1).

Fig. 1. (a) Project the Sparse Depth on the image plane. (b) Project the Sparse Depth on the image plane.

3.2 The Projection of a Sparse Depth Map Into Bounding Box and Segmentation Mask

The process of projecting a sparse depth map into a bounding box and segmentation mask entails determining the points that have accurate depth information and using their coordinates to establish the object's spatial boundaries. Initially, these points are derived from the sparse depth map. Subsequently, the sparse depth map points that are inside the YOLO bounding box rectangle or segmentation mask frame are detected. Simultaneously, the points are recorded in a list. This process enables the effective projection of sparse depth map points onto bounding boxes and segmentation mask frames. As shown in Fig. 2(a) shows the result of YOLOv8's object detection method, and (b) exhibits the projection of the Sparse Depth Map points onto the coordinates of the bounding box of the object detected by YOLO.

Fig. 2. (a) YOLOv8s object detection output. (b) Project the Sparse Depth Map points on coordinates of YOLO Bounding Box.

The coordinates of the top-left corner, which is the rectangle, are denoted by (x_1, y_1), while the coordinates of the bottom-right corner, which is the rectangle, are denoted by (x_2, y_2). Also, (x, y) represents the coordinates of the sparse depth map point that is wanted to check. The mathematical method for determining whether a point lies within a rectangle is as follows:

$$x_1 \leq x \leq x_2 \quad \text{and} \quad y_1 \leq y \leq y_2 \tag{3}$$

The Eq. 3 is used to each rectangle in an image whenever YOLOv8s identifies more than one object on the image.

Algorithm 1: Project Sparse Depth Points onto YOLOv8s BBoxes and Masks

Data: Sparse depth map array as x, y, z is P, and YOLOv8s frames are R
Result: Store array S containing sparse depth map points inside frames.
$S \leftarrow \emptyset$;   // Create an empty array for storing points
for *each r in R* do
  for *each p in P* do
    if $r_{x_{min}} \leq p_x \leq r_{x_{max}}$ *and* $r_{y_{min}} \leq p_y \leq r_{y_{max}}$ then
    | $S \leftarrow S \cup \{p\}$;   // Add p to S
    end
  end
end

The pixels that correspond to a particular object or group inside an image are identified by a segmentation mask, which is a binary image. In order to determine whether a pixel in the segmentation mask is part of the foreground or the background, a value is assigned to each pixel in the mask by the segmentation algorithm.

$$\text{Mask\_Seg}(I) = f(I, \theta) \tag{4}$$

The segmentation mask of YOLOv8 is a function that provides a mask value to every pixel present in an image. The original input image is denoted by I in the Eq. 4, the function f generates the segmentation mask, and the parameters or weights of the YOLOv8s model are represented by θ. By applying this equation to the YOLOv8s mask polynomial array, the coordinates in the Sparse depth map are projected onto the segmentation mask that YOLOv8s provides for each object.

(a) (b)

Fig. 3. (a) YOLOv8s object detection output. (b) Project the Sparse Depth Map points on coordinates of YOLOv8s Segmentation mask.

3.3 Increasing Sparse Depth Map Resolution in Bounding Box and Segmentation Mask

Linear interpolation is an essential technique for enhancing the resolution of sparse depth maps for a variety of reasons. The first reason is to increase the visual quality. The linear interpolated depth values serve to fill in the gaps between sparse measurements, which ultimately results in surfaces that are smoother and scenarios that are that much more lifelike. A second factor is to improve the accuracy. The process of linear interpolating between sparse depth measurements offers a straightforward and efficient method for estimating intermediate depth values based on the data that are already available. This method helps to reduce mistakes in the depth map. The third reason is to reduce the limitations of the sensor. The limitations in sensor resolution, occlusions, or environmental conditions that are present in depth sensors frequently result in the production of sparse readings. Through the utilization of linear interpolation is lessened the impact of these limits and acquired depth data that is both more dense and more relevant for the purpose of subsequent processing.

In linear interpolation, a straight-line connection is created between data points that are down to up. So linear interpolation determines a value x between x_0 and x_1, as well as a value y between y_0 and y_1, given two points with known values, such as (x_0, y_0) and (x_1, y_1). Below is an expression that can be used to express the formula for linear interpolation:

$$f(x) = \frac{(x_1 - x_0)}{2} \quad \text{and} \quad f(y) = \frac{(y_1 - y_0)}{2} \tag{5}$$

The Euclidean distance is a metric that quantifies the shortest distance between two places in Euclidean space, and the shortest distance between the two points is determined by its length. The formula for calculating the Euclidean distance between two points (x_0, y_0) and (x_1, y_1) in a two-dimensional space, using the Pythagorean theorem is as follows:

$$\text{Euclidean distance} = \sqrt{(x_1 - x_0)^2 + (y_1 - y_0)^2} \tag{6}$$

In order to enhance the resolution of the sparse depth map projected on the image, the Yolov8s object detection technique is initially employed to analyze the image and extract the coordinates of the bounding boxes and segmentation masks for the items present in the image. The bounding box frame has the coordinates of two points, (x0,y0) and (x1,y1), while the segmentation mask consists of polynomial points.

$$\text{Frame coord} = f(\text{YOLOv8}(image)) \tag{7}$$

Here, Frame coord represents the coordinates of the bounding box and Segmentation mask in the image, and YOLOv8($image$) represents the output of the YOLOv8 model applied to the image. For the purpose of determining the points that are reflected on the frames of the bounding box and segmentation

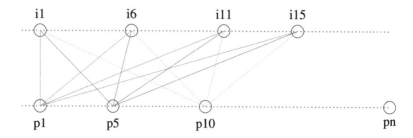

Fig. 4. The process of increasing resolution in a sparse depth map using linear interpolation along the y-axis.

mask, then generating arrays ($[p_1, p_2, p_3, \ldots, p_n]$) for the bounding box and segmentation mask by comparing the points in the sparse depth map with the coordinates indicating the object positions on the picture, as described in Eq. 3.

LiDAR devices generate distance measurements at ranges of 1, 4, 8, 16, 32, 64, and 128. While the points on each range line may vary on the x-axis, their positions on the y-axis are closely clustered together. Based on this, by calculating the absolute difference between these two points and dividing it by the range value provided by LiDAR (range-64), the estimated difference between the range lines was determined along the y-axis, and this difference was as *range_gap*. After that, as shown in Fig. 4, to increase the sparse depth map resolution speed, the linear interpolation approach was used on the two points with the shortest distance between them along the y-axis, rather than on two points along the x-axis.

The *range_gap* parameter was utilized to jump from point p_1 to point $p_n + 1$. Subsequently, the euclidean distance formula, as presented in Eq. 6, was employed to calculate the distance between point p_1 and points $i_1, i_2, i_3, \ldots, i_n]$. A list was formed to record each distance found. To create a new sparse depth point, linear interpolation was used between the point i with the smallest distance value and the point p_1.

(a) (b)

Fig. 5. (a) Projected sparse depth points onto bounding box coordinate. (b) Projected increased resolution sparse depth points onto bounding box coordinate.

The points obtained by linear interpolation are merged with the sparse depth map points and then projected onto the image. The LiDAR point cloud points are projected into the bounding box and segmentation mask frames of the objects in the image, as depicted in Fig. 5(a) and Fig. 6(a). The enhanced resolution of the Sparse depth map is illustrated in Fig. 5(b) and Fig. 6(b).

(a) (b)

Fig. 6. (a) Projected sparse depth points onto segmentation mask coordinate. (b) Projected increased resolution sparse depth points onto segmentation mask coordinate.

Algorithm 2: Increasing Sparse Depth Map Resolution and Projecting onto BBoxes and Masks

Data: Sparse depth map points P, YOLOv8s BBoxes and Masks R, Image
Result: Projected points inside BBoxes and Masks on the image
Initialize empty lists M for merged points and D for distance values
for *each point p_1 in P* **do**
  for *each point p_n in P* **do**
    Calculate euclidean distance d between p_n and p_1
    Append d to list D
  end
  Find index i of minimum value in D
  Interpolate linearly between p_1 and p_i to obtain point p_{new}
  Append p_{new} to M, and Clear D
end
Project points in S onto the image to obtain BBoxes and Masks frames

---

4 Dataset

The KITTI object detection benchmark comprises 7481 training images and 7518 test images, along with the associated point clouds data, calibration, and labeling files. The KITTI object detection benchmark dataset comprises the categories of person, bicycle, car, bus, and truck. Additionally, it has occlusions of varying difficulty levels, namely easy, moderate, and hard [4] (Table 1).

Table 1. The KITTI object detection benchmark.

Image Feature	Easy	Moderate	Hard
Min. bounding box height:	40 Px,	25 Px	25 Px
Max. occlusion level:	Fully visible	Partly occluded	Difficult to see
Max. truncation :	15	30	50
Number of Object	14550	6428	3511

5 Result and Discussion

5.1 The Projection of a Sparse Depth Map Into Bounding Box and Segmentation Mask

In this section, the results of projecting the sparse depth map points created by mapping LiDAR point cloud information onto the camera image to the coordinates of the bounding box and segmentation mask frames of the objects detected with the Yolov8s object detection method will be examined.

Fig. 7. (a) Project sparse depth onto bounding box coordinate. (b) Project sparse depth onto segmentation mask coordinate.

Figure 2 and Fig. 3 show the projecting of spars depth map points onto the coordinates of the bounding box and segmentation mask frames. These frames correspond to the object detected by the Yolov8s algorithm in the given image. The Yolov8s object detection algorithm was provided with an input image that contained multiple objects. The coordinates of all bounding box and segmentation mask frames produced by Yolov8 were used to project sparse depth map points, and the image that was obtained is shown in Fig. 7 (Table 2).

The duration for projecting limited depth points to the bounding box and segmentation mask frame varies based on the object's category (Person, Bicycle, Car, Bus, Truck) and the level of occlusion (Easy, Moderate, Hard). The projection time is shorter when the object has a small volume and a hard occlusion

Table 2. Time to project sparse depth points to bounding box and segmentation mask frame.

Occlusion	Person (Second)		Bicycle (Second)		Car (Second)		Bus (Second)		Truck (Second)	
Frame Type	BBox	Mask	BBox	Mask	BBox	Mask	BBox	Mask	BBox	Mask
Easy	1.52	1.34	1.47	1.30	1.46	1.29	1.55	1.46	1.51	1.40
Moderate	1.48	1.32	1.46	1.29	1.46	1.27	1.51	1.38	1.46	1.31
Hard	1.47	1.26	1.45	1.24	1.44	1.25	1.35	1.87	1.47	1.27

type, whereas it is slower when the object has a large volume and an easy occlusion type. It was observed that the mean duration for reflecting the sparse depth points onto the bounding box and segmentation mask frame exceeded one second. This time interval causes a sluggish implementation of our approach.

5.2 Increasing Sparse Depth Map Resolution in Bounding Box and Segmentation Mask

This section presents the outcomes of our approach technique for projecting a sparse depth map onto the bounding box and segmentation mask. The efficacy of our approach is evaluated through several comparisons. The outcomes demonstrate how well our approach performs in precisely recreating depth information and improving sparse depth map resolution onto bounding boxes and segmentation masks. The obtained results are analyzed in detail to provide insights into the strengths and limitations of our approach. The evaluations demonstrate the effectiveness and dependability of our method in processing sparse depth map data and generating meaningful results.

Fig. 8. (a Projected all sparse depth onto bounding box coordinate. (b) Projected all sparse depth onto segmentation mask coordinate.

The images in Fig. 8 were acquired by merging the depth map points generated through linear interpolation with the sparse depth map. Upon comparing the sparse depth image depicted in Fig. 7(a) with Fig. 8(a), it has been ascertained that the resolution of the sparse depth map points within the bounding box frame has improved. Nevertheless, the enhanced resolution of the sparse depth points, which are unrelated to the object in the bounding box frame, resulted in image distortion. By increasing the resolution of the sparse depth points inside the segmentation mask frame depicted in Fig. 8(b), the sharpness of the objects improved without introducing any image distortion.

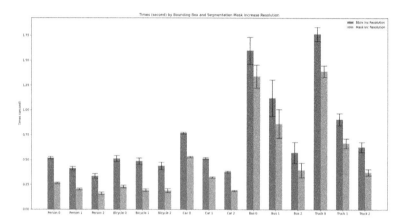

Fig. 9. (a) Times (second) by Bounding Box and Segmentation Mask Increase Resolution.

The KITTI object detection benchmark data set comprises 5 categories: person, bicycle, car, bus, and truck. It also includes three occlusion types: easy, moderate, and hard. The implementation of our approach involved separately addressing easy, moderate, and hard occlusion types for each object category. The outcomes acquired are displayed in Fig. 9. When comparing the implementation times of increasing resolution on the bounding box and segmentation mask frames, it was found that the resolution of the sparse depth map points in the segmentation mask frames increased more rapidly than the sparse depth points within the bounding box frames.

6 Conclusions

Our study on YOLOv8-LiDAR Fusion: Increasing Range Resolution in Self-Driving Vehicles via Image Guided-Sparse Depth Fusion introduces an innovative approach for improving the range resolution of autonomous vehicles. This is achieved by combining sparse depth information with YOLOv8 object detection. Through the utilization of the synergistic capabilities of LiDAR and image data, the fusion method we have put forth attains noteworthy enhancements in range resolution. This, in turn, empowers the detection and localization of objects within the vehicle's surroundings with greater precision. The KITTI object detection benchmark data set includes the person, bicycle, car, bus, and truck. It also has easy, moderate, and hard occlusions. Our method addressed easy, moderate, and hard occlusion types for each object category. The sparse depth map points in segmentation mask frames increased faster than those in bounding box frames when our approach was implemented.

For future work, investigating the incorporation of sophisticated algorithms like the Segment Anything Model (SAM) could further optimize and strengthen the effectiveness of our method. The great accuracy of SAM's ability to segment diverse object classes could enhance our fusion method, potentially enhancing detection and localization performance in different settings. Integrating these advancements into our method not only seeks to improve the current methodology but also prepares us for exploring wider applications and scalability in autonomous driving and other areas.

Acknowledgments. This work is supported in part by the Horizon Europe COVER project under grant number 101086228, via UKRI grant EP/Y028031/1, and Ahmet Serhat Yildiz's PhD is sponsored by the Ministry of National Education of Türkiye.

References

1. Benedek, C., Majdik, A., Nagy, B., Rozsa, Z., Sziranyi, T.: Positioning and perception in lidar point clouds. Digital Signal Process. **119**, 103193 (2021)
2. Dhall, A., Chelani, K., Radhakrishnan, V., Krishna, K.: Lidar-camera calibration using 3d-3d point correspondences. arxiv 2017. arXiv preprint arXiv:1705.09785
3. Dong, H., et al.: Superfusion: multilevel lidar-camera fusion for long-range HD map generation. arXiv preprint arXiv:2211.15656 (2022)
4. Geiger, A., Lenz, P., Urtasun, R.: Are we ready for autonomous driving? The kitti vision benchmark suite. In: 2012 IEEE Conference on Computer Vision and Pattern Recognition, pp. 3354–3361. IEEE (2012)
5. Hu, M., Wang, S., Li, B., Ning, S., Fan, L., Gong, X.: Penet: towards precise and efficient image guided depth completion. In: 2021 IEEE International Conference on Robotics and Automation (ICRA), pp. 13656–13662. IEEE (2021)
6. Jocher, G., Chaurasia, A., Qiu, J.: YOLO by Ultralytics (2023). https://github.com/ultralytics/ultralytics. Accessed 30 Feb 2023
7. Kaliappan, V.K., Manjusree, S., Shanmugasundaram, K., Ravikumar, L., Hiremath, G.B.: Performance analysis of yolov8, RCNN, and SSD object detection models for precision poultry farming management. In: 2023 IEEE 3rd International Conference on Applied Electromagnetics, Signal Processing, & Communication (AESPC), pp. 1–6. IEEE (2023)
8. Kumar, G.A., Lee, J.H., Hwang, J., Park, J., Youn, S.H., Kwon, S.: Lidar and camera fusion approach for object distance estimation in self-driving vehicles. Symmetry **12**(2), 324 (2020)
9. LaValle, S.M.: Planning Algorithms. Cambridge University Press, Cambridge (2006)
10. Liu, H., Wu, C., Wang, H.: Real time object detection using lidar and camera fusion for autonomous driving. Sci. Rep. **13**(1), 8056 (2023)
11. Makarov, I., Korinevskaya, A., Aliev, V.: Sparse depth map interpolation using deep convolutional neural networks. In: 2018 41st International Conference on Telecommunications and Signal Processing (TSP), pp. 1–5. IEEE (2018)

12. Nguyen, A.D., Nguyen, T.M., Yoo, M.: Improvement to lidar-camera extrinsic calibration by using 3D–3D correspondences. Optik **259**, 168917 (2022)
13. Reis, D., Kupec, J., Hong, J., Daoudi, A.: Real-time flying object detection with yolov8. arXiv preprint arXiv:2305.09972 (2023)
14. Sapkota, R., Ahmed, D., Karkee, M.: Comparing yolov8 and mask RCNN for object segmentation in complex orchard environments. arXiv preprint arXiv:2312.07935 (2023)
15. Terven, J., Cordova-Esparza, D.: A comprehensive review of yolo architectures in computer vision: from yolov1 to yolov8 and yolo-nas. arXiv preprint arXiv:2304.00501 (2023)
16. Zhao, X., Sun, P., Xu, Z., Min, H., Yu, H.: Fusion of 3D lidar and camera data for object detection in autonomous vehicle applications. IEEE Sens. J. **20**(9), 4901–4913 (2020)

Efficient 2D and 3D Corresponding Object Identification Using Deep Learning Models

Haowen Liu(✉)[iD], Mark Post[iD], and Andy Tyrrell[iD]

School of PET, University of York, York, UK
{haowen.liu,mark.post,andy.tyrrell}@york.ac.uk

Abstract. This paper proposes an Efficient Perspective-n-Point (EPnP)-based You Only Look Once (YOLO) 3D object detection algorithm that aims to be applied to the navigation and localization system of autonomous robots when they work in outdoor environments. In the 3D object detection research area, although the classic PointNet series detection methods can learn shape and structure information directly from original point cloud data without complex preprocessing or feature extraction steps, the generalisation ability of this deep neural network model is limited by the training data. The EPNP based sensor fusion detection method proposed in this paper greatly reduces the dependence on the training data set due to its accuracy mostly reliant on the geometric transformation between camera and Lidar coordinate system. By combining YOLO this real-time 2D object detection method with EPnP pose estimation and Lidar data, this algorithm will be able to provide 3D object detection in real-time with less training data dependence. The key contribution is that the EPNP-based detection algorithm will first identify and classify targets in the 2D colour frames obtained by the depth camera, and use the absolute position relationship between the camera and the Lidar to get the camera pose relative to the Lidar coordinate system. Use the 2D pixel point coordinates that have also been transformed and the camera pose as input, apply EPnP in reverse to obtain the 3D point coordinates of the objects around the robots. The goal of this novel 3D object detection method through sensor fusion is to maintain a real-time speed while reducing the difficulty of training its depth model.

Keywords: Projection geometry of point clouds · sensor fusion · EPnP · roboticis vision · localization · navigation · Lidar · YOLO · identification · classification · geometric transformation · semantic SLAM

1 Introduction

Intelligent robots need to have autonomous decision-making capabilities so that they can move in outdoor environments and replace humans in tedious tasks, such as urgently needing office documents which are forgotten somewhere far away from you or picking up express delivery. When using autonomous decision-making robots to work in such uncertain or unknown outdoor areas, humans do not need to use a remote to manually direct the robot in which direction it needs to move or at a point in time it will encounter obstacles. Humans only need a simple input to the start point and destination into their system. The navigation and localization system of the autonomous decision-making robot can automatically understand the semantic information in their surrounding environment, build a real-time and detailed three-dimensional map, and accurately plan the path to complete the task. Having excellent environmental perception, self-localization and real-time construction of three-dimensional maps are necessary to meet the needs of contemporary intelligent quadruped robot configurations. The application of Simultaneous Localization and Mapping (SLAM) can enable the robot to not only locate its own position in real-time in a complex environment, but also maps can also be constructed in real time. The research works like Lidar Odometry and Mapping (LOAM), a SLAM method first proposed in 2014 by Zhang et al. [10], is to achieve continuous self-motion estimation and map updating without relying on GPS by relying only on Lidar point cloud data.

The main contribution of the algorithm proposed in this paper to the navigation research work on autonomous robots is to enhance the robot's semantic understanding of its surrounding environment, that is, the autonomous robot can clearly know the coordinate points of objects in the three-dimensional world by performing 3D object detection. After understanding the specific location and category information of each object, decisions can be made based on this information to complete the path planning task in the outdoor environment. The 3D coordinate point information can be used to determine the height and width of the object, so that the robot can determine whether to bypass the obstacle from the left and right sides, or to determine whether to cross the obstacle if it is not enough to hinder normal walking. In this algorithm, the fusion of the two sensors of the visual sensor camera and Lidar allows the robot to complete 3D object detection in three-dimensional space, thereby identifying the specific category and location of the object. The research uses YOLO 3D object detection based on EPnP to enable the robot to obtain semantic information about the outdoor environment, so that the environment map is no longer composed of undifferentiated 2D pixels or 3D point clouds. Although there are already related studies on 3D object detection using YOLO, a 2D target detection algorithm proposed by Redmon et al. [7] in 2016. For example, Complex-YOLO proposed in 2018 by Simony et al. [8] by introducing Euler-Region-Proposal to represent 3D bounding boxes, in their network architecture YOLO framework can directly process point cloud data from 3D Lidar. However, so far, there is no relevant research using the geometric correspondence between 2D pixels and 3D point clouds to achieve the purpose of using YOLO to do 3D object detection.

Regarding the EPnP part of the proposal research, EPnP is a solution to the PnP problem proposed by Lepetit et al. [4] in 2008. They provide an algorithm that can efficiently estimate the pose of the camera when the projection in 2D image and several points in three-dimensional space are known. This method is very helpful for 3D object detection related research based on solving the correspondence between 2D pixels and 3D point cloud data. For example, in our proposed research work, YOLO will first locate objects in the 2D image acquired by the camera. Then EPnP will calculate the three-dimensional spatial coordinates and orientation of these objects in the real world based on the known 2D prediction results and the 3D point clouds information obtained from the Lidar.

The network architecture of this algorithm is based on the correspondence between 2D pixels and 3D point clouds, thereby using the geometric projection relationship between them to enable the deep neural network (DNN) model to estimate the camera pose and the 3D coordinate points of the object relative to the Lidar coordinate system. This DNN and geometric projection based research work aiming to reduce the dependence on the diversity of the training data set while maintaining the efficiency of recognition and classification in the 3D detection process. In the Convolutional Neural Network (CNN) - SLAM research proposed by Tateno et al. [9] in 2017, it is no longer the traditional SLAM algorithm that only focuses on how to extract the geometric structure from the environment and object matching. It also combines deep learning technology to use a CNN to predict depth information and semantic segmentation from monocular images, integrating semantic information into traditional SLAM technology.

In summary, the abundance of semantic information in the navigation system determines whether the robot has the ability to make intelligent autonomous decisions. This EPnP based YOLO 3D object detection technology can not only position the robots from a geometric perspective, but also has the ability to identify and classify objects in the surrounding environment, allowing robots to achieve more intelligent path planning and environmental interaction, which will be able to have potential contribution on semantic SLAM related research work.

2 The Network Architecture Background

Due to the advancement of sensor technology and the development of deep learning, research related to 3D object detection has received widespread attention in the past decade or so. In the papers PointNet and PointNet++ published by Qi et al. [5,6] in 2017, they proposed a deep learning network for 3D object detection directly from the 3D point cloud data obtained by Lidar. The original point cloud data from Lidar represented only by coordinate points will be used as the input of the PointNet series deep learning architecture. Object recognition and semantic segmentation of 3D point cloud data are performed by feature extraction for each point cloud data that has been processed by the symmetry function into order. The deep learning architecture PointNet series does not rely on data conversion and can directly and effectively process and extract object feature points, thereby having more accurate object recognition capabilities. However, when large-scale point cloud data needs to be processed quickly,

PointNet's feature extraction of each 3D point individually will lead to a sharp increase in computing intensity and memory requirements for hardware devices, making it unable to be effectively applied in the real-time applications such as the robot navigation system. As shown in the Fig. 1, the speed and accuracy of PointNet series (F-PointNet) and other 3D object detection methods such as VoxelINet and Complex-YOLO are compared when running in the embedded system (NVIDIA Titan X/Titan XP)., it can be seen that Complex-YOLO has an FPS (frame per second) of 50.4, and AVOD, as the fastest processing frame number among other algorithms, only reaches 12.5.

Method	Modality	FPS	Car			Pedestrian			Cyclist		
			Easy	Mod.	Hard	Easy	Mod.	Hard	Easy	Mod.	Hard
MV3D [2]	Lidar+Mono	2.8	71.09	62.35	55.12	-	-	-	-	-	-
F-PointNet [5]	Lidar+Mono	5.9	81.20	70.39	62.19	51.21	44.89	40.23	71.96	56.77	50.39
AVOD [7]	Lidar+Mono	12.5	73.59	65.78	58.38	38.28	31.51	26.98	60.11	44.90	38.80
AVOD-FPN [7]	Lidar+Mono	10.0	81.94	71.88	66.38	46.35	39.00	36.58	59.97	46.12	42.36
VoxelNet [3]	Lidar	4.3	77.47	65.11	57.73	39.48	33.69	31.51	61.22	48.36	44.37
Complex-YOLO	Lidar	50.4	67.72	64.00	63.01	41.79	39.70	35.92	68.17	58.32	54.30

Fig. 1. Performance comparison for 3D object detection [8]

The reason why Complex-YOLO 3D object detection algorithm can achieve real-time detection speed is that it applies and expands YOLO, an efficient deep learning framework. This enables it to not only perform 2D object detection but also obtain depth information of objects in space. The YOLO network framework transforms the object detection task into a single regression problem, that is, it does not need to first generate potential Region Proposals (RPs) and then classify these generated RPs like the traditional two-step method of object detection, such as the Region-based Convolutional Neural Network (R-CNN). Due to the fact it directly predicts multiple object bounding boxes and object categories directly on two-dimensional image pixels, the algorithm can also use global features in the image for feature extraction and object detection. The Complex-YOLO algorithm performs object detection on a 2D bird's-eye view converted from 3D point cloud data obtained only by Lidar by applying YOLO while retaining key spatial information, and then regress the length, width, height and direction parameter information of the 3D bounding box to infer the detected position and orientation in 3D space. The entire detection process does not involve 3D object detection on RGB images from camera, and the Complex YOLO algorithm directly performs 3D object detection on the bird's-eye view of 3D point cloud data.

Although the real-time nature of Complex-YOLO meets the speed requirements of the robot navigation system for 3D object detection technology, the design concept of Complex-YOLO places speed and detection efficiency in an overly important position compared to the computational complexity and model size, causing its algorithm to lose some accuracy for small objects and occluded objects in complex detection environments. As shown in the Fig. 1, the Average Precision (AP) of Complex-YOLO in the difficult pedestrian detection experiment is only 35.92.

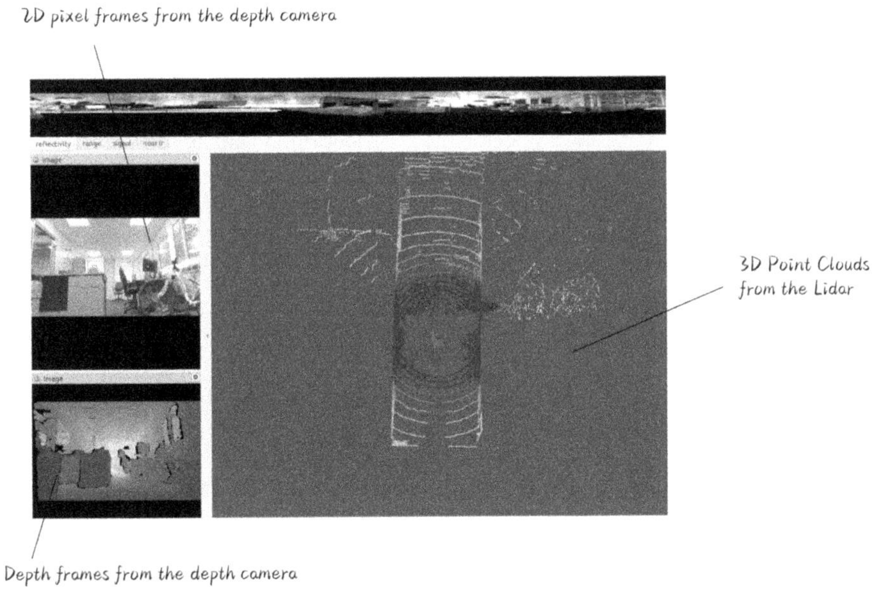

2D pixel frames from the depth camera

3D Point Clouds from the Lidar

Depth frames from the depth camera

Fig. 2. Comparison of Data format between Depth Camera and Lidar

The goal of the research work described in this paper is to develop a YOLO-based 3D object detection algorithm that maintains accuracy while maintaining speed, so it can be used in the robot's navigation system to enhance its ability to perceive the surrounding environment. Considering about the problem of loss of accuracy in the Complex-YOLO 3D object detection algorithm, the Lidar will not be used as a single sensor to perform environment perception work like the Complex-YOLO detection model, but will use a depth camera sensor and Lidar for environmental perception to improve the robustness of the 3D object detection algorithm.

Using Lidar alone, like the research work of Complex-YOLO, can indeed provide accurate distance and 3D spatial information and can save the time of sensor fusion to a certain extent and speed up the detection efficiency. However, as shown in the Fig. 2, the 3D point cloud data information obtained by Lidar is too sparse in some long-distance or complex environments. The RGB information and depth map from the camera as shown on the left side of the Fig. 2 can capture the rich color and texture information of objects in the scene in the detection area that Lidar cannot cover.

As shown in the middle of the Fig. 2, when the depth map from the depth camera cannot provide the depth value information corresponding to each two-dimensional pixel under ideal circumstances, you can rely on the 3D point cloud information obtained by Lidar and the RGB data obtained by the depth camera to apply the sensor fusion to have more accurate environment modeling capabilities.

3 Methodology

The novelty of the research method this paper proposes is to use YOLO, a 2D object detection method, to perform 3D object detection based on the correspondence between 2D projections and 3D point clouds. The difficulty lies in the geometric coordinate system transformation and how to apply EPnP reversely to efficiently estimate the camera pose to get the object 3D points based on the Lidar system instead of the camera.

3.1 EPnP Based Corresponding Relationship

In the computer vision research area, the aim of the Perspective-n-Point problem—PnP in short—is to determine the position and orientation of a camera given its intrinsic parameters and a set of n correspondences between 3D points and their 2D projections. [4] EPnP is a solution that can effectively solve the PnP problem, that is, by representing specific 3D key coordinate points as a set of predefined points as a reference basis to represent all other 3D points, so that the 3D coordinate points that originally need to be processed in large quantities in the three-dimensional scene can be obtained by calculating the position of the weighted specific 3D control coordinate point using the mathematical linear equation (1) shown below. The weight is the coefficient of the linear combination, which speeds up the estimation of the camera pose. The key to solving the PnP problem is to analyze and research how to estimate the pose information of the camera in the three-dimensional scene from the 2D pixels in the two-dimensional image and the corresponding 3D coordinate points.

$$P = w_1 C_1 + w_2 C_2 + w_3 C_3 + w_4 C_4 \tag{1}$$

As shown in the above Eq. (1), C1, C2, C3 and C4 are respectively four preset 3D coordinate points (four virtual 3D control points), and W1, W2, W3 and W4 are the weights corresponding to the four virtual control points respectively. The actual 3D coordinate point P is a linear combination of these four virtual control points.

Since the PnP problem is solved only in a non-iterative way by using mathematical means to solve a small number of constant degree equations in the process of camera pose estimation. Compared with some algorithms whose time complexity is $O(n^5)$ or $O(n^8)$, which approximate the optimal solution by repeating the same iteration steps many times, EPnP is only $O(n)$, where n is the input number of 2D-3D point pairs, and its time complexity increases linearly. In terms of the speed and stability of attitude estimation in three-dimensional scenes, the EPnP solution is very suitable for applications that have great real-time requirements, such as the robot navigation system. Because the EPnP solution has efficient calculation speed while also maintaining reasonable calculation accuracy, unlike the method of iteratively approaching the optimal solution, which relies heavily on guessing the initial value and needs to deal with some complex problems like the convergence of the solution and setting iteration termination conditions, etc., so EPnP has very good robustness characteristics whether considered from the perspective of computational complexity or time efficiency. It is the great reference for research on real-time analysis of the position of objects in three-dimensional space and environmental perception of robot vision in embedded systems.

In this research work, the depth camera will be used as a visual sensor to capture 2D pixel points obtained from the two-dimensional image data and obtain the three-dimensional coordinate points of the measured objects in the scene from the captured depth map. The correspondence between these known two-dimensional pixel points and three-dimensional coordinate points obtained by the depth sensor can be understood as the 2D pixel point is the projection of the 3D coordinate point on the two-dimensional image. The Lidar is used to obtain sparse but more accurate 3D point cloud data to complete subsequent research work on sensor fusion with data from depth camera.

3.2 Projection Geometry of Point Clouds

As shown in the network architecture Fig. 3, depth cameras and the Lidar are used as sensors that provide direct visual information and precise three-dimensional environmental information respectively to jointly improve the accuracy and efficiency of 3D object detection in the robot navigation system. Based on the working principle and data acquisition method of the depth camera, there is a corresponding relationship between the coordinate points of the two-dimensional image and the three-dimensional depth map from the same depth camera as shown in the following Eq. (2). That is, theoretically, for 2D pixel coordinate points in a two-dimensional image, the depth map generated by the same depth camera provides distance information corresponding to each coordinate point.

From this, the 3D bounding box in the depth map and the corresponding object label category in the bounding box can be generated according to the known YOLO algorithm and the three-dimensional coordinate point information corresponding to the 2D pixels of the corresponding object. So far, two conditions necessary for using the EPnP algorithm to estimate the camera's pose in three-dimensional space have been obtained, namely, the 2D-3D coordinate point pair information in the measurement scene and the depth camera internal parameters focal length (f_x, f_y) and principal point coordinates (c_x, c_y). Additionally the 2D pixel information obtained from the depth camera is combined with the Lidar coordinate system as the origin. The EPnP algorithm is used again to reversely calculate the Lidar's three-dimensional coordinate information relative

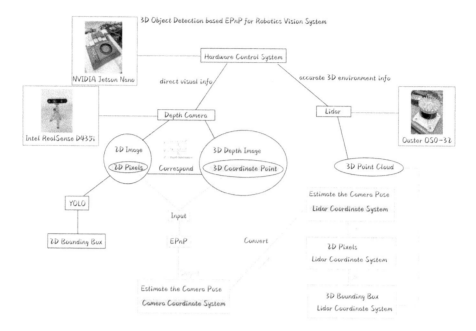

Fig. 3. The Network Architecture of the 2D to 3D object correspondence using EPnP

to the object, so that achieves the research purpose of generating 3D bounding boxes and corresponding object label categories on point cloud data obtained by Lidar.

$$X = \frac{(x - c_x) \times Z}{f_x}$$
$$Y = \frac{(y - c_y) \times Z}{f_y}$$
$$Z = \text{Depth Information} \qquad (2)$$

As shown in the above Eq. (2), when the coordinates of the 2D pixel points in the two-dimensional image are known, the corresponding 3D coordinate points (X, Y, Z) can be obtained by mathematical calculation with the internal parameters of the depth camera. Z depth value is obtained directly from the depth camera and represents the straight-line distance from the 2D pixel point in the two-dimensional image to the camera. (x, y) is the coordinate point of the pixel in the two-dimensional image. (f_x, f_y) are the focal lengths in the x-axis and y-axis of the depth camera's internal parameters, which represent the length of the imaging plane from the center of the lens. (c_x, c_y) are the principal point coordinates of the depth camera in the x-axis and y-axis, which represent the focus of the imaging plane and the lens optical axis.

In an ideal situation, the 2D pixels in the captured 2D image of each frame will be assigned their corresponding depth values by the same depth sensor. According to the projection relationship between pixels, which is the smallest unit of the two-dimensional image and corresponding depth information, the depth camera can theoretically directly create its corresponding 3D coordinate point for each 2D pixel point. Unlike Lidar environment modeling systems that use laser pulses to measure the distance between objects in the scene and the Lidar sensor to independently scan the position information of objects in three-dimensional space. The four virtual control points in the core step of solving the PnP problem in the EPnP algorithm are calculated from the correspondence between 2D-3D point pairs and the internal parameters of the depth camera used. The accurate estimation of the camera pose depends to a large extent on the quality of the 2D-3D point pairs mentioned above that can be calculated by the depth sensor. As shown in the following Eq. (3), $\mathbf{M}$ is a 12×12 matrix constructed from the three coordinate values (X, Y, Z) of each of the four virtual control points in the linear combination equation, $\mathbf{v}$ is the eigenvector, and λ is the eigenvalue corresponding to the eigenvector. By solving the eigenvector $\mathbf{v}$ of the 12×12 matrix $\mathbf{M}$, the optimal solution to the PnP problem, that is, the camera position and direction, is obtained through mathematical calculations.

$$M\mathbf{v} = \lambda \mathbf{v} \tag{3}$$

The correspondence between pixels and depth values in the depth camera not only provides the technology of a visual sensor to capture objects in the environment and generate two-dimensional images, but also can theoretically obtain 3D point cloud data from the perspective of the depth camera. The Lidar scanning system is independent of the two-dimensional plane image, so the 3D point cloud information it generates has no direct correspondence with the 2D pixels in the two-dimensional image generated by the depth camera. However, the breaking point of the problem is that the relative positions of the depth camera and Lidar remain unchanged in the robot. This EPnP based detection and classification research work on reference coordinate system conversion and sensor calibration in the book Multiple View Geometry in Computer Vision [3], which provides fundamental theoretical support for the research on sensor data fusion between depth camera and the Lidar.

3.3 Coordinate System Transformation

Since the relative position between the camera and the Lidar is fixed in the robots, the camera pose relative to the depth camera obtained when using EPnP for the first time can be converted through the absolute position relationship between the camera and Lidar coordinate systems. When applying EPnP in the secondary time as shown on the right side of the Fig. 4 below, the 3D coordinate points of the object based on the Lidar's initial position as the reference origin by reversely inputting to EPnP the 2D pixels with the coordinate-converted Lidar as the reference origin and the camera pose, and then generating the corresponding 3D bounding boxes.

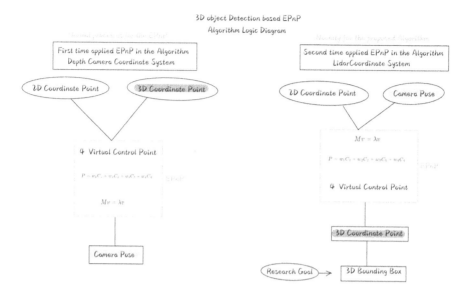

Fig. 4. The Logic Diagram of the Algorithm while using EPnP

4 Experiments

4.1 Current Results

The Fig. 5 below, the video frames generated from intel realsense D435i [1] and the outputs from this algorithm are shown. The left side is the output window of the colour frame and the right side is the corresponding depth map. In the output window of the colour frame on the left, the 2D object detection technology YOLO identified the vase, bowl, teddy bear and mouse respectively and its confidence. These objects are marked by a green rectangular border around them through using OpenCV [2]. The 2D pixel centre points of each bounding box is also represented in the Fig. 5 in the colour frame, and the distance from the camera optical centre to the object plane is also marked after the 2D pixel.

The output in the lower part of Fig. 5 shows the 3D point coordinates of the object in the depth map corresponding to the 2D pixel point in the color frame. For example, the 2D pixel point coordinates of the bounding box of the vase are (266, 374) in the colour frame, the corresponding 3D point coordinates of this point in the depth map are (−0.0754937, 0.173762, 0.798). The Z value 0.798 in this 3D point matches the distance information 0.80 m of the 2d pixel point displayed in the color frame. The outputs also have the camera pose in the world coordinate system, that is, camera's rotation matrix and translation vector. The world coordinate system of the camera's external parameters obtained in the existing results of this algorithm is the optical center of the camera, which is why its rotation matrix and translation vector are nearly 0. Please note in Fig. 5 below that the green bounding boxes around the objects recognized in the left

2D colored frames are generated for this algorithm, the remaining colored lines are just drawn for better explanation.

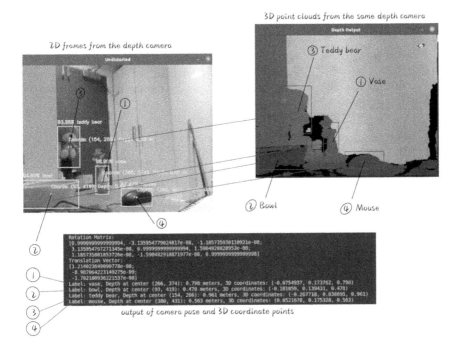

Fig. 5. Camera pose and 2D bounding boxes with its correspondence 3D points

4.2 Future Work

In the next stage of the experiment, point cloud library (PCL) will be embed in this existing algorithm to display the corresponding bounding boxes from the 2D RGB frame in the depth map obtained from the realsense depth camera. In order to make the depth camera and Lidar work better together, robot operating system (ROS) will also be used in subsequent work to maintain the synchronization of the sensors.

5 Conclusion

This paper proposes a YOLO 3D real-time object detection algorithm based on the EPnP. In the network structure of the algorithm, EPnP is used in reverse to obtain the corresponding 3D coordinate points relatively to the Lidar coordinate system, so that the corresponding 3D bounding boxes of the objects can be obtained on the 3D point clouds. This research also uses sensor fusion to complement the deficiencies between sensors. Like the pixel data of the depth camera can give more intuitive visual information, while the 3D point cloud obtained by the Lidar is too sparse to support the accuracy of 3D object detection. However, When the depth cameras are affected by lighting and cannot function properly, Lidar can provide accurate environmental modelling information. The proposed work can improve the environmental perception and navigation capabilities of autonomous robots while reducing reliance on training data sets, and has potential applications for the use of lightweight robots for unmanned automated delivery of express deliveries and so on.

References

1. Intel RealSense SDK. https://dev.intelrealsense.com
2. OpenCV: Open source computer vision library. https://opencv.org/
3. Hartley, R., Zisserman, A.: Multiple View Geometry in Computer Vision. Cambridge University Press, Cambridge (2003)
4. Lepetit, V., Moreno-Noguer, F., Fua, P.: Ep n p: an accurate o (n) solution to the p n p problem. Int. J. Comput. Vis. **81**, 155–166 (2009)
5. Qi, C.R., Su, H., Mo, K., Guibas, L.J.: Pointnet: deep learning on point sets for 3D classification and segmentation. In: Proceedings of the IEEE Conference on Computer Vision and Pattern Recognition, pp. 652–660 (2017)
6. Qi, C.R., Yi, L., Su, H., Guibas, L.J.: Pointnet++: deep hierarchical feature learning on point sets in a metric space. In: Advances in Neural Information Processing Systems, vol. 30 (2017)
7. Redmon, J., Divvala, S., Girshick, R., Farhadi, A.: You only look once: unified, real-time object detection. In: Proceedings of the IEEE Conference on Computer Vision and Pattern Recognition (CVPR) (2016)
8. Simony, M., Milzy, S., Amendey, K., Gross, H.M.: Complex-yolo: an Euler-region-proposal for real-time 3D object detection on point clouds. In: Proceedings of the European Conference on Computer Vision (ECCV) Workshops (2018)
9. Tateno, K., Tombari, F., Laina, I., Navab, N.: CNN-slam: real-time dense monocular slam with learned depth prediction. In: Proceedings of the IEEE Conference on Computer Vision and Pattern Recognition, pp. 6243–6252 (2017)
10. Zhang, J., Singh, S.: LOAM: lidar odometry and mapping in real-time. in: robotics: science and systems X. Robot. Sci. Syst. Found. (2014). https://doi.org/10.15607/RSS.2014.X.007, http://www.roboticsproceedings.org/rss10/p07.pdf

WeedScout: Real-Time Autonomous Blackgrass Classification and Mapping Using Dedicated Hardware

Matthew Gazzard[1], Helen Hicks[2], Isibor Kennedy Ihianle[1], Jordan J. Bird[1], Md Mahmudul Hasan[3], and Pedro Machado[1(✉)]

[1] Computational Intelligence Applications Research Group, Nottingham Trent University, Clifton Campus, Nottingham NG11 8NS, UK
n0932346@my.ntu.ac.uk, {isibor.ihianle,jordan.bird, pedro.machado}@ntu.ac.uk

[2] School of Animal, Rural and Environmental Sciences, Nottingham Trent University, Brackenhurst, Southwell NG25 0QF, UK
helen.hicks@ntu.ac.uk

[3] Mediprospects AI, 5-7 High Street, London E13 0AD, UK
m.mahmudul@mediprospects.ai
https://www.ntu.ac.uk/research/groups-and-centres/groups/computational-intelligence-applications-research-group,
https://www.ntu.ac.uk/study-and-courses/academic-schools/animal-rural-and-environmental-sciences, https://www.mediprospects.ai/

Abstract. Blackgrass (Alopecurus myosuroides) is a competitive weed that has wide-ranging impacts on food security by reducing crop yields and increasing cultivation costs. In addition to the financial burden on agriculture, the application of herbicides as a preventive to blackgrass can negatively affect access to clean water and sanitation. The Weed-Scout project introduces a Real-Rime Autonomous Black-Grass Classification and Mapping (RT-ABGCM), a cutting-edge solution tailored for real-time detection of blackgrass, for precision weed management practices. Leveraging Artificial Intelligence (AI) algorithms, the system processes live image feeds, infers blackgrass density, and covers two stages of maturation. The research investigates the deployment of You Only Look Once (YOLO) models, specifically the streamlined You Only Look Once (YOLO)v8 and YOLO-NAS, accelerated at the edge with the NVIDIA Jetson Nano (NJN). By optimising inference speed and model performance, the project advances the integration of AI into agricultural practices, offering potential solutions to challenges such as herbicide resistance and environmental impact. Additionally, two datasets and model weights are made available to the research community, facilitating further advancements in weed detection and precision farming technologies.

Keywords: SDG12 · SDG13 · SDG15 · YOLOv8 · YOLO-NAS · Blackgrass · Robotics · NVIDIA Jetson nano

1 Introduction

Blackgrass (Alopecurus myosuroides) is a problematic weed that has potential impact on food security due to its interruption of oilseed rape and several winter cereals. As weed species became increasingly resistant to herbicides, managing weed populations became more difficult. Varah et al. [17] concluded that herbicide resistant blackgrass resulted in £400 million lost gross profit and 0.8 million tonnes lost yield in England alone. Furthermore, Varah et al. [17] predicted that in the absence of herbicide control, losses could reach £1 billion and 3.4 million tonnes annually. In this paper, the WeedScout project presents an Artificial Intelligence (AI) solution tailored to detect blackgrass in wheat crops, streamline farmers' identification processes, and resource allocation. The innovative approach not only facilitates precision weed management, but also holds promise in reducing herbicide costs and hindering the advancement of herbicide resistance. The contributions of the paper include the compilation and sharing of a diverse blackgrass dataset, employing state-of-the-art Deep Learning (DL) algorithms, and the deployment of the models on dedicated hardware. Initially, our focus was on exploring the use of AI in object detection, with particular emphasis on real-time implementations, given the challenge of accurately mapping blackgrass. Furthermore, our work contributes annotated two blackgrass datasets [6,7] to support future work. The WeedScout delivers Real-Rime Autonomous Black-Grass Classification and Mapping (RT-ABGCM) powered by the NVIDIA Jetson Nano (NJN). By leveraging the computational prowess of the NJN, AI inference is accelerated at the edge, enabling swift detection and analysis of blackgrass. Moreover, the proposed approach not only maps the blackgrass density but also provides real-time geolocation, obviating the necessity for a continuous internet connection. Thus, WeedScout offers a comprehensive and efficient means of managing blackgrass infestations in wheat fields.

The paper is organised into the following sections: Sect. 2 reviews relevant works; Sect. 3 elaborates on the proposed methodology; Sect. 4 conducts an analysis of the results; and, Sect. 5 presents the concluding remarks and outlines future research directions.

2 Literature Review

Weeds are commonly recognised as the most significant biotic factor affecting crop production [15]. However, it is noteworthy that low densities of weeds can also provide agronomic and ecological benefits [15]. Therefore, weeds should be viewed as agricultural entities that require management rather than outright elimination. Historically, herbicides have been the most widely employed method for weed management, aimed at maximising crop yield while minimising human effort and cost [15,17]. However, extensive use of herbicides has led to the evolution of resistance on a large scale, jeopardising yield benefits as herbicide efficacy decreases [4,9,15,17]. Resistance to herbicides presents a significant problem, as resistance to one class of herbicides can indicate resistance to multiple types, and the use of herbicide mixtures in hopes of reducing specialist resistance can paradoxically lead to more general resistance [4,9]. Such reliance on herbicides as the primary weed management strategy is unsustainable, given the expense and time required to formulate new weed-killing chemicals, coupled with the inevitable development of resistance rendering them obsolete [9]. Additionally, it is crucial to acknowledge the adverse environmental impacts of herbicides, which harm biodiversity, water, and soil quality [17].

The abundance of A. myosuroides weeds is currently mediated by herbicide resistance [9], a phenomenon supported by Varah et al. [17]. The consequential impact of A. myosuroides on yield stands out as the most economically significant weed in Western Europe due to its high yield impact [9]. Comont et al. [4] discuss how glyphosate has emerged as one of the final lines of defence in controlling A. myosuroides, yet A. myosuroides is already developing resistance to the herbicide. Resistance to herbicides mirrors the broader trend of resistance in agricultural sectors, posing a substantial challenge. Given the increasing difficulties in herbicide-based weed management, it becomes imperative to explore more diverse methods of weed management that integrate targeted site management, an area where AI proves invaluable.

Ahmed et al. [2] outline challenging environments like occlusion and poor lighting, where object detection using AI has struggled. Crop fields are such environments, as noted by Hasan et al. [8]. Two-stage detectors, despite higher accuracy, are unsuitable for real-time applications due to slower processing speeds [21]. For embedded or edge devices in blackgrass detection, lightweight models are essential [2], discounting two-stage detectors as per Zaidi et al. [21]. You Only Look Once (YOLO) is ideal for real-time detection, balancing speed and accuracy [3,14,16]. It offers various model scales for specific needs [3,14,16] and has been effectively implemented in agricultural edge devices [3,14,16].

In agriculture, AI deployment for weed detection and mapping of weeds is increasingly prevalent, as evidenced by studies such as those conducted by Krähmer et al. [12], Xu et al. [19], Jabir and Falih [10], Hasan et al. [8], Wu et al. [18], Xu et al. [20], and Dainelli et al. [5]. Krähmer et al. [12] further specify that herbicide resistance distribution can also be effectively mapped. In the context of wheat crops, Xu et al. [19] elaborate on the tillering stage as the opportune time for weed detection due to reduced occlusion between wheat and weeds, a factor alleviating some challenges encountered in challenging environments [2,8]. However, despite the availability of numerous datasets, challenges persist in training models specifically for weed detection within crop fields. Dainelli et al. [5] highlight limitations in data collection methods and availability, stressing the costliness and time-consuming nature of creating custom datasets. Krähmer et al. [12] underscore the difficulty in identifying growth stages and the challenge detection algorithms face in distinguishing between similar weeds and a plethora of unknown species [11]. Nonetheless, finding an automated solution for weed detection remains imperative, as manual methods prove economically unviable [11]. While the hardware for weed monitoring may incur initial expenses, the resulting reductions in herbicide usage offset costs and offer environmental benefits. Moreover, the inefficiency of manual weed detection underscores the potential of technologies such as Unmanned Aerial Vehicle (UAVs) or robots to supplement or replace manual labor. UAVs equipped with high-resolution cameras are particularly effective in detecting low-density weed patches [18]. However, limitations exist, as most UAV applications involve capturing images only above the crop canopy, potentially missing crucial information [12]. Thus, a solution that closely monitors the wheat fields is desirable. Examples of such solutions already exist, contributing to Integrated weed Management (IWM) and Site-Specific Weed Management (SSWM). For instance, BonnBot-I [1] serves as a precision weed management and field monitoring platform, utilising ground-based rovers for weed detection and herbicide application. Moreover, Mattivi et al. [13] exemplify the successful deployment of cost-effective UAVs equipped with open-source software for implementing smart farming techniques, even on small-scale farms. These systems play pivotal roles in advancing Agriculture 4.0 initiatives [18,20]. The localised application of herbicides can be carried out using weed maps or real-time sensors [11]. The WeedScout project extends upon the literature reviewed in the present section by perform RT-ABGCM, independent of internet connectivity, and leveraging edge acceleration with the NJN.

3 Methodology

The WeedScout system receives live image feeds, executes AI inference on the NJN, and produces an output depicting the density of detected blackgrass, which could discern various stages of maturation. The chapter provides a detailed description of the RT-ABGCM module which is the core of WeedScout.

As discussed in the previous section, YOLO models are well-known for their efficacy in weed detection. However, due to limited computational resources on edge devices, larger AI models, despite their higher accuracy compared to smaller ones, are not feasible for deployment. Moreover, larger YOLO models exhibit increased latency, making them unsuitable for real-time detection systems. To determine the veracity of the effectiveness of YOLO-NAS, YOLOv8 models were also trained on the same two custom datasets [6,7] and benchmarked to determine which type of model works best in the problem domain.

The developed models were deployed on the cost-effective NJN. The compact computer is purpose-built for AI applications and leverages Compute Unified Device Architecture (CUDA)[1] for accelerating AI operations. However, given the edge deployment's limited resources, training the model directly on the device is impractical. Therefore, the models will undergo training on a separate, more powerful machine. The models performed inference on the blackgrass video input, generating video output with overlays of the model's detections.

---

[1] Available online, https://developer.nvidia.com/cuda-zone, last accessed: 08/05/2024.

Fig. 1. NVIDIA Jetson Nano.

The NJN (see Fig. 1) was flashed with JetPack SDK 2[2] (see Fig. 2), running Robotic Operating System 2 (ROS2) Dashing[3] powered by Ubuntu 18.04.06 LTS[4]. The JetPack SDK provides all the NVIDIA libraries necessary for AI edge applications and ROS2 is necessary for compatibility with mobile robotic platforms (Fig. 3).

Fig. 2. NVIDIA JetPack SDK 2.0.

Fig. 3. NJN AI inference.

3.1 WeedScout Platform Workflow

A meticulously designed and implemented prototype for RT-ABGCM is depicted in Fig. 4. The system integrates an NJN equipped with an Intel RealSense D435i depth sensor. The NJN serves as the computational backbone for accelerating the YOLOv8 and YOLO-NAS inference at the edge. Specifically tailored for a myriad of tasks, the NJN handles image identification, object detection, data segmentation, and speech processing. It boasts a robust configuration, featuring a 128-core Maxwell Graphics Processing Unit (GPU), a quad-core ARM A57 Application Processor Unit (APU) operating at 1.43 GHz, and 4 GB of 64-bit LPDDR4 memory running at 25.6 GB/s. Moreover, the system is bolstered by the inclusion of a pre-trained DL models, enhancing its detection capabilities. While semantic segmentation would be beneficial for clearly delineating boundaries between blackgrass and wheat, annotating a sufficiently large dataset would be exceedingly time-consuming due to the likelihood of images containing

---

[2] Available online, https://developer.nvidia.com/embedded/jetpack, last accessed: 08/05/2024.
[3] Available online, https://docs.ros.org/en/dashing/index.html, last accessed 08/05/2024.
[4] Available online, https://releases.ubuntu.com/18.04/, last accessed: 08/05/2024.

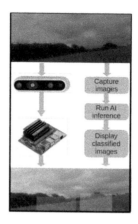

Fig. 4. The proposed WeedScout architecture workflow.

numerous individual plants. Annotating images for object detection is comparatively simpler as it does not require drawing complex polygons around the desired objects. Moreover, it was determined that employing object detection would not adversely affect the performance of the generated models. Therefore, object detection was chosen as the computer vision methodology.

Object detection necessitates annotating a given dataset by drawing bounding boxes around objects within images and assigning each box to the corresponding object class. Roboflow[5] was used to increase productivity. Leveraging Roboflow proved advantageous as it facilitated the easy management and storage of datasets; annotated images could be categorised into training, validation, and test sets before undergoing augmentation and export.

The dataset for mature blackgrass was collected using an Intel RealSense D435i[6] at a wheat field in England, UK. The Intel RealSense D435i was used to capture Red, Green, Blue and depth (RGBd) data and stored in a bag file. Consequently, it was necessary to extract standard video footage from the file using Intel RealSense SDK[7]. The process yielded thousands of PNGs of standard images alongside thousands of images containing depth data. Since the depth data was unnecessary for the proposed work, it was discarded, and a random selection of 780 PNGs was extracted using a PowerShell script to form an initial dataset.

Following the loading of the 780 images into Roboflow, 156 (20%) were allocated for validation, and 78 (10%) for testing, leaving the remaining 546 (70%) for training. Subsequently, all images were annotated using Roboflow's annotation tool, categorising blackgrass into three different density categories: low, medium, and high. To augment the size and diversity of the initial dataset, data

---

[5] Available online, https://roboflow.com/, last accessed: 10/05/2024.
[6] Available online, https://www.intelrealsense.com/depth-camera-d435i/, last accessed: 10/05/2024.
[7] Available online, https://www.intelrealsense.com/sdk-2/, last accessed: 10/05/2024.

Table 1. Dataset augmentations

Type	Applied to	Option
Flip	Image, Bounding Box	Horizontal, Vertical
90° Rotate	Image, Bounding Box	Clockwise, Counter-Clockwise, Upside, Down
Crop	Image, Bounding Box	0% Minimum Zoom, 20% Maximum Zoom
Rotation	Image, Bounding Box	Between $-15°$ and $+15°$
Shear	Image, Bounding Box	$\pm15°$ Horizontal, $\pm15°$ Vertical
Hue	Image	Between $-25°$ and $+25°$
Saturation	Image	Between -30% and $+30\%$
Brightness	Image, Bounding Box	Between -25% and $+25\%$
Exposure	Image, Bounding Box	Between -14% and $+14\%$
Blur	Image	Up to 2.5px
Noise	Image, Bounding Box	Up to 1.99% of pixels
Cutouth	Image	3 boxes with 10% size
Mosaic	Image	Applied

augmentation techniques were applied to decrease the likelihood of overfitting and enhance model performance. Augmentations were applied to both the images and the bounding box annotations (refer to Table 1). Additionally, a version of the dataset without augmentations was preserved for training, enabling performance comparison. A separate approach was also explored on Roboflow utilising blackgrass seedling dataset obtained from the Plant seedling datasets available on Kaggle[8]. The images of the plant seedling dataset were annotated, and data augmentation was applied due to the limited size of the dataset (see Table 1). The seedling dataset comprised only one class, blackgrass as it was not intended for detecting weed density. Instead, it was envisaged that any resulting model could be used for precision weed management in newly planted wheat fields.

Developing YOLO-NAS models required the utilisation of the open-source model library SuperGradients[9]. In addition to facilitating the training of regular models, the library enables the SuperGradients libraries to quantitise models through its post-training quantisation and quantization aware training functionalities. SuperGradients also provides the ability to connect third-party monitoring software to display the metrics tracked during model training and validation. Given the iterative nature of training AI models, where adjustments to the dataset or hyperparameters may be necessary based on training outcomes, it is crucial to monitor performance at both training and validation stages to assess the impact of any modifications. Weights and Biases (WnB) serves as a platform

---

[8] Available online, https://www.kaggle.com/datasets/vbookshelf/v2-plant-seedlings-dataset, last accessed: 10/05/2024.
[9] Available online, https://github.com/Deci-AI/super-gradients, last accessed: 10/05/2024.

for storing AI model data, facilitating the comparison and evaluation of model performance. Any monitoring data collected during model training was hosted on WnB, enabling the automatic generation of graphs for essential metrics such as mean Average Precision (mAP), precision, and recall for use in model comparison. The use of WnB also served as the project's test suite, as the performance metrics of the models are intrinsically linked to the successful video inference. Although training AI models is resource intensive, maximising the use of the hardware used for training is crucial to ensuring models are trained within an acceptable timeframe. Although it is possible to train AI models using a computer's Central Processor Unit (CPU), the method is not the most efficient. Leveraging CUDA with an NVIDIA GPU can significantly reduce training time by accelerating processing through GPU-acceleration. Thus, training with GPU-acceleration is the preferred methodology, either by using available resources or using cloud solutions such as Google Collab.

Using a GPU offers benefits beyond accelerated training; it also enables faster model inference. This is essential as any recorded footage of blackgrass should be inferred at the fastest possible speed to maintain usability. The chosen training machine utilised an NVIDIA GeForce RTX 3080 10GB GPU with CUDA version 12.4 (see Fig. 13). Before starting training, it was necessary to prepare the training machine by downloading the CUDA Toolkit, as it does not come pre-installed. SuperGradients utilises PyTorch, a Python-based machine learning framework. To enable PyTorch to utilise CUDA, it is necessary to download versions of it that come packaged explicitly with CUDA capabilities. Without these specific packages, training and inference using NVIDIA GPUs would be impossible.

Before deploying models to the NJN, it was essential to devise a design for presenting the blackgrass detections to users. By default, the prediction function of SuperGradients draws labelled bounding boxes around the detections, indicating the confidence threshold of the prediction (see Fig. 5). However, the objective of blackgrass inference is to identify areas with varying densities of blackgrass weeds, and SuperGradients does not offer a density overlay option. Additionally, SuperGradients cannot be used to infer YOLOv8 models, necessitating the cre-

Fig. 5. Blackgrass detection overlay initial design. Yellow is being used to demonstrate low density and orange is being used to demonstrate medium density. High density patches would have a red overlay. (Color figure online)

ation of a custom overlay. Inspiration was drawn from heatmaps, where higher densities are represented by increasingly warmer colours. For instance, lower densities may be highlighted in yellow, while higher densities can be depicted in orange. Areas with the highest densities could have a red overlay to denote their severity. Figure 5 was devised to provide guidance for designing the output of model inference returned to users.

4 Results Analysis

In this section, we will discuss the diverse results obtained for the two datasets tested against both YOLOv8 and YOLO-NAS.

4.1 Blackgrass Seedling Dataset

To evaluate the suitability of the seedlings dataset (see Table 2) tested against the YOLOv8.

Table 2. Number of images within augmented blackgrass seedling datasets

Train	Validation	Test	Total
2820	168	112	3100

For the seedling dataset, there was only the blackgrass class to be detected. From the results obtained, the True Positive (TP) value of 0.81 indicated the model's success, as the vast majority of blackgrass in the positive samples was detected. While metrics for mAP quickly converged within approximately 20 epochs, the values failed to reach 1.0 The validation loss gradually decayed as expected, but training loss exhibited greater stability than its validation counterparts. Nevertheless, the results from creating a YOLO-NAS model using the blackgrass seedling dataset provided clearer insights into overfitting compared to the YOLOv8 model. Convergence was remarkably swift for both mAP and recall during training and validation. The final values for mAP and recall metrics during training were also strikingly close to 1. While validation precision remained high, it was noticeably lower compared to its training counterpart, which tended towards 1.0 like training mAP and recall. Therefore, an interesting trend was observed when comparing the training loss and validation loss, where training loss decreased while validation loss increased. The results indicate that the model was overfitting. Combined with other trends observed during both YOLO-NAS and YOLOv8 training and validation, it was concluded that overfitting negatively impacted the performance of models trained on the blackgrass seedling dataset. Unfortunately, beyond the existing data augmentations, there was no effective way to further increase the size and diversity of the blackgrass seedling dataset. Therefore, the focus shifted to prioritising the detection of mature blackgrass.

4.2 Mature Blackgrass Dataset

The mature blackgrass dataset (see Table 3 was generated using the same augmentations described in Table 1.

Table 3. Dataset Details

Dataset	Train	Validation	Test	Total
Blackgrass mature	6552	624	312	7488

Given the state-of-the-art performance typically associated with YOLO-NAS models, it was expected that similar models would surpass YOLOv8 in later epochs. However, the mAP for the YOLOv8 model demonstrated a slight improvement compared to its YOLO)-NAS counterpart. Furthermore, the recall for the YOLOv8 model outperformed YOLO-NAS when running on the NVIDIA Jetson nano, although both precision and recall are similar. The observation is intriguing considering that the YOLOv8 nano architecture is smaller than the architecture of the YOLO-NAS small. The discrepancy in size makes it obvious that the YOLOv8 model has a better performance than the YOLO-NA model. Additionally, the absence of loss escalation during validation with the YOLOv8 model contrasts with the trend observed in the YOLO-NAS models, where validation loss increases.

The results in Table 4 indicate that the YOLOv8 model significantly outperforms the YOLO-NAS model across various density levels of blackgrass. YOLOv8 achieves the highest metrics for the "blackgrass high density" class with recall, specificity, precision, accuracy, and F1-score all around 0.94, reflecting its strong performance. For both "blackgrass low density" and "blackgrass mid density" classes, YOLOv8 maintains similar high performance with recall, specificity, precision, accuracy, and F1-score around 0.91. In contrast, YOLO-NAS shows a meaningful drop in performance, particularly for "blackgrass low density" where its metrics hover around 0.64, and for "blackgrass high density" with scores around 0.77. To assess the performance at runtime of the mature

Table 4. Results per class for each of the Convolutional Neural Network CNN architectures.

Class	Model	Recall	Specificity	Precision	Accuracy	F1-Score
Blackgrass high density	**YOLOv8**	**0.937**	**0.945**	**0.946**	**0.941**	**0.942**
Blackgrass low density	YOLOv8	0.910	0.917	0.918	**0.913**	**0.914**
Blackgrass mid density	YOLOv8	0.910	0.917	0.918	**0.913**	**0.914**
Blackgrass mid density	YOLO-NAS	0.833	0.839	0.840	**0.836**	**0.836**
Blackgrass high density	YOLO-NAS	0.768	0.772	0.774	**0.770**	**0.771**
Blackgrass low density	*YOLO-NAS*	*0.642*	*0.644*	*0.646*	***0.643***	***0.644***

blackgrass models, a brief 5-second video was used for inference. Upon examining the inference times of various models listed in Table 5, it became evident that the YOLOv8 model, trained on the mature blackgrass dataset, exhibited notably faster inference speeds compared to the YOLO-NAS models. The discrepancy can be attributed to the variance in model size and possibly to differences in the efficiency of prediction functions between Ultralytics[10] and SuperGradients.

In order to utilise GPU, the NJN was configured to run the inference using the pre-trained YOLOv8 weights. The runtime for inference on the same video used for training on the testing machine is presented in Table 5). Although the inference time is considerably longer, the results were anticipated as YOLO-NAS runs on the CPU while YOLOv8 utilises GPU. Given the significantly longer inference time compared to the supplied source video, optimisations to the model and code are imperative to reduce inference time and potentially enable real-time detection.

Table 5. Mean average time for mature blackgrass models to inference a 5 s long video on the CPU only, and CPU and GPU accelerated.

Model	Inference 1 [s]	Inference 2 [s]	Inference 3 [s]	Mean inference [s]
YOLOv8 (CPU and GPU)	82.98	86.25	82.69	83.97
YOLO-NAS (CPU only)	229.71	223.75	217.93	223.80

5 Conclusions and Future Work

This paper introduces a novel RT-ABGCM designed to operate effectively at the edge. Following multiple iterations of dataset augmentation and annotation, two types of models were developed capable of successfully detecting blackgrass. Given constraints with the blackgrass seedling dataset, the emphasis shifted towards detecting mature blackgrass. The results highlight YOLOv8's consistency and reliability in identifying blackgrass across different densities. The challenge arose from the mature blackgrass dataset's significant bias towards high-density patches, making it challenging to generate annotated datasets containing an adequate number of images depicting lower densities for robust training. Rectifying the challenge requires obtaining a more diverse dataset and conducting additional experimentation to identify the best mix of training hyperparameters and dataset augmentations for refining density detection models. Despite these challenges, models were successfully deployed on the NJN, demonstrating the efficacy of the model production pipeline. While the NJN enables inference, it operates at sub-optimal GPU acceleration, resulting in slower inference times that preclude real-time performance.

---

[10] Available onine, https://docs.ultralytics.com/, last accessed: 12/05/2024.

Inference speed of the current solution could be improved by investigating other methods of model inferencing. Other lightwight AI frameworks such as TensorRT and DeepStream SDK could be utilised to improve the frame rate. Improving the Frames Per Second FPS is necessary for real-time inference. If real-time inference at a stable FPS was achieved, it would be possible to create an automated sprayer or electric discharge combined with more research into detecting blackgrass seedlings so that blackgrass could be tackled at the earliest possible stage.

Disclosure of Interests. The authors have no competing interests to declare that they are relevant to the content of this article.

References

1. Ahmadi, A., Halstead, M., McCool, C.: BonnBot-I: a precise weed management and crop monitoring platform. In: 2022 IEEE/RSJ International Conference on Intelligent Robots and Systems (IROS), pp. 9202–9209. IEEE (2022)
2. Ahmed, M., Hashmi, K.A., Pagani, A., Liwicki, M., Stricker, D., Afzal, M.Z.: Survey and performance analysis of deep learning based object detection in challenging environments. Sensors **21**(15), 5116 (2021)
3. Brandenburg, S., Machado, P., Lama, N., McGinnity, T.: Strawberry detection using a heterogeneous multi-processor platform. arXiv preprint arXiv:2011.03651 (2020)
4. Comont, D., et al.: Evolution of generalist resistance to herbicide mixtures reveals a trade-off in resistance management. Nat. Commun. **11**(1), 3086 (2020)
5. Dainelli, R., et al.: Recognition of weeds in cereals using AI architecture. In: Precision Agriculture 2023, pp. 401–407. Wageningen Academic Publishers (2023)
6. Gazzard, M., Machado, P.: Annotated black-grass seedling dataset (2024). https://doi.org/10.5281/zenodo.10981308
7. Gazzard, M., Machado, P.: Annotated mature black-grass dataset (2024). https://doi.org/10.5281/zenodo.10981327
8. Hasan, A.M., Sohel, F., Diepeveen, D., Laga, H., Jones, M.G.: A survey of deep learning techniques for weed detection from images. Comput. Electron. Agric. **184**, 106067 (2021)
9. Hicks, H.L., et al.: The factors driving evolved herbicide resistance at a national scale. Nat. Ecol. Evol. **2**(3), 529–536 (2018)
10. Jabir, B., Falih, N.: Deep learning-based decision support system for weeds detection in wheat fields. Int. J. Electr. Comput. Eng. **12**(1), 816 (2022)
11. Kavhiza, N.J., Vvedenskiy, V., Behzad, A., Bayat, M., Kargar, M.H., Zargar, M.: Weed mapping technologies in discerning and managing weed infestation levels of farming systems. Res. Crops **21**(1), 93–98 (2020)
12. Krähmer, H., et al.: Weed surveys and weed mapping in Europe: state of the art and future tasks. Crop Prot. **129**, 105010 (2020)
13. Mattivi, P., et al.: Can commercial low-cost drones and open-source GIS technologies be suitable for semi-automatic weed mapping for smart farming? A case study in NE Italy. Remote sensing **13**(10), 1869 (2021)
14. Pomykala, J., de Lemos, F., Ihianle, I.K., Adama, D.A., Machado, P.: Deep learning approach for classifying trusses and runners of strawberries. arXiv preprint arXiv:2207.02721 (2022)

15. Scavo, A., Mauromicale, G.: Integrated weed management in herbaceous field crops. Agronomy **10**(4), 466 (2020)
16. Terven, J., Cordova-Esparza, D.: A comprehensive review of yolo: from YOLOv1 to YOLOv8 and beyond. arXiv:2304.00501 (2023)
17. Varah, A., et al.: The costs of human-induced evolution in an agricultural system. Nat. Sustain. **3**(1), 63–71 (2020)
18. Wu, Z., Chen, Y., Zhao, B., Kang, X., Ding, Y.: Review of weed detection methods based on computer vision. Sensors **21**(11), 3647 (2021)
19. Xu, K., Li, H., Cao, W., Zhu, Y., Chen, R., Ni, J.: Recognition of weeds in wheat fields based on the fusion of RGB images and depth images. IEEE Access **8**, 110362–110370 (2020)
20. Xu, K., et al.: Precision weed detection in wheat fields for agriculture 4.0: a survey of enabling technologies, methods, and research challenges. Comput. Electron. Agric. **212**, 108106 (2023)
21. Zaidi, S.S.A., Ansari, M.S., Aslam, A., Kanwal, N., Asghar, M., Lee, B.: A survey of modern deep learning based object detection models. Digit. Signal Process. **126**, 103514 (2022)

What Criteria Define an Ideal Skeletonisation Reference in Object Point Clouds?

Qingmeng Wen[1(✉)], Seyed Amir Tafrishi[1], Ze Ji[1], and Yu-Kun Lai[2]

[1] School of Engineering, Cardiff University, Cardiff CF24 3AA, UK
{wen1,tafrishisa,jiz1}@cardiff.ac.uk
[2] School of Computer Science and Informatics, Cardiff University, Cardiff CF24 4AG, UK
laiy4@cardiff.ac.uk

Abstract. The skeletonisation methods of 3D object models gained noteworthy attention in recent years and have been extensively investigated, inspired by their wide-range applications, including computer graphics and robotics. However, there remains a gap that the quantitative reference of high-quality skeletonisation results is hardly given by the previous studies. Building upon previous research on point cloud skeletonisation, this paper explores the inherent characteristics of the skeletonisation process across objects of varying shapes. This analysis aims to provide intuitive insights into the quality of the resulting desirable skeletons. Additionally, we introduce a new concept of stable convergence of contraction based on distributions of geometric curvature and vectorial normal changes.

Keywords: Reference skeleton · Point cloud · Computer vision · Geometric methods

1 Introduction

The skeleton of a shape is a contracted and medial structure that intuitively represents the shape both topologically and geometrically [23]. For 3D shapes captured from point clouds, there are two known shape skeletons, the medial surface and curve skeletons. Accordingly, skeletonisation is the process that generates the desirable skeletal descriptors of the object model by a set of computations [8]. This skeletonisation technique is highly valuable for robotics applications, ranging from object reconstruction and manipulation [25,27] to localization and navigation [8,19]. Although the algorithms for skeletonisation from point clouds have been extensively researched by previous studies, quantitative analysis of desirable skeleton is rarely found [23], and there has been no approach to understanding the contraction convergence state from a control theory point of view.

Supported by China Scholarship Council (No. 202006760092).

© The Author(s), under exclusive license to Springer Nature Switzerland AG 2025
M. N. Huda et al. (Eds.): TAROS 2024, LNAI 15051, pp. 422–433, 2025.
https://doi.org/10.1007/978-3-031-72059-8_35

The medial surface, typically known as the medial axis transform (MAT), has been extensively researched by previous studies. Either by simplifying Delaunay triangulation or computing a number of balls, MAT can be estimated with the information of Voronoi diagram [1,2]. Chazal et al. [6] proposed λ axis transform that is calculated by estimating the center point sets of the maximal inscribed sphere. Extended by Giesen et al. [10], a scale factor is used to generate MAT with fewer irrelevant spikes. In addition, the surface skeleton, obtained through Laplacian-based contraction (LBC) applied to either a point cloud or a mesh model, serves as an alternative approximation of the medial surface [4,5]. In recent years, the learning-based method has been applied to predict the "skeletal mesh" of the shapes [12]. While the medial surface faithfully corresponds to the original shape with intricate structure, the curve skeleton offers a more simplified representation that facilitates modeling and manipulation [5,9,23].

Due to its simplicity and versatile applications, curve skeletons have attracted significant interest from researchers. The rotational symmetry axis (ROSA) is defined by a set of geometric centers of cross-section points, achieved through the design of an optimal "cutting plane". [24]. Huang et al. [11] considered the L1-medians as the pivotal points of the skeletons, resulting in accurate 1D curves. Derived from the idea of curve skeletonisation from meshes [4], Cao et al. [5] abstract the curve skeleton from contracted point clouds with a Laplacian-based contraction (LBC) method. Based on this, our previous work reduces the computation cost of LBC by an in-loop local point cloud reduction strategy [26]. Besides, Wu et al. [28] adopted the resampling strategy to calibrate the resultant skeleton from LBC to refine the skeletons of maize plants. In a recent study by Meyer et al. [13], the quality of Laplacian-based contraction (LBC) applied to cherry tree point clouds is enhanced by discriminatively weighting the Laplacian matrix, taking into account the semantic segmentation of the points.

The application of curve skeletons might cover many areas, including virtual navigation, computer graphics, medical imaging, and more [7,20,23]. Recent studies suggest that curve skeletons hold promise for advancements in robotic research. Given that topological connections often encapsulate the essential structure of objects with hole-like features, researchers have begun applying topological curves in the planning of grasps for such objects [14,22]. As argued by Przybylski et al. [15–17], the excellent geometrical representations of shapes may contribute to aiding robots in planning stable grasps. Besides, through the design of a tailored grasping strategy, it has been demonstrated that curve skeletons can be effectively utilized for efficient and high-quality grasping planning [25].

The evaluation of skeletonisation methods is of significant importance to the validation of applicability. However, evaluating and validating the shape skeletonisation results remains a challenging task. Most works on skeletonisation simply discussed their experimental results by visually comparing resultant skeletons [4,11,18]. As for the remaining quantitative analysis, Arcelli et al. [3] compared the results with recoverabilities. Sobiecki et al. [21] gave the definition of "centeredness" of various types of skeletons to evaluate skeletonisation performance. Comparing resultant skeletons with ground truth is one approach, but establishing a universally accepted definition of the ground truth presents another challenging issue [13]. Currently, defining what constitutes a high-quality skeleton remains a significant challenge [20,23]. However, the evaluation of skeletonisation results is indeed crucial for advancing the field, particularly in terms of its practical applications.

Based on the previous work, this paper investigates the geometrical properties of Laplacian-based contraction and the quality of the resultant skeleton. Firstly, the methodology of evaluation metrics and the definition of the high-quality skeleton is discussed in Sect. 2. In Sect. 3, we demonstrate our experimental results and observations. Lastly, we conclude our work in Sect. 4.

2 Methodology

In this section, we will look into the meaningful geometrical properties of stable contraction and skeletonisation, addressing the challenge of evaluating the skeletonisation results, after the brief introduction of Laplacian-based skeletonisation (LBC).

2.1 Overview of Laplacian-Based Skeletonisation

The Laplacian-based skeletonisation is a pipeline that abstracts the skeleton vertices and connections from the points contracted by Laplacian weights. The Laplacian-based skeletonisation pipeline generally consists of 4 processes as illustrated in Fig. 1.

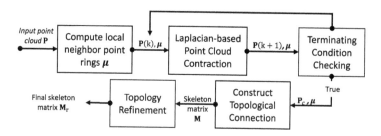

Fig. 1. General structure of Laplacian-based contraction of point clouds

Given an input point cloud $\mathbf{P} \in \mathbb{R}^{n \times 3}$, the pipeline starts with computing the point-wise neighbour-rings of the point cloud. For each point, its neighbour points can be queried by the k-nearest neighbours (KNN) algorithm. After obtaining the Delaunay triangulation of the neighbor points, the dimensions of the neighbor point coordinates data are reduced to 2D using Principal Component Analysis (PCA), and the ring points of the anchor point are derived from the triangulation results.

With the information of neighbour ring points, the Laplacian matrix $\mathbf{L} \in \mathbb{R}^{n \times n}$ is defined as

$$L_{ij} = \begin{cases} \omega_{ij} = \cot \alpha_{ij} + \cot \beta_{ij}, & \text{if } \mathbf{p}_j \in \boldsymbol{\mu}_i; \\ \sum_{k \in \mu_i}^{k} -\omega_{ik}, & \text{if } i = j; \\ 0, & \text{Otherwise;}, \end{cases} \quad (1)$$

where $\boldsymbol{\mu}_i$ ring point set of set of point $\mathbf{p}_i$, and α_{ij} and β_{ij} are two opposite angles of edge (i, j) in triangles of the point ring. Following that, the contracted cloud points in iteration $k + 1$ can be obtained by solving the system below

$$\begin{bmatrix} \mathbf{W}_L \mathbf{L} \\ \mathbf{W}_H \end{bmatrix} \mathbf{P}_{k+1} = \begin{bmatrix} 0 \\ \mathbf{W}_H \mathbf{P}_k \end{bmatrix}, \quad (2)$$

where $\mathbf{W}_L, \mathbf{W}_H \in \mathbb{R}^{n \times n}$ are contraction and attraction weights respectively. The contraction weights and the attraction weights work jointly to balance the process of pushing the points toward the medial axis. Please note, $\mathbf{W}_L, \mathbf{W}_H$ must be updated after each iteration to make sure the contraction process continues [5]. The contraction process in the pipeline runs iteratively until the resultant point clouds meet the terminating condition. The terminating condition can be defined either based on global or local features of the cloud points [5,26].

After the contraction process, the approximation of the medial surface, comprised of the contracted point set known as the surface skeleton, is obtained. This surface skeleton can be abstracted to skeleton vertices using farthest point sampling. The connections among different vertices are derived from the point-wise neighbor ring information. Lastly, the final skeleton is acquired after necessary refinement.

2.2 Characteristics of the Stable Contraction and Skeletonisation

Although Laplacian-based contraction can generate both the surface skeleton and the final curve skeleton, evaluating the quality of the resultant skeleton remains a considerable challenge. To investigate the characteristics of stable contraction, we examine deformation and distortion induced by the Laplacian contraction process by analysing the changes in surface normal vectors and curvatures.

To calculating distances among the surface normal vectors, the cosine similarity is chosen for measuring the similarity. Compared to other metrics, the cosine similarity is more indicative of the directions [30]. The general definition for m-dimensional vectors $\mathbf{n}_x$ and $\mathbf{n}_y$ is given by

$$S_C(\mathbf{n}_x, \mathbf{n}_y) = \cos(\theta) = \frac{\mathbf{n}_x \cdot \mathbf{n}_y}{\|\mathbf{n}_x\| \|\mathbf{n}_y\|} = \frac{\sum_{i=1}^{m} n_{x,i} n_{y,i}}{\sqrt{\sum_{i=1}^{m} n_{x,i}^2} \sqrt{\sum_{i=1}^{m} n_{y,i}^2}}, \quad (3)$$

where θ is the included angle between $\mathbf{n}_x$ and $\mathbf{n}_y$. Since the range of the cosine function is nonlinear to the angle, the cosine distance is normalised by

$$D_\theta = \frac{\arccos(S_C)}{\pi}, \quad (4)$$

where D_θ is the normalised angular distance, and $0 < D_\theta < 1$. For each point $\mathbf{p}_i$, we compute the distance of the normal vectors between the original and the contracted point cloud using

$$D_{\theta,k,i} = \frac{\arccos(S_C(n_{k,i}, n_{0,i}))}{\pi}, \tag{5}$$

where $n_i(k), n_i(0)$ are normal vectors of the point of the original positions and the new positions after k-th contraction iteration, respectively. As illustrated in Eq. (5), the normal vectors are required before computing their distance. Similar to computing local neighbor rings, a number of nearest neighbor points around each cloud point are used for normal vector estimation. This is achieved by calculating the covariance matrix $\mathbf{C}$ of the neighbor points, defined by

$$\mathbf{C} = \frac{1}{n} \sum_{i=1}^{n} (\mathbf{p}_i - \mathbf{c})(\mathbf{p}_i - \mathbf{c})T, \tag{6}$$

where n is the number of the neighbours used for estimation, $\mathbf{c}, \mathbf{p}_i$ are the coordinates of the anchor point and its corresponding i-th closest neighbour point, respectively. After calculating the eigenvectors $\mathbf{v}_1, \mathbf{v}_2, \mathbf{v}_3$ and the corresponding eigenvalues $\lambda_1, \lambda_2, \lambda_3$ satisfying $\lambda_1 < \lambda_2 < \lambda_3$ of that covariance matrix, the normal vector of the point is obtained as $\mathbf{v}_1$.

To obtain the curvature differences between the results of contraction and the original point cloud, we define the curvature differences here. With the covariance matrix $\mathbf{C}$ mentioned above, the curvature of the point can be estimated and normalised by

$$\kappa_n = \frac{\lambda_1}{\sum_{i=1}^{3} \lambda_i}. \tag{7}$$

Since the curvatures are obtained as scalars, we can easily obtain the differences of curvatures between the input point cloud and the contracted one by

$$\Delta \kappa_{n,k} = \kappa_{n,k} - \kappa_{n,0}, \tag{8}$$

where $\kappa_n(k)$ and $\kappa_n(0)$ represent the curvature at the point position after k-th contraction iteration and the original point position, respectively.

Here, we propose a hypothesis on the stable convergence of contraction and outline how it is qualitatively evaluated based on the convergence of the contracted surface. It is important to note that to the best of our knowledge, there has not been research discussing the desirability or stability of convergence of contraction. Thus, these propositions are developed and verified based on the obtained real object results from multiple skeletonisation strategies. Let's assume, we have the original point set $\mathbf{P}_o = \{\mathbf{p}_i\}$ of the object and the contracted point cloud obtained in k-th contraction iteration $\mathbf{P}_k = \{\mathbf{p}_{k,i}\}$. Here, we use i and to denote the i-th point in the point set. The surfaces of the point clouds $\mathbf{P}_o$ and $\mathbf{P}_k$ are denoted as U_o and U_k respectively. The stability of contraction is assessed locally with respect to the boundary region of set points $\mathbf{X}_o \in \mathbf{P}_o$. If the iterative contraction results, denoted as $\mathbf{X}_k \ll \mathbf{P}_k$, satisfy the condition $X_{k+1} < X_k \ll X_o$, where the contracted surface is constrained to $U_{k+1} < U_k \ll U_o$, we define this condition as boundedness to assess the stability of contraction. Moreover, the stable contraction convergence is checked through the difference of the curvature $\Delta\kappa_{n,k}$ and the difference of the normal vectors $D_{\theta,k,i}$ at the local surface areas. For the process of point cloud contraction with stable convergence, the distribution of $D_{\theta,k,i}$ and $\Delta\kappa_{n,k}$ are expected to be symmetric and unimodal, resembling "bell shape". Also, the average of $D_{\theta,k,i}$ and $\Delta\kappa_{n,k}$ are expected to converge towards 0.5 and 0 values respectively. Apart from the shared features, the unique pattern of the distribution might indicate the special geometrical structures varying in different types of shapes or unexpected shape handling. Since the contracted results are the approximation of the surface skeleton, that pattern is also applicable for evaluation of the quality of the finalised surface skeleton. Besides, the Laplacian-based curve skeletonisation also relies on stable contraction results. Thus, the stable and converged contraction pattern also contributes to generating high-quality curve skeletons.

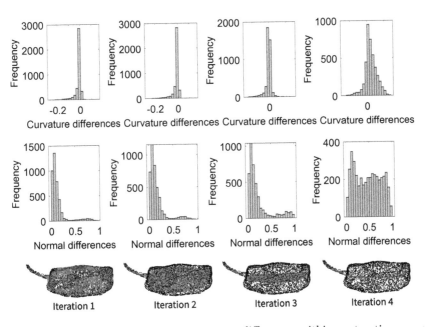

Fig. 2. Changes of normal vector and curvature differences within contraction process. The selected point cloud object is of a chilli.

3 Results and Discussion

In this section, we will present the observations of normal vector and curvature difference patterns, along with the results of various object skeletonisations.

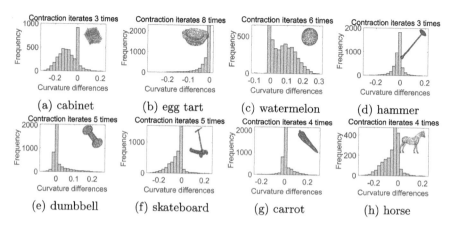

Fig. 3. Curvature differences between the input and point cloud after final contraction. The original point cloud (grey) with the contraction results (red) are put at the top right corner of each histogram (Color figure online)

The point cloud models utilized in this section are sourced from OmniObject3D dataset [29]. The point clouds from the OmniObject3D dataset are derived from real-world scans. Consequently, they contain noise and imperfections. The skeletonisation method employed by default is the baseline Laplacian-based skeletonisation [5]. Additionally, we tested our findings with GLSkeleton [26], a skeletonisation method proposed in our previous work, to validate the hypothesis. Multiple object point clouds from the dataset were tested in the experiment, and the results demonstrated similar patterns. Therefore, we selected only topologically different objects to present here. In other words, we will discuss these patterns using selected indicative objects.

To show the geometrical property changes within the contraction iterations, we generate the histogram of the differences of curvature and the normal vectors between each contraction iteration output and the original point cloud. As illustrated in Fig. 2, the histogram pattern changes demonstrate that the pattern of the curvature differences and the normal vector differences can indicate the geometrical convergence in the contraction process. While the shape of the object is deformed and pushed inward, the pattern of the mentioned geometrical characteristics are changed correspondingly along the iterations. Overall. the distribution of differences between the normal vectors and curvatures gradually becomes symmetric during contraction, suggesting stable convergence. As for the curvature differences shown in the first row of Fig. 2, the curvature differences are minimal. Its distribution remains a "bell shape" and the normality of the distribution keeps increasing. It means that the contraction on the curvature aspect converges stably in the whole process. As shown in the second row of Fig. 2, the change of the pattern of the normal vector differences are quite different in comparison with the curvature difference change. The distribution of surface normal vectors differences are skewed left at the beginning, after which

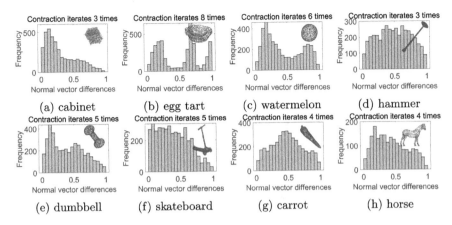

Fig. 4. Normal vector differences between the input and point cloud after final contraction. The original point cloud (grey) with the contraction results (red) are put at the top right corner of each histogram (Color figure online)

the normal vector differences gradually increased and the pattern is approaching a symmetric and unimodal distribution. It means that the contraction process is unstable at the beginning. But with the control of contraction and attraction weights (2) updated in iterations, the distribution is pushed to a stable stage. The changes are also reflected on the shape, which is demonstrated by the third row of the figure. Start with the contracted surface being non-parallel to the original surface and out of the boundary, the surface is gradually pushed inward and forming the surface skeleton at the end.

The assessment of final contraction outcomes relies on a comparative analysis of curvature and normal vectors between the contracted and original point clouds, as illustrated in Figs. 3 and 4. Notably, the contraction outcomes for certain objects, such as the cabinet, egg tart, and watermelon, deviate from the original topology, indicating lower quality. Conversely, other objects exhibit well-contracted forms with distributions that align more closely with the desired stability criteria outlined in Sect. 2.2.

As for the curvature differences, the distributions for the cabinet, egg tart, and watermelon display either skewed or spiky distributions, with average curvature differences diverging significantly from the expected value of zero. In contrast, the curvature distributions for other objects tend towards a bell-shaped curve centered around zero, indicating better adherence to the original shape.

For the normal vector differences, distributions of those with worse contraction also demonstrate less symmetric histogram patterns. Since the contraction process of the cabinet terminated unexpectedly, the normal vector differences are skewed left. Similarly, the unexpected peak on the left in the histogram graph of the dumbbell reflects the incomplete contraction on the left side of the contracted dumbbell point cloud. Here, we assume only the peaks of the frequency that drop off on both sides by at least 2% of the number of the point samples

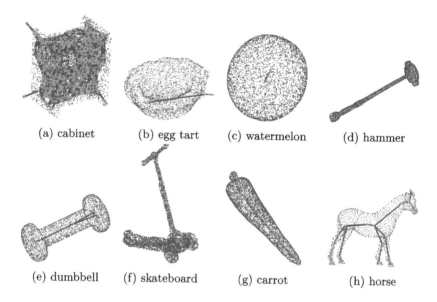

Fig. 5. Resultant curve skeletons

are meaningful, as there might be surface normal vector estimation errors and data noises. Regarding the distribution of the egg tart, the multiple spikes of the contracted surfaces are also reflected as peaks on the histogram. As for the skateboard, carrot, and horse, the histograms are closer to 0.5 on average and more symmetric in comparison to other objects, aligning with their better contraction results. It is important to note the presence of normal vector differences in skeleton surfaces, leading to variations in distributions and resulting in multiple peaks in histograms. These differences could provide valuable insights into the topology of the contracting surface, such as the segments that are contracted improperly.

Since the curve skeletons are derived from the contraction results while using the Laplacian-based skeletonisation method, the quality of the resultant curve skeletons is heavily dependent on the outcomes of the contraction. Therefore, the patterns of geometric changes observed during the contraction process effectively reflect the performance of the resultant curve skeletons. As illustrated in Fig. 5. The curve skeletons generated for objects such as the hammer, skateboard, carrot, and horse exhibit the highest quality, consistent with the quality of the contraction results. In contrast, the curve skeletons generated for the cabinet, egg tart, and watermelon lack significant topological information about the object shape, mirroring the deficiencies observed in the contraction process. Besides, the observed incomplete contraction handling of the dumbbell object is reflected in the curve skeleton results as well as the bifurcation at the undesirable position.

Additionally, we applied these findings to GLSkeleton approach [26], which selectively reduces the points while shrinking the point cloud. As points are removed by the algorithm iteratively, and the mentioned differences are computed point-wise, we decide to use only the points that are consistently retained throughout the iterations for contraction pattern analysis. As illustrated in Fig. 6, similar patterns are demonstrated by the results of the contraction process of GLSkeleton. It means that the surface skeletons generated by GLSkeleton hold similar performance which confirms our propositions.

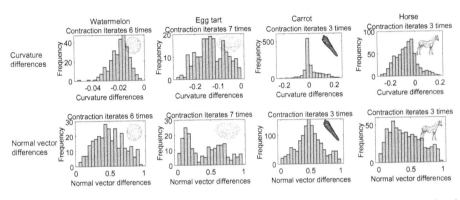

Fig. 6. Curvature vector and normal differences between the input and point cloud after contraction by GLSkeleton method. The original point cloud (grey) with the contraction results (yellow) are put at the top right corner of each histogram. (Color figure online)

Overall, the stable convergence of the contraction and the Laplacian-based curve skeletonisation process are decided by the distributions of the curvature differences and the surface normal vector differences and their changes within the contraction iteration. Besides, with the control by the contraction and attraction weights of Laplacian-based contraction, the distribution patterns can be corrected within the contraction process.

4 Conclusions

This work presents how we can determine whether skeletonisation by contraction methods e.g., Laplacian-based contraction can be understood about stable convergence. To develop our new proposition, we demonstrate the pattern of characteristics of curvature and normal vector changes during stable contraction and provide explanations. It is noteworthy that the defined metrics, surface normal vector and curvature difference changes, illustrate the correction of stability of contraction. However, the distribution pattern remains hypothetical and requires further quantitative analysis before we can establish a clear definition of high-quality skeleton.

In the future, we will concentrate on developing a robust framework for quantitatively analyzing changes in geometric properties. Our goal is to establish a comprehensive definition of high-quality skeleton and stable convergence of contraction for both surface skeletons and curve skeletons.

References

1. Amenta, N., Bern, M.: Surface reconstruction by voronoi filtering. In: Proceedings of the Fourteenth Annual Symposium on Computational Geometry, pp. 39–48 (1998)
2. Amenta, N., Choi, S., Kolluri, R.K.: The power crust. In: Proceedings of the Sixth ACM Symposium on Solid Modeling and Applications, pp. 249–266 (2001)
3. Arcelli, C., Di Baja, G.S., Serino, L.: Distance-driven skeletonization in voxel images. IEEE Trans. Pattern Anal. Mach. Intell. **33**(4), 709–720 (2010)
4. Au, O.K.C., Tai, C.L., Chu, H.K., Cohen-Or, D., Lee, T.Y.: Skeleton extraction by mesh contraction. ACM Trans. Graph. **27**(3), 1–10 (2008)
5. Cao, J., Tagliasacchi, A., Olson, M., Zhang, H., Su, Z.: Point cloud skeletons via laplacian based contraction. In: 2010 Shape Modeling International Conference, pp. 187–197 (2010)
6. Chazal, F., Lieutier, A.: The "λ-medial axis". Graph. Models **67**(4), 304–331 (2005)
7. Cornea, N., Demirci, M., Silver, D., Shokoufandeh, Dickinson, S., Kantor, P.: 3D object retrieval using many-to-many matching of curve skeletons. In: International Conference on Shape Modeling and Applications 2005 (SMI 2005), pp. 366–371 (2005)
8. Cornea, N., Silver, D., Min, P.: Curve-skeleton applications. In: VIS 2005. IEEE Visualization 2005, pp. 95–102 (2005)
9. Cornea, N.D., Silver, D., Min, P.: Curve-skeleton properties, applications, and algorithms. IEEE Trans. Vis. Comput. Graph. **13**(3), 530–548 (2007)
10. Giesen, J., Miklos, B., Pauly, M., Wormser, C.: The scale axis transform. In: Proceedings of the Twenty-Fifth Annual Symposium on Computational Geometry, pp. 106–115 (2009)
11. Huang, H., et al.: L1-medial skeleton of point cloud. ACM Trans. Graph. **32**(4) (2013)
12. Lin, C., Li, C., Liu, Y., Chen, N., Choi, Y.K., Wang, W.: Point2Skeleton: learning skeletal representations from point clouds. In: Proceedings of the IEEE/CVF Conference on Computer Vision and Pattern Recognition (CVPR), pp. 4277–4286 (2021)
13. Meyer, L., Gilson, A., Scholz, O., Stamminger, M.: CherryPicker: semantic skeletonization and topological reconstruction of cherry trees. In: Proceedings of the IEEE/CVF Conference on Computer Vision and Pattern Recognition, pp. 6243–6252 (2023)
14. Pokorny, F.T., Stork, J.A., Kragic, D.: Grasping objects with holes: a topological approach. In: 2013 IEEE International Conference on Robotics and Automation, pp. 1100–1107. IEEE (2013)
15. Przybylski, M., Asfour, T., Dillmann, R.: Unions of balls for shape approximation in robot grasping. In: 2010 IEEE/RSJ International Conference on Intelligent Robots and Systems, pp. 1592–1599 (2010)
16. Przybylski, M., Asfour, T., Dillmann, R.: Planning grasps for robotic hands using a novel object representation based on the medial axis transform. In: 2011 IEEE/RSJ International Conference on Intelligent Robots and Systems, pp. 1781–1788 (2011)
17. Przybylski, M., Wächter, M., Asfour, T., Dillmann, R.: A skeleton-based approach to grasp known objects with a humanoid robot. In: 2012 12th IEEE-RAS International Conference on Humanoid Robots (Humanoids 2012), pp. 376–383. IEEE (2012)

18. Qin, H., Han, J., Li, N., Huang, H., Chen, B.: Mass-driven topology-aware curve skeleton extraction from incomplete point clouds. IEEE Trans. Vis. Comput. Graph. **26**(9), 2805–2817 (2020)
19. Rezanejad, M., Samari, B., Rekleitis, I., Siddiqi, K., Dudek, G.: Robust environment mapping using flux skeletons. In: 2015 IEEE/RSJ International Conference on Intelligent Robots and Systems (IROS), pp. 5700–5705 (2015). https://doi.org/10.1109/IROS.2015.7354186
20. Saha, P.K., Borgefors, G., di Baja, G.S.: A survey on skeletonization algorithms and their applications. Pattern Recogn. Lett. **76**, 3–12 (2016)
21. Sobiecki, A., Jalba, A., Telea, A.: Comparison of curve and surface skeletonization methods for voxel shapes. Pattern Recogn. Lett. **47**, 147–156 (2014)
22. Stork, J.A., Pokorny, F.T., Kragic, D.: A topology-based object representation for clasping, latching and hooking. In: 2013 13th IEEE-RAS International Conference on Humanoid Robots (Humanoids), pp. 138–145. IEEE (2013)
23. Tagliasacchi, A., Delamé, T., Spagnuolo, M., Amenta, N., Telea, A.C.: 3D skeletons: a state-of-the-art report. Comput. Graph. Forum **35** (2016). https://api.semanticscholar.org/CorpusID:5740454
24. Tagliasacchi, A., Zhang, H., Cohen-Or, D.: Curve skeleton extraction from incomplete point cloud. ACM Trans. Graph. **28**(3), 1–9 (2009)
25. Vahrenkamp, N., Koch, E., Waechter, M., Asfour, T.: Planning high-quality grasps using mean curvature object skeletons. IEEE Robot. Autom. Lett. **3**(2), 911–918 (2018)
26. Wen, Q., Tafrishi, S.A., Ji, Z., Lai, Y.K.: GLSkeleton: a geometric laplacian-based skeletonisation framework for object point clouds. IEEE Robot. Autom. Lett. 1–7 (2024). https://doi.org/10.1109/LRA.2024.3384128
27. Wu, L., Falque, R., Perez-Puchalt, V., Liu, L., Pietroni, N., Vidal-Calleja, T.: Skeleton-based conditionally independent gaussian process implicit surfaces for fusion in sparse to dense 3D reconstruction. IEEE Robot. Autom. Lett. **5**(2), 1532–1539 (2020)
28. Wu, S., Wen, W., Xiao, B., Guo, X., Du, J., Wang, C., Wang, Y.: An accurate skeleton extraction approach from 3D point clouds of maize plants. Front. Plant Sci. **10**, 248 (2019)
29. Wu, T., et al.: OmniObject3D: large-vocabulary 3D object dataset for realistic perception, reconstruction and generation. In: Proceedings of the IEEE/CVF Conference on Computer Vision and Pattern Recognition (CVPR), pp. 803–814 (2023)
30. Xia, P., Zhang, L., Li, F.: Learning similarity with cosine similarity ensemble. Inf. Sci. **307**, 39–52 (2015)

Advancements in 3D X-Ray Imaging: Development and Application of a Twin Robot System

Seemal Asif[1]($^{\boxtimes}$), Yuliya Hryshchenko[1], Martin Holden[3], Matteo Contino[3], Ndidiamaka Adiuku[2], Bryn Hughes[3], Angelos Plastropoulos[2], Nico Avdelidis[2], and Phil Webb[1]

[1] Centre for Robotics and Assembly, Cranfield University, Cranfield MK43 0AL, UK
`s.asif@cranfield.ac.uk`
[2] Centre for Integrated Vehicle Health Management, Cranfield University, Cranfield MK43 0AL, UK
[3] Adaptix Ltd., Oxford University Begbroke Science Park, Oxford OX5 1PF, UK

Abstract. The development of a novel twin robot system for 3D X-ray imaging integrates advanced robotic control with mobile X-ray technology to significantly enhance diagnostic accuracy and efficiency in both medical and industrial applications. Key technical aspects, including innovative design specifications and system architecture, are discussed in detail. The twin robots operate in tandem, providing comprehensive imaging capabilities with high precision. This novel approach offers potential applications ranging from medical diagnostics to industrial inspections, significantly improving over traditional imaging methods. Preliminary results demonstrate the system's effectiveness in producing detailed 3D images, underscoring its potential for wide-ranging uses. Future research will focus on optimizing image quality and automating the imaging process to increase utility and efficiency. This development signifies a step forward in integrating robotics and imaging technology, promising enhanced outcomes in various fields.

Keywords: Mobile Xray · Robotic Imaging · Robotics Control · Image Processing · Image Stitching

1 Introduction

Two-dimensional (2D) radiography represents a prevalent imaging modality within the medical field, primarily utilized for a variety of clinical diagnoses. Despite its widespread application, 2D radiography encounters significant limitations due to the superimposition of anatomical structures within the imaging plane, which can obscure critical diagnostic details. Conversely, tomographic three-dimensional (3D) imaging techniques offer a substantial advantage by facilitating the reconstruction of cross-sectional images. This method effectively eliminates the issue of overlapping structures, thereby enhancing clarity and diagnostic accuracy.

In clinical practice, the acquisition of 2D and 3D images typically involves the use of distinct systems situated in separate facilities. This separation can affect workflow efficiency and patient experience due to the need for multiple appointments or transfers between different imaging suites. Despite the clear benefits of 3D imaging, its integration into routine clinical practice has been more readily achieved in the medical field compared to industrial applications (Strobel et al., 2009). Robot-assisted medical imaging uses robots to acquire medical images with high precision and accuracy. Examples include robot-assisted endoscopic camera imaging, ultrasound imaging with a robot-held transducer, X-ray imaging with robot-positioned sources and detectors, and actuated capsule endoscopy. This technology allows for controlled trajectories, increased aperture, volumetric or tomographic imaging, tracking of medical instruments, and real-time adjustments. Intraoperative robotic imaging provides valuable information to physicians and aids in registering preoperative images to patients (Salcudean et al., 2022). In the latter, the adoption of advanced imaging techniques such as 3D tomography remains uncommon, possibly due to higher costs, the specific technical expertise required, or the lack of immediate applicability in standard industrial processes (Salcudean et al., 2022).

X-ray Non-Destructive Testing (NDT) is a widely used technique in both medical and aerospace fields (Joseph et al., 2018). The high-resolution capability of low-power X-rays allows for the identification of small defects in critical aerospace components. To transfer this technology into aerospace applications, the feasibility of applying medical X-ray NDT methods directly to aerospace contexts involves considerations of material density and safety standards.

While medical X-ray systems are designed for imaging soft tissue and bones and are certified for safe radiation doses for occasional human exposure, aerospace X-ray NDT/NDE systems tend to require slightly higher X-ray source energy to penetrate higher density materials such as composites, aluminum, and titanium structures. These systems often involve larger and more complex geometries and must adhere to stringent safety protocols to protect operators in the workplace (Towsfyan et al., 2020).

The novelty of this ongoing research is to investigate the integration of portable, rapid low-power X-ray scanning (10–11 W, at 35–75 kV) with a fixed-to-the-floor robotic solution for flexible non-destructive evaluation and testing (NDE/NDT) of aircraft components, including both metallic and composite materials of varying small geometries and curvatures. This research, conducted at the Aerospace Integration Research Centre (AIRC) at Cranfield University in collaboration with Adaptix, focuses on developing a research framework to understand the robustness of system integration, robotic control for process automation opportunities, health and safety (H&S) clearance to safely operate X-rays in compliance with the UK Ionizing Radiations Regulations 2017 (IRR17), (Mridul Gupta Muhsin Ahmad Khan & Singari, 2022) image data acquisition and processing, and defect detection capability analysis to meet aerospace NDT standards for accuracy, reliability, and repeatability.

The rest of this paper is structured as follows: First, the system description section provides an overview of the robotic cell setup and describes its various elements. Both a virtual digital mock-up (DMU) of the robotic cell and its physical integration are presented to demonstrate the initial results of this ongoing research. Showcasing both the DMU and the real setup aims to highlight the potential applications of this developed system and its prospects for future development. The safety system considerations are described, including novel approaches for validating safe and compliant work with ionizing radiation in a research and development (R&D) environment, with potential implications for production environments.

2 System Description

The experimental cell setup included two industrial robots mounted at a fixed tooling height with floor mounting, an Adaptix low-power X-ray device, and fixed tooling designed for the precise positioning of a 2-m longitudinal composite aircraft component. The system utilized articulated robotic arms, each with a 20 kg payload capacity, and both robots were safety-rated for remote operation, necessitating additional safety measures to prevent human proximity during operation.

The Adaptix low-power X-ray device, originally designed and certified for medical and veterinary applications, features a low-power X-ray source (10 W at 70 kV). This generator is specifically designed for use in mobile X-ray instruments and has demonstrated potential applications in scanning relatively small composite aircraft components, offering high-quality image acquisition with low-power X-rays.

In this system setup, the original design of the Adaptix X-ray device was modified by mounting the source and detector separately onto two end-effectors, each attached to one of the robotic arms, as illustrated in Fig. 1. This modification allowed for precise and coordinated movements of the X-ray source and detector, ensuring comprehensive scanning coverage of the aircraft component.

Furthermore, during this initial testing phase, the fixed tooling design concept was developed with flexibility in mind to accommodate relatively small, curved longitudinal aircraft composite parts. This approach aimed to simplify the design and expedite the initial integration process, ensuring readiness for physical demonstration. Future tooling design considerations will incorporate automated positioning driven by tooling and part metrology.

2.1 Process Flow for X-ray Scanning

In X-ray radiography, high-energy X-ray photons (short wavelength electromagnetic radiation) are used to penetrate different materials and create a shadowgraph image of the object being tested. With increased material density, the X-ray absorption also increases, leading to greater attenuation of the X-rays as they pass through the object toward the detector. The attenuation of X-rays with a specific energy as they interact with matter is described by the Beer-Lambert law, represented by the following formula:

$$I = I_0 e^{-\mu x}$$

where:

- I is the intensity of the transmitted X-ray beam,
- I_0 is the initial intensity of the X-ray beam,

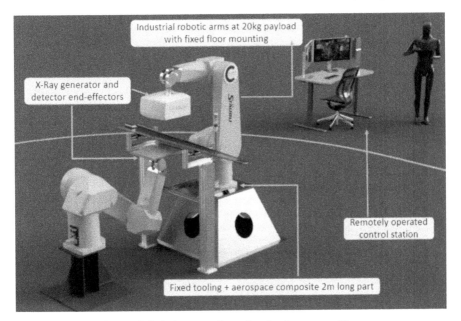

Fig. 1. Cell design setup for twin robot low power X-ray

- μ is the linear attenuation coefficient of the material,
- x is the thickness of the material.

The non-absorbed energy, captured by the detector, results in variations in image darkness, which is a measure of the density of the scanned material. For composite materials, low-power X-rays are capable of characterizing defects such as solid inclusions, fiber misalignment, and matrix cracking. The orientation of the composite internal layup affects the characterization effectiveness.

The choice of scanned part material and thickness is carefully considered for this initial setup, and there is a certain degree of freedom in determining these parameters, as they are crucial for investigating the system's effectiveness and capabilities to match aerospace accuracy standards for non-destructive testing (NDT) and non-destructive evaluation (NDE). In this experimental setup, the X-ray device capability plays a major role in defining the robotics process flow and kinematics. The generated X-ray beam power, projection angle, and the relative distance between the X-ray source, detector, and part within its workspace are critical for acquiring high-quality and accurate images necessary for effective processing software results. Therefore, the concept of separated movement for the source and detector end-effectors robots was integrated.

This concept's flexibility allows for accommodating complex geometry parts for scanning within the same working envelope (Sattar & Brenner, 2009). The robotic kinematics were defined so that different sections (A, B, C, etc.) can be scanned sequentially or repeated on-demand with pre-defined and pre-programmed robot positions for active X-ray scanning, as shown in Fig. 2 below.

Different local areas can be scanned with sequential stitching to accomplish full part scan in flexible an on-demand approach for automating rapid 3D X-ray process flow. In contrast to conventional NDT methods, such as ultrasonic and eddy current testing where typically a constant probe contact is required to contoured components surface, 3D X-Ray process allows for contactless scanning of complex geometries, therefore allowing a rapid inspection process with less concern robots position accuracies (Ajman & Abdullah, 2024).

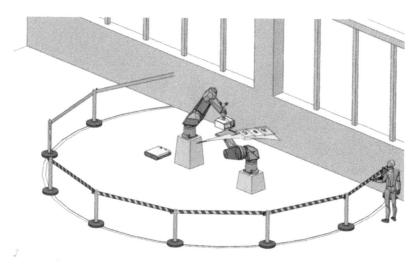

Fig. 2. Rapid X-ray Process Flow

2.2 Safety System Considerations

In-built safety system approach had been followed to consider both the robotic safe control and X-ray safety and legal considerations working environment with potential Ionizing Radiation exposure, as summarized in Fig. 3 below.

Fig. 3. H&S clearance approach for robotic open cell concept with low power X-Ray

This research health and safety (H&S) approach is designed to comply with the annual dose limit exposures to ionizing radiation in the workplace across industry and research fields. Additionally, it ensures that doses follow the ALARP principle (As Low As Reasonably Possible). More detailed work on the methodology will be published following approval from the Health and Safety Executive (HSE).

Given the demanding safety margins for proximity and the additional measures required for the X-ray safety system, the global cell safety in-built design is driven by the X-ray system and is currently under approval by the UK HSE regulator.

Our robotic 3D X-ray concept involves a clear line of sight (LOS) and, therefore, does not use a cabinet for the X-ray equipment or a physical enclosure for the robotic cell. Consequently, scatter radiation monitoring within the controlled area and across the 5-m perimeter is the primary factor driving the safety system interlocks. For this purpose, a designated Controlled Area has been established, with a safety perimeter of 5 m to prevent human (or foreign object) access while the system is in operation, as shown in the Fig. 4 below.

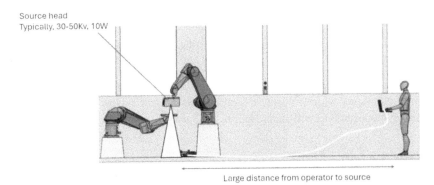

Fig. 4. Remotely operated robotics system with clear LOS

A breach of the programmed 5-m perimeter will be automatically detected by the robotics safety proximity sensors, which will trigger an immediate power-off of the X-ray (if it was on) and stop the robots from moving as illustrated in Fig. 5. All safety interlocks, including proximity sensors and emergency stops (E-stops), are wired through the safety Programmable Logic Controller (PLC). Additionally, the X-ray generator beam angle and the synchronized movements of the robots (source and detector) are preprogrammed to operate in safe positions at all times while the X-ray is armed and powered on (scanning in progress).

Both robots' controllers and the X-ray data acquisition system are wired into the safety PLC to enable the relevant safety triggers to be lifted automatically, preventing unsafe system operation. Moreover, the process status is indicated through traffic lights and an audible alarm during the brief process step when the X-ray is on, as shown in the figure below. This integrated safety system ensures that any breach of the controlled area is promptly addressed, maintaining a safe environment for personnel and equipment. The use of proximity sensors, E-stops, and safety PLC wiring, combined with preprogrammed safe positions for the X-ray beam and synchronized robotic movements, provides a robust safety framework. The visual and audible alarms further enhance

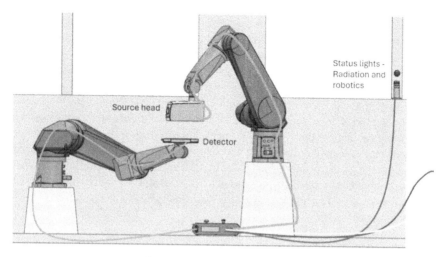

Fig. 5. X-Ray system data acquisition wiring, visual and audible signage.

the safety measures by alerting personnel to the operational status of the X-ray system, ensuring compliance with stringent safety protocols and minimizing the risk of exposure to ionizing radiation (Great Britain. Health and Safety Executive, 2017).

3 Xray System

The Adaptix 3D X-ray system employs a technique known as tomosynthesis. Unlike computed tomography, which captures images through a 360° rotational motion around the subject, tomosynthesis involves acquiring a series of images from a limited range of angles along a line or, in the case of the Adaptix system, a grid pattern. In the Adaptix approach, a source head held in a stationary position above the subject, emitting a sequence of X-rays from many different positions within the source head. A detector is placed underneath the subject, as close as possible to the unit being examined. A control module powers and synchronizes the interaction between the source and the detector, ensuring operational safety as the operator remains at a safe distance. The X-ray device is synchronized with robotic systems and safety protocols, enabling operation only when the surrounding area is clear and the robots are stationary.

A rectangular array of emission positions provides enhanced depth resolution compared to a linear sweep of emitter positions, as it captures projection images over a two-dimensional sweep. Additionally, a rectangular array allows for a reduction in the source-to-image distance (SID). Halving this distance decreases the required beam current by a factor of four, potentially reducing the size, weight, and cost of the overall system, albeit with an increase in the angular width of the focal spot. Both simulations and physical experiments indicate that depth resolution significantly improves with a rectangular array compared to a linear array. When utilizing a rectangular array at half the SID and a quarter of the power, this benefit is maintained despite the larger angular width of the focal spot. This approach facilitates enhanced imaging from lower-cost, smaller devices compared to conventional systems, making practical-sized robot cells feasible for production lines and non-destructive testing areas. Moreover, this reduces the weight of the source head, ensuring it remains within the load limits of smaller robotic arms.

3.1 X-Ray Image Acquisition

During a tomosynthesis acquisition, a series of images are captured with the X-ray source positioned at various locations, while the sample and detector remain stationary. Consider two objects placed at different heights but centered on a flat-panel X-ray detector. As illustrated in Fig. 6, when illuminated by an X-ray source positioned such that it aligns with both objects (Position 2), the detector displays an overlapping shadow formed by the shadows of each object.

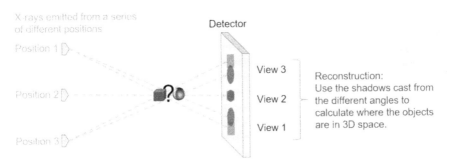

Fig. 6. Shadow formation on a detector during tomosynthesis.

Objects that created overlapping shadows with 2D X-ray can then be seen in isolation. Acquiring images from multiple positions allows for the reconstruction of the 3D positions of each element along the path between the X-ray source and the detector.

3.2 Image Reconstruction

Once all X-ray images are collected, the 3D information is extracted using the method described in (Soloviev et al., 2020). In digital tomosynthesis, a stack of 2D planes is generated, each corresponding to a different height from the detector, rather than rendering a 3D volume as in computed tomography.

The approach can be briefly described without delving into the mathematical details of the reconstruction algorithm. The algorithm is akin to the filtered back-projection method used in traditional CT scans. For each acquired image, the source position and the X-ray cone angle of emission are recorded, and the image is propagated back to the original source position. As illustrated in Fig. 7: (a) Pixels' backprojection at the specified height along rays connecting pixel's corners to the emitter. (b) Backprojected pixels mapping onto the reconstruction slice grid. As illustrated in Fig. 7, the same pixel on the acquired image can correspond to different spatial positions at a given height from the detector, necessitating a "pixel remapping" strategy (Evangelista et al., 2020).

In traditional CT algorithms, acquired images are filtered in the Fourier space, and a 3D mesh is built to remap the projection into the same coordinate system. In this method, the images are back-propagated to the emitter position while applying an offset based on the source position for each image. A spatial frequency filter is then applied, and the average resulting image is calculated. This process, repeated for each reconstructed slice, creates an "out-of-focus effect" for objects at heights different from the reconstructed plane. Objects in the selected plane appear brighter and with sharper edges due to the averaging effect.

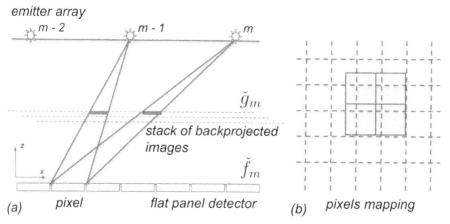

Fig. 7. (a) Pixels' backprojection at the specified height along rays connecting pixel's corners to the emitter. (b) Backprojected pixels mapping onto the reconstruction slice grid.

Some artifacts or impressions of all the back-projected images may still be present in the reconstructed slices, but these become less prominent as the number of images increases.

4 Image Processing

Image processing encompasses exploration and enhancement of X-ray scans and other types of images, for improved insight and decision making in technical operations across various fields including material science, medical diagnostics and robot platforms. In the context of X-ray scans, this includes acquiring multiple x-ray image measurements around a sample from different angles to provide a comprehensive view of x-rays attenuation throughout the sample. The x-ray penetrates the material, revealing details such as hidden flaws or irregularities. This produces 2D X-ray images of the object from different perspective and are used to facilitates image reconstruction providing detail information about the sample's density and material composition. Different challenges limit effective reconstruction and 3D visualization of the 2D slices which include variation in x-ray beam and material's absorption capability (Ou et al., 2021). These affects noise rate, defect detection ability and general image quality. These requires relevant processing techniques that can reduce noise, enhance contrast, quality and uniformity in the acquired 2D images slices. Figure 8 shows samples of pre-processed images slices acquired from composite material with the plot of average intensity value showing significant changes in the image's structures and components. This can guide further adjustment and enhancement to achieve visual improvements and facilitate accurate reproduction of image structures and properties. Image processing techniques like image stitching are among the commonly applied methods and have shown significant improvement in data visualization picking up defect profile and structures with clarity and high fidelity.

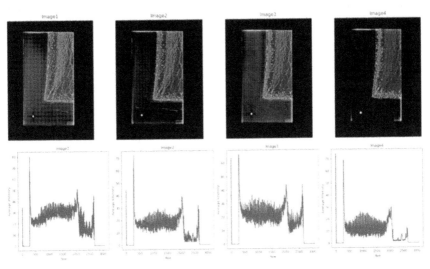

Fig. 8. Image slices and plot of average intensity value.

4.1 Image Stiching

Image stitching methods combines 2D image slices to create comprehensive and high-resolution image representation. This requires identifying and matching corresponding key points or features between the overlapping images using algorithms like SIFT (Scale-Invariant feature transform) (Burger & Burge, 2016) and SURF (speeded-Up Robust Features) (Oyallon & Rabin, 2015). These algorithms align and merge multiple overlapping images based on the matched features to produce a single seamless image. Techniques like feathering and blending(Gu & Rzhanov, 2006) help achieve smooth transitions, reducing visible seams and resulting to high quality and high-resolution visualization of 3D model structures. Leveraging image stitching approaches has shown significant improvement and application in domains like robotics, composite material science and medical imaging leading to enhanced analysis and better decision-making in technical operations.

5 Discussion and Conclusion

The development of the twin robot system for 3D X-ray imaging represents a significant advancement in both medical diagnostics and industrial inspections. The collaborative efforts between Cranfield University and Adaptix Ltd. have resulted in a system that not only enhances imaging capabilities but also improves the accuracy and efficiency of diagnostic procedures. The integration of advanced robotic control with portable X-ray technology has shown promising results in initial tests, particularly in producing high-quality 3D images for NDE environments.

One of the critical aspects of this system is its ability to operate in tandem, providing comprehensive imaging from multiple angles. This dual-robot approach ensures that the subject is thoroughly scanned, reducing the likelihood of missed anomalies and improving diagnostic confidence. The system's precision and reliability are bolstered by its sophisticated control algorithms and image processing techniques, which enable precise positioning and consistent image quality.

The potential applications of this technology are vast. In the medical field, it can be used for more accurate and detailed imaging of complex structures, such as bones and organs, which is crucial for diagnosing conditions that are difficult to detect with traditional 2D X-ray. In industrial settings, the system can be employed for non-destructive testing of materials and components, ensuring safety and integrity in critical infrastructure.

However, the system is not without its challenges. The complexity of coordinating two robots and the associated computational requirements for real-time image processing and stitching are significant. Future research will focus on optimising these processes to enhance the system's performance further. Additionally, exploring automation in the imaging process can reduce the need for manual intervention, making the system more user-friendly and accessible for various applications.

In conclusion, the twin robot system for 3D X-ray imaging developed by Cranfield University and Adaptix Ltd. represents a promising step change in imaging technology. The system's ability to provide detailed and accurate 3D images has the potential to revolutionize both medical diagnostics and industrial inspections. Preliminary results have demonstrated the system's effectiveness, and future research will aim to refine its capabilities and expand its applications.

The successful integration of advanced robotics and portable X-ray imaging highlights the potential for interdisciplinary collaboration to drive technological innovation. By addressing the current challenges and optimizing the system's performance, this technology could become a standard tool in various fields, offering significant benefits in terms of diagnostic accuracy, efficiency, and overall outcomes. The continued development and refinement of this system will be crucial in realizing its full potential and ensuring its widespread adoption.

Acknowledgment. This work was supported by ATI funding for advanced manufacturing innovation - ATI ROBOT-MOUNTED 3D X-RAY INSPECTION. We express our gratitude to Jamie Rice and Daniel Oakley, Lab Technicians, and James Fowler, Senior Technical Officer at the Intelligent Automation Lab, for their invaluable support and technical expertise, which were essential for the project's success. We also thank Dr. Charles Antrobus and Jessica Evans from Adaptix for building the systems, and James Whittle and Linde Van Parijs from the Adaptix software team for their contributions.

References

Ajman, M.A.A., Abdullah, E.J.: Design of a robotic arm for inspecting curved surface in aerospace non-destructive testing. J. Aeron. Astron. Aviat. **56**(1), 501–511 (2024). https://doi.org/10.6125/JoAAA.202403_56(1S).37

Burger, W., Burge, M.J.: Scale-invariant feature transform (SIFT). In: Burger, W., Burge, M.J. (eds.) Digital Image Processing, pp. 609–664. Springer, London (2016). https://doi.org/10.1007/978-1-4471-6684-9_25

Evangelista, D., Terreran, M., Pretto, A., Moro, M., Ferrari, C., Menegatti, E.: 3D mapping of X-ray images in inspections of aerospace parts. In: 2020 25th IEEE International Conference on Emerging Technologies and Factory Automation (ETFA), vol. 1, pp. 1223–1226 (2020). https://doi.org/10.1109/ETFA46521.2020.9212135

Great Britain. Health and Safety Executive: Work with ionising radiation: ionising radiations regulations 2017 (2017)

Gu, F., Rzhanov, Y.: Optimal image blending for underwater mosaics. In: OCEANS 2006, pp. 1–5 (2006). https://doi.org/10.1109/OCEANS.2006.307037

Joseph, L., Padois, V., Morel, G.: Towards X-ray medical imaging with robots in the open: safety without compromising performances. In: 2018 IEEE International Conference on Robotics and Automation (ICRA), pp. 6604–6610 (2018). https://doi.org/10.1109/ICRA.2018.8460794

Gupta, M., Khan, M.A., Butola, R., Singari, R.M.: Advances in applications of Non-Destructive Testing (NDT): a review. Adv. Mater. Process. Technol. **8**(2), 2286–2307 (2022). https://doi.org/10.1080/2374068X.2021.1909332

Ou, X., et al.: Recent Development in X-Ray Imaging Technology: Future and Challenges. Research (2021). https://doi.org/10.34133/2021/9892152

Oyallon, E., Rabin, J.: An analysis of the SURF method. Image Process. Line **5**, 176–218 (2015). https://doi.org/10.5201/IPOL.2015.69

Salcudean, S.E., Moradi, H., Black, D.G., Navab, N.: Robot-assisted medical imaging: a review. Proc. IEEE **110**(7), 951–967 (2022). https://doi.org/10.1109/JPROC.2022.3162840

Sattar, T.P., Brenner, A.-A.: Robotic system for inspection of test objects with unknown geometry using NDT methods. Ind. Robot. **36**(4), 340–343 (2009). https://doi.org/10.1108/01439910910957093

Soloviev, V.Y., Renforth, K.L., Dirckx, C.J., Wells, S.G.: Meshless reconstruction technique for digital tomosynthesis. Phys. Med. Biol. **65**(8) (2020). https://doi.org/10.1088/1361-6560/ab7685

Strobel, N., et al.: 3D imaging with flat-detector C-arm systems. In: Reiser, M.F., Becker, C.R., Nikolaou, K., Glazer, G. (eds.) Multislice CT, pp. 33–51. Springer, Heidelberg (2009). https://doi.org/10.1007/978-3-540-33125-4_3

Towsfyan, H., Biguri, A., Boardman, R., Bluemensath, T.: Successes and challenges in non-destructive testing of aircraft composite structures. Chin. J. Aeron. **33**(3), 771–791 (2020). https://doi.org/10.1016/j.cja.2019.09.017

Author Index

A
Adetoro, Mayo II-184
Adiuku, Ndidiamaka I-434
Adorno, Bruno Vilhena II-143
Agrane, Joseph I-3
Ahmed, Abdullahi II-266
Ahmed, Sarfraz I-194
Al Assam, Hisham I-132
Alboraei, Youssef G. II-130
Aldhafeeri, Bandar I-283
Ali, Omar I-249
Alispahić, Ibrahim II-333
Althoefer, Kaspar I-296, II-242
An, Xiangyu II-164, II-184
Andrew, Jeremy I-270
Angelmahr, Martin I-296
Aragon-Camarasa, Gerardo I-3
Arezzo, Alberto I-296
Arnaud, Soumo Emmanuel I-249
Arvin, Farshad II-281, II-320
Asif, Seemal I-434, II-33
Athar Ali, Mohammad I-132
Attenborough, Eden I-249
Avdelidis, Nico I-434

B
Badyal, Arjun I-169
Bahaidarah, Mazen II-281
Baker, Daniel H. II-27
Bao, Lingfan II-118
Batty, David I-302
Benali, Khairidine II-14
Bettles, Joshua I-270
Bin Firoz, Hasan I-169
Bin Hassan, Mohd Norhakim I-234
Bird, Jordan J. I-409
Biswas, Mriganka I-119
Boé, Laurence I-28
Borja, Pablo II-254
Bosilj, Petra I-53
Braghin, Francesco I-313

Brown, James M. I-53
Buckow, Pia II-266
Buffet, Olivier II-3

C
Caleb-Solly, Praminda II-14
Caliskanelli, Ipek I-302
Camara, Fanta I-182, II-27
Campbell, Adam I-28
Carrasco, Joaquin I-283
Carty, Jim I-3
Cielniak, Grzegorz I-15, I-355
Clark, Alex II-266
Clement, Benoit II-242
Colas, Francis II-3
Colombo, Gualtiero B. II-85
Conn, Andrew T. II-171
Contino, Matteo I-434
Costi, Leone I-327
Craenen, Bart I-28

D
Darby, Iain I-270
Das, Gautham P. II-294
Davies, George I-249
Dawood, Abu Bakar I-296
Delfaki, Andromachi Maria II-118
Denzel Nyasulu, Tawanda I-256
Dillikar, Sairaj R. II-93
Dong, Jiale II-58
Du, Shengzhi I-256

E
Edan, Yael II-157
Elliott, Harrison II-164
Elstner, Laurenz II-223
Espejel Flores, Juan Pablo I-15

F
Faris, Omar I-222

Felicetti, Leonard   II-93
Fox, Charles   I-222, I-249
Friedl, Werner   I-313

G

Gazzard, Matthew   I-409
Ghalamzan, E. Amir   I-209
Godary-Dejean, Karen   I-72
Gofton, Hannah   II-27
Goodsell-Carpenter, Liam   II-266
Grech, Raphael   I-84
Groves, Keir   II-143
Guo, Zhihao   I-367

H

Handelzalts, Shirley   II-157
Hardin, Benjamin   I-169
Hasan, Md Mahmudul   I-409
He, Yuanzhi   II-85
Hernández Vega, Juan David   I-155
Hernndez, Juan D.   II-85
Heselden, James R.   II-294
Hewett, Daniel   II-266
Hicks, Helen   I-409
Hickson, Henry   II-171
Hodge, Victoria   I-169
Holden, Martin   I-434
Hou, Ningzhe   I-327
Hryshchenko, Yuliya   I-434
Hu, Junyan   II-320
Huchard, Marianne   I-72
Hughes, Bryn   I-434
Humphreys, Joseph   II-118

I

Ihianle, Isibor Kennedy   I-339, I-409
Imam, Adil   II-266

J

James, Katherine Margaret Frances   I-355
Ji, Ze   I-84, I-422
Jones, Charlotte   I-182
Jose Lopez Pulgarin, Erwin   I-283

K

Kaleel, Danyaal   II-242
Kanoulas, Dimitrios   II-118
Khan, Fahad   II-33
Klüpfel, Leonard   I-313

Krakovski, Maya   II-157
Krysov, Mark   II-195
Kyrkjebø, Erik   II-223

L

Lai, Yu-Kun   I-422
Lascheit, Kai   I-3
Lee, Alexandra   II-48
Lennox, Barry   I-270
Leslie, Cameron   II-93
Leslie-Dalley, Seyonne   II-143
Lewinska, Paulina   I-169
Li, Jizhang   I-194
Liang, Yichen   I-41
Lindsay, Alan   I-28
Liu, Haowen   I-397
Lu, Qi   II-307
Luna, Ryan   II-307

M

Machado, Pedro   I-339, I-409
Maiolino, Perla   I-327
Manna, Soumya K.   II-266
Marjanovic, Ognjen   II-281
Martin, Ana Elvira Huezo   I-313
Mawle, Richard   I-107
McGhan, Fraser   I-84
McGloin, Helen   I-107
Meng, Hongying   I-383
Merchant, Catherine   I-222
Moroni, Michele   I-313
Mosquera-Maturana, Juan Sebastian   I-155
Munafò, Andrea   I-28
Murray, John   I-119

N

Nahri, Syeda Nadiah Fatima   I-256
Naz, Nabila   II-266
Nazari, Kiyanoush   I-209
Nazmul Huda, Md   I-194
Neill, Oliver   I-3
Neubert, Peer   I-94
Ngom, Fama   I-72
Nicholls, Benjamin   I-222

O

O'Dowd, Paul   II-130
Oikonomou, Andreas   I-339

Author Index

P
Paoletti, Paolo I-302
Paparas, Dimitris I-222
Parris, Matthew Marlon Gideon I-132
Parsons, Simon I-209
Peng, Tianhu II-118
Perrett, Andy I-53
Petrick, Ronald P. A. I-28
Philamore, Hemma II-171
Plastropoulos, Angelos I-434
Polydoros, Athanasios I-60
Pontin, Marco I-327
Post, Mark I-397
Protzel, Peter I-94

R
Rafiq Swash, Mohammad I-383
Rahvar, Faraz II-106
Rajendran, S. Vishnu I-209
Ramírez-Duque, Andrés A. I-28
Raper, Rebecca I-144
Reed, Darren I-182
Rekabi-bana, Fatemeh II-281
Ren, Hanchi II-320
Roa, Máximo A. I-313
Robb, Jamie I-3
Robinson, Breeshea II-266
Romero Cano, Victor I-155
Ryan, Philippa I-169

S
Schubert, Stefan I-94
Serhat Yildiz, Ahmet I-383
Si, Weiyong II-58
Sirithunge, Chapa II-236
Skilton, Robert II-3
Smith, Elliot I-222
Soriano Avendaño, Luis Arturo I-15
Stancu, Alexandru I-41
Stevenson, Robert I-222
Stimson, Christina E. I-144
Stoelen, Martin F. II-223
Studley, Matthew I-107, II-48
Sun, Jingcheng II-73
Sundaram, Ashok M. I-313
Suthar, Kedar II-236
Swann, Michael I-339
Swindell, Jacob I-249
Sze, Samuel I-169
Sznaidman, Yael II-157

T
Tafrishi, Seyed Amir I-422, II-195, II-208
Tahirović, Adnan II-333
Tang, Gilbert II-93
Thidrasamee, Chayada II-130
Thomas, Vincent II-3
Tyrrell, Andy I-397

U
Upadhyay, Saurabh II-93

V
V. Adorno, Bruno I-283
van Wyk, Barend J. I-256

W
Wallbridge, Christopher D. II-85
Wang, Mingfeng II-164, II-184, II-236
Wang, Ning II-58
Wang, Peng I-367
Watson, Michael II-320
Watson, Simon I-234
Webb, Phil I-434, II-33
Weißflog, Markus I-94
Wen, Qingmeng I-422
West, Andrew I-270, I-302
White, Francis II-266
Williams, Emlyn I-60
Wiltshire, Ollie II-208
Winfield, Alan I-107
Winter-Glasgow, Ted II-254

X
Xu, Xiaoxian I-222

Y
Yahaya, Salisu I-339
Yang, Chenguang II-58
Yilmaz, Abdurrahman I-15, II-106
You, Yang II-3

Z
Zhang, Cheng I-234
Zhang, Huaxi Yulin I-72
Zhang, Lei I-72
Zhang, Zhenyu I-296
Zhang, Zhixin I-41
Zhou, Chengxu II-73, II-118
Zhou, Liyou I-249
Zorriassatine, Farbod I-339

www.ingramcontent.com/pod-product-compliance
Lightning Source LLC
Chambersburg PA
CBHW072016120125
20267CB00006B/77